MW00464660

WILDLIFE SCIENCE

CONNECTING RESEARCH
WITH
MANAGEMENT

WILDLIFE SCIENCE
CONNECTING RESEARCH
WITH
MANAGEMENT

Edited by

Joseph P. Sands, Stephen J. DeMaso,
Matthew J. Schnupp, and Leonard A. Brennan

CRC Press
Taylor & Francis Group
Boca Raton   London   New York

CRC Press is an imprint of the
Taylor & Francis Group, an **informa** business

CRC Press
Taylor & Francis Group
6000 Broken Sound Parkway NW, Suite 300
Boca Raton, FL 33487-2742

Printed in the United States of America on acid-free paper
Version Date: 20120424

International Standard Book Number: 978-1-4398-4773-2 (Hardback)

Library of Congress Cataloging-in-Publication Data

Wildlife science : connecting research with management / editors: Joseph P. Sands ... [et al.].
    p. cm.
Includes bibliographical references and index.
ISBN 978-1-4398-4773-2 (hardcover : alk. paper)
    1. Wildlife management. 2. Wildlife research. 3. Wildlife conservation. I. Sands, Joseph P., 1982-

SK355.W58 2012
639.9--dc23   2012014554

Visit the Taylor & Francis Web site at
http://www.taylorandfrancis.com

and the CRC Press Web site at
http://www.crcpress.com

This book is for all biologists, naturalists, students of ecology, and proponents of conservation in the spirit of Roosevelt and Leopold.

Contents

SECTION I Introduction

SECTION II Research and Management Entities

SECTION III   Species Case Studies

SECTION IV Management and Policy Case Studies

SECTION V Conclusions and Future Directions

Foreword

Fred C. Bryant
Caesar Kleberg Wildlife Research Institute
Texas A&M University-Kingsville
Kingsville, Texas

Research, the pursuit of discovery, is a long, tedious, and sometimes thankless process. This is especially true for ecological systems. It gets complicated because there are many variables that are difficult to control in a research setting: weather, past land use practices, varying soil types and topography, elusive and shy wildlife species, and wide-ranging wildlife species, among others. So research that applies to the real world of a manager is hard to obtain to begin with. It is especially hard when the scientist does not consider how his or her research can or will be applicable to a field biologist, state or federal agency, or land manager. The cultural abyss, and the reason for this book, is addressed in the Preface—it is real and is widening in my opinion because no one today seems to ask the *"so what?"* question of the scientist. We do that every day at the Caesar Kleberg Wildlife Research Institute because our constituents are private landowners. They want new information they can apply or that at least reinforces what they have observed as they manage a piece of ground vital to the wildlife about which they care. Moreover, most of the ones I know care deeply about all the wild species that inhabit their vital ground. I applaud the editors and the authors for their penetrating and visionary look at this important topic. If it creates a forum for communication between the research community and the manager, or if it generates new thinking in how we approach or conduct research, it will have served a valuable and beneficial purpose.

Preface

As we begin the second decade of the twenty-first century, the field of natural resource conservation is faced with many new challenges and continues to struggle with challenges from the past. Conservation achieves its highest level of success when research results and management applications of these results intersect. Despite the potential synergy that can result from basing management applications on results from research, there is currently a polarization of cultures between wildlife managers and wildlife researchers. The polarization persists even though there is a consensus that linking management with sound ecological theory will improve both management and research over the long term. Whether one's work in natural resource conservation involves the direct management of resources or the pursuit of new knowledge through the scientific process, the importance of understanding and improving the relationship and interaction between management and research will be paramount to achieving critical conservation goals in the twenty-first century.

The idea for this book evolved from a symposium we organized for The Wildlife Society's 15th Annual Conference in Miami, Florida during November 2008. The purpose of the symposium was to investigate what we and other biologists have viewed as a widening gap between wildlife research and wildlife management and to provide suggestions for bridging this gap. To achieve our objectives, we compiled a panel of experts from universities, state and federal agencies, and the private sector, and asked participants to provide their perspectives on the interrelationships between management and research. Within this context, we asked speakers to present case studies highlighting the role of state and federal agencies and private organizations in management and research, the lingering disconnects between grassland birds, quail, and deer research and management, as well as the development of management techniques from field research, and rangelands and ranch management. After the symposium, we held several conversations about the possibility of compiling material from the presentations into an edited volume. We identified several important topics not covered during the symposium, such as greater sage-grouse (*Centrocercus urophasianus*) management in the western United States, white-tailed deer (*Odocoileus virginianus*) management in the northeastern United States, private entity financing of applied research to facilitate improvements in management, and the emerging issues between wind-energy development and wildlife conservation, to round out the scope of the book.

The objective of the book is to investigate the differing cultural priorities between wildlife research and wildlife management and to provide suggestions for bridging this gap. Section I (Chapters 1 and 2) introduces and characterizes the issue and serves as the thesis for the rest of the volume. Section II (Chapters 3–8) investigates the roles of research and management from the perspective of the institutions upon which the wildlife profession is composed: universities, state governments, the federal government, and private organizations. Section III (Chapters 9–14) presents species-oriented case studies focusing on the lingering disconnects between northern bobwhite (*Colinus virginanus*), sage grouse, grassland birds, waterfowl, and deer research and management. Section IV (Chapters 15–17) presents place-based and conceptual case studies involving rangelands and ranch management, the role of science in the development of natural resource law and policy, wind energy, and the development of management techniques from field research. Section V (Chapter 18) provides strategies and tactics for bridging the research and management disconnect and makes recommendations for improving wildlife research and management in order to achieve conservation objectives in the twenty-first century.

The book provides examples of how wildlife research and management policies have succeeded and failed, and represents the first comprehensive collaboration and review of this important topic compiled into a single volume. The lessons learned from these examples are then used to develop a synthesis of how we can bridge the cultural and communication gaps between the wildlife management and research communities. The case studies highlight past and ongoing controversies from throughout North America and provide a platform for discussing the underlying issues embedded within each of these topics.

This book is intended to serve as a resource for field biologists, university researchers, and especially students who are in the process of defining their career goals. All of the topics covered are designed to address relevant issues related to natural resource conservation and provide a platform for vigorous discourse and debate among readers of this book. Identifying factors that help connect wildlife research with management, or work to form barriers between wildlife research and management, more often than not boil down to opinions and interpretations. Our goal was to provide a wide range of examples that illustrate ways to overcome these barriers and forge a stronger connection between wildlife research and management applications based on scientific research. Therefore, the opinions and interpretations presented in the individual chapters of this book are exclusively those of the authors who wrote each chapter.

Joseph P. Sands
Caesar Kleberg Wildlife Research Institute
Texas A&M University-Kingsville
Kingsville, Texas

Stephen J. DeMaso
Caesar Kleberg Wildlife Research Institute
Texas A&M University-Kingsville
Kingsville, Texas

Matthew J. Schnupp
King Ranch, Inc.
Kingsville, Texas

Leonard A. Brennan
Caesar Kleberg Wildlife Research Institute
Texas A&M University-Kingsville
Kingsville, Texas

Acknowledgments

This book is the result of more than four years of work. Initial discussions about this topic arose through conversations among the editors in the fall of 2007, and work on this project has continued since then. Throughout this process, we received help from numerous individuals. We have listed many here and apologize for any omissions. Thanks to Dr. Barry Dunn at King Ranch Institute for Ranch Management (now Dean of the College of Agricultural and Biological Sciences at South Dakota State University) for supporting the initial concept development and Dr. Fred Bryant and the Caesar Kleberg Wildlife Research Institute for providing us with the flexibility to pursue this project both as a symposium for The Wildlife Society's 15th Annual Conference in Miami, Florida, and also as a compiled volume. Thank you to James Gallagher, Dave Godwin, John Haufler, Andrew Tri, Samantha Wisniewski, and John Young for candid suggestions, constructive criticism, and creative ideas during the developmental stages of the book. A special thank you to Jennifer Ahringer and Randy Brehm at CRC Press, whose guidance and unending support for this project was instrumental to its successful completion. Thank you to each of the contributing authors; without your knowledge, expertise, and experience, this book would not have been possible.

JPS—I would like to thank my wife, Jessica, and son, Morgan, for their love and support throughout this process, and my parents, Kevin and Emily Sands, long-time natural resource professionals who encouraged me to pursue a career in wildlife science and management.

SJD—I would like to thank the Caesar Kleberg Wildlife Institute for the time and environment to work on this project. Also, being able to work with friends and colleagues like Joe Sands, Lenny Brennan, and Matthew Schnupp has made this project a delightful endeavor.

MJS—I am fortunate to be employed by King Ranch, Inc., one of the few profit-oriented companies that encourage and support nonprofit activities such as this book project. Without the support of the family shareholders and management, this project would not have been possible. Additionally, I would also like to thank my family for their unwavering encouragement.

LAB—The peaceful activities of writing and editing scientific publications require space and time. A person needs quiet, uninterrupted space and significant amounts of time in order to be an effective writer and editor. I am fortunate that my position at the Caesar Kleberg Wildlife Research Institute provides me with the space and allows me the time to work on projects such as this book. Furthermore, I am fortunate to be supported by the C.C. Charlie Winn Endowed Chair for Quail Research at the Richard M. Kleberg, Jr. Center for Quail Research. Without such support, this book would simply not be possible.

About the Editors

Originally from Days Creek, Oregon, **Joseph P. Sands** received an Honors Bachelor of Science from Oregon State University (2004), and an M.S. (2007) and a Ph.D. (2010) from Texas A&M University–Kingsville. Joseph worked as a Research Assistant and Research Associate for the Caesar Kleberg Wildlife Research Institute at Texas A&M University–Kingsville from June 2004 to September 2011. At the time of publication, Joseph was the Migratory Game Bird, Small Game and Wild Turkey Program Coordinator for the New Mexico Department of Game and Fish in Santa Fe.

Stephen J. DeMaso is the monitoring coordinator for the Gulf Coast Joint Venture in Lafayette, Louisiana. Prior to moving to Louisiana, he worked as a research scientist at the Caesar Kleberg Wildlife Research Institute at Texas A&M University–Kingsville in Kingsville, Texas; upland game bird program leader for the Texas Parks and Wildlife Department in Austin, Texas; and the Oklahoma Department of Wildlife Conservation where he served as the lead researcher on the nationally recognized Packsaddle quail research project. He is a member of the Louisiana, Texas, and National Chapters of The Wildlife Society. Steve has served as the book review editor for the *Journal of Wildlife Management* and previously served as the program chairman and editor for the *Proceedings of the Fifth National Quail Symposium*. Steve was raised in southern Michigan and received his B.S. from Michigan State University, M.S. from Texas A&M University–Kingsville, and his Ph.D. from Texas A&M University.

Matthew Schnupp was born and raised in Wheeling, West Virginia. He attended West Virginia University where he graduated in 2005 with a degree in Wildlife and Fisheries Science. Matthew has been part of various research projects in West Virginia, South Carolina, Montana, and Wyoming. In the fall of 2009, he graduated with a Masters in Wildlife and Range Science from the Caesar Kleberg Wildlife Research Institute (CKWRI) at Texas A&M University–Kingsville. During his tenure at CKWRI, Matthew studied abroad at the Centre for Research into Ecological and Environmental Modelling (CREEM) and received the Clarence Cottam Award for his outstanding student research efforts. Matthew was hired in October 2008 as the King Ranch Santa Gertrudis Heritage Society Biologist. Matthew is currently the wildlife biologist for over 200,000 acres of King Ranch and the wildlife research liaison between King Ranch and various universities, private organizations, and government entities.

Leonard A. Brennan grew up in Connecticut and earned a B.S. from The Evergreen State College in Olympia Washington, an M.S. from Humboldt State University in Arcata, California, and a Ph.D. from the University of California, Berkeley. He was Editor-in-Chief (2001–2002) and is currently Associate Editor (2005–present) of the *Journal of Wildlife Management*. He chaired the Technical Support Committee for the Texas Quail Council (2003–2006). He is editor of the book, *Texas Quails: Ecology and Management* published by Texas A&M University Press, which received the 2008 Outstanding Edited Book Award from The Wildlife Society, the 2007 Special Category Publication Award from the Texas Section of the Society for Range Management, and the 2007 Outstanding Edited Book Award from the Texas Chapter of The Wildlife Society. Lenny has authored or co-authored more than 130 scientific publications and more than 100 extension and magazine articles. He is currently Editor-in-Chief of *Wildlife Society Bulletin*, an online journal dedicated to integrating wildlife science and management.

Contributors

Edward B. Arnett
Bat Conservation International
Austin, Texas

Robert A. Askins
Department of Biology
Connecticut College
New London, Connecticut

Chris Bauman
Georgia Wildlife Resources Division
Fitzgerald, Georgia

Laura Bies
The Wildlife Society
Bethesda, Maryland

William M. Block
U.S. Forest Service
Rocky Mountain Research Station
Flagstaff, Arizona

John Bowers
Georgia Wildlife Resources Division
Social Circle, Georgia

Chad Boyd
Eastern Oregon Agricultural Research
  Station
U.S. Department of Agriculture
Burns, Oregon

Leonard A. Brennan
Caesar Kleberg Wildlife Research
  Institute
Department of Animal and Wildlife
  Science
Texas A&M University–Kingsville
Kingsville, Texas

Paul M. Castelli
New Jersey Division of Fish and
  Wildlife
Nacote Creek Research Station
Port Republic, New Jersey

David deCalesta
Wildlife Analysis Consulting
Hammondsport, New York

David S. DeLaney
King Ranch, Inc.
Kingsville, Texas

Stephen J. DeMaso
Caesar Kleberg Wildlife Research
  Institute
Department of Animal and Wildlife
  Science
Texas A&M University–Kingsville
Kingsville, Texas

Stephen DeStefano
U.S. Geological Survey
Massachusetts Cooperative Fish and
  Wildlife Research Unit
University of Massachusetts
Amherst, Massachusetts

Jennifer N. Duberstein
Sonoran Joint Venture
U.S. Fish and Wildlife Service
Tucson, Arizona

James J. Giocomo
Oaks and Prairies Joint Venture
American Bird Conservancy
Temple, Texas

Mary Gustafson
Rio Grande Joint Venture
American Bird Conservancy
Mission, Texas

James R. Heffelfinger
Arizona Game and Fish Department
University of Arizona
Tucson, Arizona

Steven G. Herman
The Evergreen State College
Olympia, Washington

Michael Hutchins
The Wildlife Society
Bethesda, Maryland

Paul R. Krausman
Boone and Crockett Program in
    Wildlife Conservation
Wildlife Biology
University of Montana
Missoula, Montana

Ronald E. Masters
Tall Timbers Research Station and
    Land Conservancy
Tallahassee, Florida
and
Oklahoma State University
Stillwater, Oklahoma

John F. Organ
U.S. Fish and Wildlife Service
Hadley, Massachusetts

William Palmer
Tall Timbers Research Station and
    Land Conservancy
Tallahassee, Florida

Leonard Ruggiero
U.S. Forest Service
Rocky Mountain Research Station
Missoula, Montana

Victoria A. Saab
U.S. Forest Service
Rocky Mountain Research Station
Montana State University Campus
Bozeman, Montana

Joseph P. Sands
Caesar Kleberg Wildlife Research
    Institute
Texas A&M University–Kingsville
Kingsville, Texas

Matthew J. Schnupp
King Ranch, Inc.
Kingsville, Texas

Reggie Thackston
Georgia Wildlife Resources Division
Forsyth, Georgia

Jack Waymire
Oklahoma Department of Wildlife
    Conservation
Clayton, Oklahoma

Christopher K. Williams
Department of Entomology and Wildlife
    Ecology
University of Delaware
Newark, Delaware

Section I

Introduction

1 A History of the Disconnect between Wildlife Research and Management

Stephen J. DeMaso*
Caesar Kleberg Wildlife Research Institute
Texas A&M University-Kingsville
Kingsville, Texas

CONTENTS

First, a game man, even if not engaged in research, must be by nature a scientific investigator, because the greater part of the facts he uses he must find for himself.

—Aldo Leopold (1933:413)

Scientific training should prepare men for professional research and teaching work. It should take graduate biologists and teach them how to use their biology in solving game management problems.

—Aldo Leopold (1933:415)

* Current address: U.S. Fish and Wildlife Service, Gulf Coast Joint Venture, c/o National Wetlands Research Center, 700 Cajundome Blvd., Lafayette, LA 70506.

ABSTRACT

Many wildlife professionals perceive that the disconnect between wildlife management and research has happened during the last few decades. I review the wildlife literature and construct a timeline of wildlife management and research to determine if and when the disconnect initially occurred. There appears never to have been a strong relationship between wildlife management and research, even though our profession has strived to link the two prior to the publication of *Game Management*, by Aldo Leopold in 1933. This weak connection, at best, has lurked in the shadows as the wildlife profession has evolved from a game management-oriented profession into the broad field of wildlife conservation we see today. Wildlife managers manipulate systems to achieve a management objective rather than to find out how the system works. Wildlife researchers generate knowledge by testing hypotheses. This knowledge is then used to guide and support wildlife management. Romesburg (1981) states that some of the principles of wildlife management were based on unreliable knowledge and that most proposed hypotheses are not tested, but become dogma through verbal repetition. Since the mid-1980s, there have been several papers on how to improve the rigor in wildlife science. Recommendations include improving the quantitative rigor in both undergraduate and graduate wildlife programs, more emphasis on study designs, the formulation of hypotheses, and the testing of those hypotheses. As landscapes become more fragmented, habitat patches become smaller and smaller, and more species are threatened with extinction, wildlife managers and researchers will have to work together on some very complex, dynamic problems. There may not be another chance to get it right.

INTRODUCTION

Prior to 1905, the philosophy toward game was to perpetuate, rather than to improve or create hunting opportunities (Leopold 1933:16). The thought was that hunting restrictions could "string out" the remnant game species and make them last a longer time. Later in this decade, Theodore Roosevelt began to promote the idea of "conservation through wise use." Wildlife, forests, ranges, and waterpower were conceived by Roosevelt to be renewable organic resources, which might last forever if they were harvested scientifically, and not faster than they were reproduced. This was the foundation for the Roosevelt doctrine of conservation (Leopold 1933:17–18). Part of the doctrine was Roosevelt's idea that science should be used as a tool for conservation of natural resources, a very new concept in 1910. Roosevelt probably imagined science being used for management of bag limits and season lengths of game species. Leopold (1933:20) states, "I do not know who first used science creatively as a tool to produce game crops in America." However, we do know that the first, partially management-oriented research project conducted in America was the "Cooperative Quail Study Investigation" conducted by Herbert L. Stoddard from 1924–1929 (Stoddard 1931). In the hiring letter to Stoddard, Dr. E. W. Nelson, Director of the U.S. Biological Survey, wrote that one of the objectives of the investigation was "to definitely determine methods whereby the quail can be increased and maintained in numbers far beyond these at present" (Komarek 1978:iii–iv).

Leopold (1933:3) defined game management as "The art of making the land produce sustained annual crops of wild game for recreational use." However, this definition has evolved over time. Leopold described management as an art, but did not mention science in his definition. As the profession of game management evolved into a broader field, the use of terms like ecology and conservation became more common. Eventually, science was added to the evolving definition of wildlife management, so that science would be the foundation and the guide for the management activities that were applied to the habitats and populations being managed.

THE DISCONNECT

There seems to be the perception among wildlife professionals that the disconnect between wildlife management and research is something that just happened during the last couple of decades. This is especially true among young professionals and graduate students who may have limited experience and are not familiar with the beginnings and the evolution of their profession. It makes sense that to know where you are going, you have to know where you have been. What is and what should be are usually two different things. It seems that students in wildlife programs get more of how things should be and not as much of how things are in the real world concerning the relationship of wildlife management and research (see Figure 1.1).

Even before the publication of *Game Management* in 1933 by Leopold, there were serious debates over the role of science in management (Table 1.1). About 1913–1914, William T. Hornaday was credited with saving the American bison (*Bison bison*) and the Alaskan fur seal (*Callorhinus ursinus*) from extinction. Despite the fact that he was a member of the scientific community, Hornaday was contemptuous of the rudimentary wildlife research effort that was developing throughout his career, although he readily accepted any findings that supported his own prejudices (Trefethen 1975:178).

FIGURE 1.1   A popular topic at early wildlife conferences was the relationship between wildlife research and management. Figure shows the plenary session of the Fifth North American Wildlife Conference in 1940 at the Mayflower Hotel in Washington, DC. Photo courtesy of the Wildlife Management Institute.

TABLE 1.1

Timeline of Events That Have Contributed to the Development of the Wildlife Research Field in the United States

Year	Event
1862	Morrill Land Grant Act allowed for the creation of land grant colleges.
1885	Formation of the Section of Economic Ornithology in the Department of Agriculture, later renamed the Biological Survey.
1914	The Ecological Society of America was formed.
1920	First issue of *Ecology* published.
1924	First game management research project, Stoddard's (1931) northern bobwhite work in South Georgia.
1930	American Game Policy published.
1933	First university wildlife program at the University of Wisconsin-Madison.
1935	First Cooperative Fish and Wildlife Research Units formed at land-grant universities.
1936	First North American Wildlife Conference held in Washington, DC
1937	The Wildlife Society was formed.
1937	First issue of the *Journal of Wildlife Management* published.
1937	Federal Aid in Wildlife Restoration (Pittman-Robertson) Act is passed by Congress.
1939	U.S. Fish and Wildlife Service is created.
1958	First *Wildlife Monograph* published.
1972	First issue of the *Wildlife Society Bulletin* published.
1985	The Society for Conservation Biology is formed.
1987	First issue of *Conservation Biology* published.
1994	First annual Wildlife Society conference.

During the late 1920s, one of the weaknesses of the proponents of the wildlife (i.e., waterfowl) refuge bills before the U.S. Congress was the dearth of scientific information on the status and movements of the various segments of the continental waterfowl population (Trefethen 1975:192). This lack of information was cause for the U.S. Bureau of Biological Survey to implement waterfowl banding projects and a primitive annual waterfowl survey (Trefethen 1975:178). In 1934, passage of the Migratory Bird Hunting Stamp Act provided funds to purchase wetland habitat to be protected in the National Wildlife Refuge System.

From about 1910 until the early 1930s, the Kaibab deer issue on the Kaibab Plateau in northwestern Arizona was a contentious issue among hunters and federal and state agencies (Trefethen 1975:195–201). After about two decades of applying different strategies to keep the deer population at levels that the habitat could support and then through scientific research, the Arizona Game and Fish Commission and the U.S. Forest Service developed a kit of management tools designed to maintain the deer population at a level that would ensure the health of the animals and the habitat on which they depended (Trefethen 1975:200).

In 1935, the Cooperative Wildlife Research Units were established to educate wildlife biologists and conduct wildlife research (Peek 1986:13). In 1937, The

Wildlife Society was formed to advance competency of wildlife biologists and advance the art and science of wildlife management. Leopold (1937:104) makes a case for increasing wildlife research funding, stating "Half a dozen New Deal bureaus are spending a score of millions on wildlife work, but not a red penny for research." Also in 1937, the enactment of the Federal Aid to Wildlife Restoration Act was legal recognition of the research into problems of wildlife management were needed, as well as habitat acquisition (Peek 1986:14). However, (Leopold, Taylor, Bennitt, and Chapman, 1938) cautioned that even after fifteen years of research on how wildlife can be increased by means of management, early wildlife management had failed except for a few species.

From the 1940s to the 1970s, the field of wildlife management continued to evolve. In 1979, the North American Wildlife and Natural Resources Conference hosted a session entitled "Wildlife and Fisheries Research Needs." Conclusions of the session were: (1) wildlife and fisheries research programs are in trouble, and (2) programs were in trouble because researchers, managers, administrators, and politicians do not understand the research process; because the research process is misunderstood, its past and potential contributions to management are underappreciated and undervalued; and because research is undervalued, it is underfunded (Gill 1985). Romesburg (1981) stated that some of the principles of wildlife management were based on unreliable knowledge and that most proposed hypotheses are not tested, but become dogma through verbal repetition. MacNab (1983) argued that wildlife management should be used as an experiment and that management treatments could be designed as experiments to answer questions. Since the mid-1980s, there have been several papers on how to improve the rigor in wildlife science (Anderson 2001; Anderson et al. 2003; Romesburg 1991; Sinclair 1991; White 2001; Wolff 2000). These recommendations include improving the quantitative rigor in both undergraduate and graduate wildlife programs, more emphasis on study designs, the formulation of hypotheses, and the testing of those hypotheses. White (2001:380) states, "The past record of the wildlife profession does not speak well for our ability to demonstrate rigor in our management, as evidenced by the increasing rate at which management decisions are being made by courts and ballot initiatives."

THE ROLE AND LIMITATIONS OF MANAGERS

Wildlife managers manipulate systems to achieve a management objective rather than to find out how the system works (MacNab 1983). Wildlife management is a difficult, respectable job (Bailey 1982). However, often the best wildlife students are groomed by academics for research positions, while the remaining students are written off to become wildlife managers in state or federal agencies. Like research, management requires original thinking. However, many university departments have changed their names, removing the word "management," and making science more inferred (Bailey 1982). Even though wildlife management is not a science, it is both complicated and important. Wildlife managers deserve equal continuing education opportunities, equal compensation, and respect regarding researchers.

THE ROLE AND LIMITATIONS OF RESEARCHERS

Wildlife researchers generate knowledge by testing hypotheses. This knowledge is then used to guide and support wildlife management. Bennitt in Leopold et al. (1938) summarizes criticisms of wildlife researchers as, first, highly trained investigators doing things that appear to have little or no value. Second, researchers cannot guarantee results quickly, and third, the results may not be the ones the sponsors of the investigation had hoped for (Leopold et al. 1938).

Franklin (1995) points out some of the limitations of researchers. Researchers tend to think in terms of a single solution to a problem, researchers do not like to base proposals on incomplete information, and when faced with incomplete information, researchers are usually conservative. In addition, researchers lack training or experience in policy analysis, have difficulties in communication, and often suffer from hubris (Franklin 1995).

Gill (1985) gives several reasons why wildlife research is underfunded and undervalued in the wildlife management profession. First, research is undervalued because the quality of the product has been poor. Political correctness is preferred to constructive criticism. Rarely does one hear debate about hypotheses, experimental design, data analysis, or interpretation of results at wildlife conferences. Second, it is undervalued because the contributions of wildlife research are taken for granted by wildlife managers. Research conducted today is aimed at the questions of tomorrow. As the findings from research are incorporated into management programs, everyone has forgotten where those findings (i.e., knowledge) came from. Third, wildlife research fails to solve wildlife management problems. This is partly because of the misunderstanding of the differences between problem solving and research, and often the problem lies in the misapplication of the knowledge by the wildlife manager. Fourth, wildlife research is undervalued because knowledge is viewed as a means to an end, and not as a valued end unto itself. Knowledge is a marketable commodity (Gill 1985). There is a growing demand by the public, especially non-consumptive users of the wildlife resource for information about wildlife. These are untapped revenue streams for funding the management of wildlife. As these potential sources are passed over, the contribution of wildlife research is undervalued. Lastly, wildlife research is undervalued each time its contribution to the vigor of the entire wildlife profession and wildlife organizations and institutions is overlooked (Gill 1985). Wildlife research serves as a catalyst for the wildlife profession because it generates new knowledge and tools for the profession to use in managing wildlife populations.

THE CHANGE FROM GAME MANAGEMENT
TO CONSERVATION BIOLOGY

From 1930 to the 1950s, there was a close association and collaboration between applied wildlife programs, both university and government, and ecology and behavior programs in university biology departments (Wagner 1989). Near the end of the 1950s, a new phase of ecology and animal behavior appeared that sent academic ecology and applied wildlife management down different paths and dissolved the close association of previous decades (Wagner 1989). During the 1960s, 1970s, and

1980s, community ecology and behavioral ecology became more mainstream in American ecology. Population ecology became more mathematical and abstract, and assumed an evolutionary dimension emphasizing life-history patterns (Wagner 1989). This shifted to a strong orientation to theory, emphasizing mathematical modeling, elegant experimental tests, and new statistical techniques. Scalet (2007) comments that university academic programs are now scaled more toward ecology/ conservation, while federal and state agencies still average to the management side of the scale. He argues that this represents a disconnect because university academic programs conduct research for these agencies, educate future agency employees, and provide technical services to agencies (Scalet 2007).

Wagner (1989) points to a class dichotomy among people who should be working together for the management, conservation, and preservation of all wildlife resources. Some traditional wildlifers look at conservation biologists as too theoretical, ivory-towerish, and unwilling to deal with the political, economic, and applied management of on-the-ground parts of wildlife management (Wagner 1989). Some conservation biologists look at wildlifers as too focused on consumptive uses, and not sophisticated enough scientifically (Wagner 1989). However, Murphy (1990) claimed that rigorous application of the scientific method in conservation biology is essential if conservation biology is to be respected.

SIMILARITIES BETWEEN THE MEDICAL FIELD AND WILDLIFE MANAGEMENT

Romesburg (1981) suggested that the medical profession was a good role model for the wildlife management profession because the medical profession has already sorted out the relationships between medical practice and medical science (Gill 1985). Munson's (1981:180) thesis was " … that, although medicine can be scientific, it is not and cannot become a science" (Gill 1985). Whereas medicine uses science as a means toward its end, the end product of medicine is decidedly different from the end product of science. Gill (1985) considered the ideal role of wildlife research in the practice of wildlife management was to substitute the words *wildlife management* for medicine or medical practice and *wildlife research* for science in Munson's (1981:195) summary of the relationship between medicine and science. The revised wildlife summary would read:

It should be no surprise that contemporary *wildlife management*, despite its commitment to *wildlife research*, continues to employ practical success or control as its basic criterion for evaluating rules, procedures, and casual claims. *Wildlife management* is an eminently practical enterprise that must attempt to meet immediate and urgent demands. It cannot afford to wait for the acquisition for the appropriate *wildlife research* knowledge, but must do the best it can in the face of ignorance and uncertainty. We do not look to *wildlife management* to tell us what the world is like. Rather, we count on *wildlife management* to conserve wildlife populations.

Wildlife research, by contrast is a leisured pursuit. It may be prodded by external demands to solve practical problems, but its internal standard of success continues to be truth. It can afford to wait and work until this is met. For *wildlife management*, the constant external demand to promote conservation is identical with *wildlife management's* internal aim, and as a result, practical success is both the external and internal standard. What we expect of *wildlife management* is what it expects of itself.

CONCLUSIONS

It appears that the disconnect between wildlife management and research is nothing new. It has been lurking in the shadows of our profession since the 1920s, as well as in other professions. Professionals are beginning to see that agencies and universities are facing budgetary shortfalls and reductions in staff, and as a result managers and researchers will have to do a better job of working together in the future (Huenneke 1995, Kessler, Salwasser, Cartwright, and Caplan, 1992, Moorman 2000).

If you look at the references for this chapter, the majority of publications are from university professionals. To me this is disturbing. Why don't we hear more from the management side of this partnership? I understand that different jobs have different priorities and publishing is a higher priority for researchers than managers. However, I do believe that we need to hear from the managers and that their experiences and insights are important regarding what students are taught in university programs. As a profession, we need to encourage dedicated managers in the field to publish their management-oriented work in outlets such as the recently resurrected *Wildlife Society Bulletin*. As landscapes become more fragmented, habitat patches become smaller and smaller, and more species are threatened with extinction, wildlife managers and researchers will have to work together on some very complex, dynamic problems. There may not be another chance to get it right.

REFERENCES

Anderson, D. R. 2001. The need to get the basics right in wildlife field studies. *Wildlife Society Bulletin* 29:1294–1297.

Anderson, D. R., E. G. Cooch, R. J. Gutiérrez, C. J. Krebs, M. S. Lindberg, K. H. Pollock, C. A. Ribic, T. M. Shenk. 2003. Rigorous science: suggestions on how to raise the bar. *Wildlife Society Bulletin* 31:296–305.

Bailey, J. A. 1982. Implications of "muddling through" for wildlife management. *Wildlife Society Bulletin* 10:363–369.

Franklin, J. F. 1995. Scientists in wonderland: experiences in development of forest policy. *Bio Science Supplement* 74–78.

Gill, R. B. 1985. Wildlife research—an endangered species. *Wildlife Society Bulletin* 13:580–587.

Huenneke, L. F. 1995. Involving academic scientists in conservation research: perspectives of a plant ecologist. *Ecological Applications* 5:209–214.

Kessler, W. B., H. Salwasser, C. W. Cartwright, Jr., J. A. Caplan. 1992. New perspectives for sustainable natural resources management. *Ecological Applications* 2:221–225.

Komarek, E. V. 1978. Preface *in* H. L. Stoddard, author. *The bobwhite quail: its habits, preservation, and increase*, 3rd ed. David A. Avant, III, Quincy, FL.

Leopold, A. 1933. *Game management*. Charles Scribner's Sons, New York.

Leopold, A. 1937. The research program. *North American Wildlife Conference* 2:104–107.

Leopold, A., W. P. Taylor, R. Bennitt, H. H. Chapman. 1938. Wildlife research—is it a practical and necessary basis for management? *North American Wildlife Conference* 3:42–55.

MacNab, J. 1983. Wildlife management as scientific experimentation. *Wildlife Society Bulletin* 11:397–401.

Moorman, C. E. 2000. Designing and presenting avian research to facilitate integration with management. *Studies in Avian Biology* 21:109–114.

Munson, R. 1981. Why medicine cannot be a science. *Journal of Medicine and Philosophy* 6:183–208.

Murphy, D. D. 1990. Conservation biology and scientific method. *Conservation Biology* 4:203–204.

Peek, J. M. 1986. *A review of wildlife management*. Prentice-Hall, Englewood Cliffs, NJ.

Romesburg, H. C. 1981. Wildlife science: gaining reliable knowledge. *Journal of Wildlife Management* 45:293–313.

Romesburg, H. C. 1991. On improving the natural resources and environmental sciences. *Journal of Wildlife Management* 55:744–756.

Scalet, C. G. 2007. Dinosaur ramblings. *Journal of Wildlife Management* 71:1749–1752.

Sinclair, A. R. E. 1991. Science and the practice of wildlife management. *Journal of Wildlife Management* 55:767–773.

Stoddard, H. L. 1931. *The bobwhite quail: its habits, preservation, and increase*. Charles Scribner's Sons, New York.

Trefethen, J. B. 1975. *An American crusade for wildlife*. Winchester Press, New York.

Wagner, F. H. 1989. American wildlife management at the crossroads. *Wildlife Society Bulletin* 17:354–360.

White, G. C. 2001. Why take calculus? Rigor in wildlife management. *Wildlife Society Bulletin* 29:380–386.

Wolff, J. O. 2000. Reassessing research approaches in the wildlife sciences. *Wildlife Society Bulletin* 28:744–750.

2 Strengthening the Ties between Wildlife Research and Management

Paul R. Krausman
Boone and Crockett Program in Wildlife Conservation,
Wildlife Biology
University of Montana
Missoula, Montana

CONTENTS

All factions, whatever their other differences, should unite to make available the known facts, to promote research to find the additional facts needed, and to promote training of experts qualified to apply them.

—Aldo Leopold (1933:412)

ABSTRACT

In the wildlife profession, research and management are the cornerstones that should be forever linked to maintain and enhance wildlife and the habitats they depend on for survival. Unfortunately, research and management often work against each other for an array of reasons (e.g., management has to proceed without all the necessary data, land conflicts, and economic concerns). I present the major causes of disagreements between research and management (i.e., poor communication, narrow focus of universities, lack of cooperation, differing objectives, research that is not relevant to management, competition for resources and funding, and uncooperative administrations). This discussion is followed with mechanisms to solidify the important cornerstones (i.e., recognize the gap; encourage cooperation between researchers, managers, and the public; use information from other disciplines for management; provide incentives for cooperation between researchers and managers; educate students on the importance of cooperation). Cooperation is critical. Our profession and the wildlife we serve depend on it.

INTRODUCTION

Ever since Leopold (1933) opened the doors to modern wildlife management in North America, science has been the cornerstone leading to sound management of wildlife and the habitats they depend on. Game laws in the United States were essentially a mechanism to divide the resources nature produced; " ... a game system based on an equally distributed citizenship, rather than, as in Europe, on an unequally distributed landownership" (Leopold 1933:17). Unfortunately, the mechanisms employed to divide game resources for greater public good, however, did not stop the decline in game supply. Thus, better law enforcement and the elimination of market hunting were two of the next steps that followed implementation of game laws. The game literature of the closing century is saturated with these two ideas. They became the personal dogma and public law throughout the United States and Canada. Game protection became a cause. The game hog and the market hunter were duly pilloried in press and banquet hall, and to some extent in field and wood, but the game supply continued to wane.

Into this situation came Theodore Roosevelt, with the idea of "conservation through wise use" (Leopold 1933:17). At this point, management and science were linked forever as Roosevelt argued that our natural resources would last forever if they were managed scientifically and not harvested faster than they reproduced. This was a new philosophy to which the American public had not been exposed; it led to the Roosevelt doctrine of conservation that was based on three conceptual pillars (Leopold 1933).

1. Natural resources were not independent and their interactions with each other had to be considered; this idea ultimately led to the philosophy of ecosystem management proposed in the 1990s.
2. Conservation through wise use was a public responsibility and a public trust.
3. Science was recognized as a key tool for discharging the responsibility of public trust.

The wildlife profession has been dealing with these issues for over a century. We still struggle with a question asked by Leopold (1933:19) decades ago: "How shall we conserve wild life without evicting ourselves?" To address this question, the task of science is " ... not only to furnish biological facts, but also to build on them a new technique by which the altruistic idea of conservation can be made a practical reality" (Leopold 1933:20). This vision combining science and management has been the major philosophy of The Wildlife Society, the society for professional wildlife biologists since its inception in 1937 (see Chapter 1).

The Code of Ethics for The Wildlife Society states that each member should " ... recognize research and scientific management of wildlife and their environments as primary goals...." With such a broad goal, how can research and management be at odds, especially since our founding fathers instilled both into the profession? Perhaps the discussion should begin with the establishment of clear definitions of research and management. Philosophically, one could argue that management includes research, biology, ecology, and conservation, related to the scientific management of wildlife and its habitats; without a scientific basis, management would not be effective. Moreover, until data were generated, there was no scientific management; the two go hand in hand and cannot be separated. Definitions related to the management of natural resources dating back to the Roosevelt doctrine include science as an integral part of conservation. They cannot be divorced. This is especially true as anthropogenic activities continue to cause unnaturally high rates of species extinction and habitat loss (Vitousek, Mooney, Lubchenco, and Melillo 1997; Czech et al. 2005). Unfortunately, research often is time consuming, underfunded, and short-termed, resulting in many management decisions being made without as much information as managers would like. Decisions have to be made, however, and politics dictate that they will often be made with or without all the data that could be acquired if there were more time, more money, more personnel, and more cooperation among relevant stakeholders.

In this chapter, I consider wildlife research as the studious inquiry or examination through experimentation to discover facts to answer questions. It depends on carefully designed experiments, comparisons, and models. Sound wildlife management is built on solid scientific investigation that produces objective and relevant information. Thus, research and management go hand in hand as wildlife research is the accumulation of information and wildlife management is the application of the data to meet objectives (i.e., to make populations increase, decrease, harvest for continuing yield, or only monitoring; Caughley 1977). Obtaining the data (e.g., research) often takes considerable time and objectives (e.g., management) often have to be met sooner than the research can be completed. As a result, management is forced to proceed without all of the needed data. Even when the data are available, they will not always be applied to management because of economic conflicts: the cost of applying the data, conflicts in land use, or other economic factors. Other reasons research and management can be at odds include research being conducted with a narrow focus or without considering management objectives, limited exposure to human dimensions by researchers, managers unfamiliar with relevant research, limited education tying research to conservation issues, and a host of other limitations. There will always be management actions that occur without all the desired information but these situations should not be common if wildlife is to be managed scientifically. The practice of management is a test to live in harmony with wildlife and each other. "[T]wenty centuries of 'progress' have brought the average citizen a vote, a national anthem, a Ford, a bank account, and a high opinion of himself, but not the capacity to live in high density without befouling and denuding his environment, nor a test of whether he is civilized. The practice of game management may be one of the means of developing a culture which will meet this test" (Leopold 1933:423). Wildlifers need to strengthen the relationship between research and management to ensure the test is passed.

Unfortunately, the disconnect between research and management is common in many disciplines including conservation biology, ecology, ecosystem management,

environmental psychology, landscape ecology, management and organizational science, and restoration ecology (Knight et al. 2008), many of which are directly related
to wildlife management. Leopold (1933:182) outlined " … three essential steps in
diagnosis of game productivity:

1. Visualizing the mechanism as it is, and as it should be.
2. Making an intelligent guess as to what is wrong.
3. Testing whether the guess is correct without too heavy a risk of time, funds,
   or damage.

If the test verifies the diagnosis, its findings may then be applied on a larger scale.

This simple diagnosis has been applied to many management projects beyond
game productivity (Krausman 2002), but the third essential step is often ignored.
Monitoring and following up research projects are often considered a luxury that
many agencies cannot afford, and some organizations do not even have regular and
systematic mechanisms to monitor the success or failure of their actions (Ehrenfeld
2000). The wildlife profession is built upon the triad of a solid understanding of
the wildlife, its habitat, and the human dimensions that influence both (Giles 1978).
Fortunately, the human dimension is being recognized as a key component of the
triad and is emphasizing the importance of incorporating the public into management decisions and the evaluation of progress along the wildlife management front.

Our wildlife resources are too valuable for their management to be disrupted by
unnecessary conflicts among wildlifers. The resources that have contributed to habitat loss and species extinction caused by anthropogenic actions (Czech et al. 2005)
are significantly greater than the resources that have been allocated to reverse the
trend (James, Gaston, and Balmford 1999, Knight et al. 2008). Part of the problem
lies in not understanding the conflicts between research and management.

My objectives in this chapter are to address the various reasons that exist for
the conflicts between wildlife research and management and to present solutions to
minimize the hiatus.

TOP CAUSES SEPARATING RESEARCH AND MANAGEMENT

Anyone who has been involved in research or management likely has a list of topics
that have led to conflicts between them. A good portion of this book will examine
how the dichotomy in views is expressed in universities, at the state and federal
levels, on private lands, and within similar professional societies, followed with
numerous examples. Although individuals have different experiences in relation to
research and management, I suspect that most could find the reasons they consider
responsible for the conflicts between wildlife research and wildlife management in
the following list.

1. *Communication or lack thereof.* I think this is one of the most important
   issues leading to the disconnect between wildlife management and wildlife
   research. Too often researchers do not talk to managers about their research
   (or even know the managers in the area in which they are working) and

managers do not let researchers know what type of information is necessary for them to be more efficient in the management of natural resources (Finewood and Porter 2010). This lack of communication also extends to the public. Wildlife in North America belongs to the public and they are ultimately the main stakeholders that will dictate how resources are managed. Therefore, it would behoove both researchers and managers to have excellent communication with the stakeholders that their efforts will influence. Leopold (1933) emphasized the importance of incorporating the public into management decisions but only in the past few decades have human dimensions been recognized as a critical aspect of wildlife management (Decker, Brown, and Siemer 2001). Since the 1980s, there has been an infusion of research to better understand the views, attitudes, and behaviors of different stakeholders so management can be more effective and meet the needs of the public (McCleery, Ditton, Sell, and Lopez 2006). With better communication skills on the part of everyone involved in the management of our natural resources, many of the other problems identified herein would disappear or be minimized. Making working and communicating with people a priority in research and management will greatly enhance the conservation of our nation's resources (Bonar 2007).

2. *The narrow focus of many universities.* Too often, students are limited to a few courses during college other than their major and few are exposed to classes that enhance their environmental literacy (Moody and Hartel 2007). This has led to a serious deterioration of graduating students knowledgeable about natural resources. Universities are not producing environmentally literate students (National Wildlife Federation 2001). Even students in natural resource programs are seeking change (e.g., building a foundation for critical thinking in applying statistics; Butcher et al. 2007) so their work can be more relevant. To meet the challenges facing wildlife professionals, universities will need to embrace change (Brown and Nielson 2000) and recognize that the wildlife profession is one of lifelong learning (Matter and Steidl 2000). Without constant modifications and acquisition of new data by both researchers and managers, neither will be good bedfellows.

3. *Many universities do not promote cooperation between research and management.* "The future of our wildlife resources is tied directly to solid education, both in and out of the classroom, involving wildlife, their habitats, and all of the anthropogenic forces that threaten their future" (Krausman 2000:491). Some have described a shift in universities away from traditional wildlife and fisheries management toward basic ecology and conservation (Scalet 2007). Causes for the shift within universities include the expansion of research that avoids linkages to management and pressure from administrators to obtain funds that pay more overhead than the traditional funding sources (e.g., state and federal natural resource agencies). In some cases, wildlife and fisheries programs have gone from being state-supported to state-assisted to only state-located (Scalet 2007). Other causes for the shift in universities are the impression that studying ecology and conservation is more prestigious than management, and the holistic approach that is often

taken leaves management out of the mix (Bleich and Oehler 2000). This can lead to shifts in faculty views that eventually cause rifts within programs. There are also fewer faculty and students who are involved with fishing and hunting, and there are fewer students entering programs with backgrounds in wildlife and fisheries. In addition, there is pressure from the legislature for programs to show economic returns on their investment resulting in fewer interactions with management and a shift from organismal biology toward the cellular and molecular level. Finally, many universities are combining wildlife programs and fisheries programs with other natural resource programs, which can lead to an identity crisis (Scalet 2007).

Similar issues within state and federal agencies have occurred over the past three decades, and there is less money from agencies for research into management, fewer personnel to conduct the research, and with the increased costs of doing research at universities the problem is exacerbated (Scalet 2007). In addition, most state and federal agencies have not embraced the certification programs of The Wildlife Society or the American Fisheries Society. Both of these programs require management-oriented class work and by ignoring them, agencies are eroding a tie that binds them to universities. Leadership also needs to be stronger at all levels if management is to be scientifically based (Scalet 2007).

Cooperation between research in universities and management within agencies (e.g., internships) is critical for college students majoring in fisheries and wildlife. Their professional careers are shaped, in part, by their experiences in universities and colleges, and if the importance of management is not instilled during their university years, it is unlikely to be considered important in future years.

4. *There are different missions for research and management within state and federal agencies and universities.* Natural resource management is complex and simply trying to describe a wildlife professional can be complicated because of the myriad involvement of people, animals, and habitat. Likewise, laws dictate what agencies can and cannot do, and compromise is a common companion of management when numerous agencies are involved in planning research and management. Even the establishment of clear objectives can be daunting as different agencies may have different objectives, which conflict with others. Conducting research or management on a wildlife refuge that is funded by the U.S. Fish and Wildlife Service may sound like a straightforward proposition, but it rapidly becomes complicated as researchers and managers have to address concerns of state agencies, adjoining federal agencies, Native American issues, local sportsmen organizations, university regulations, local "Friends of ..." groups, and other stakeholders. Neither research nor management is conducted in a vacuum, and opposing views can often widen the chasm between management and research.

5. Researchers often work on issues not relevant to management and when they do, the guidelines that are recommended for management are not feasible financially, logistically, or socially. Some published papers dealing with

conservation assessments are translated into conservation action (Knight et al. 2008). When they are, researchers bemoan the fact that there is a lack of representativeness of study areas, try to improve the efficiency of theory, and experimentally test data, but are not often doing conservation (Knight et al. 2006, 2008). This problem is compounded further as the activities of many conservation organizations are directed by in-house documents instead of peer-reviewed literature (Knight et al. 2008).

6. *Competition for the land.* As society expands in numbers and subsequently increases its use of resources and wildlife habitat, there are multiple demands for the available habitat for fish and wildlife that have to be managed. Conflicts can be moderate such as deciding how best to use rangelands (e.g., for livestock, wildlife and fisheries versus hunting leases, photography, and day trips; Rios, Thompson, and Hellickson 2007), to the removal of forestry projects due to political and administrative issues (Davis 2009), use of coastlines (Bhat and Stamatiades 2003), and altering habitats for oil and gas development and other anthropogenic desires. These conflicts often give the impression that scientists and managers do not get along but others have proposed that the disconnect stems not from different ecological views but from economic development pressures (Finewood and Porter 2010) that are not often addressed.

7. *Funding.* Funding influences the system and in natural resources, financial resources are especially limited. There is not enough from the public, state agencies, or federal agencies (Brown and Nielsen 2000). As a result, researchers often view themselves as academic beggars following the money for their research, which often moves them away from management and contributes to the disconnect (Scalet 2007).

8. *Unrealistic administrative policies.* Examine the expectations of any wildlife professional whether in management or research. Very few have the luxury of concentrating on a single topic. For example, university professors at land grant universities are evaluated (and retained or fired) based on their teaching, research, and service, all of which take considerable time. Preparing lectures and keeping up with the literature can be a full-time job but obtaining grants and publishing the results of research is also a requirement. On top of teaching and research, faculty are expected to " ... serve on departmental, college, and university committees; be active in their professional societies; write popular articles; advise student clubs; get involved in distance and continuing education; referee journal articles; attend recruiting efforts; judge local science fairs; give tours of the building; and do their own typing and copying" (Brown and Nielsen 2000:499–500). For these contributions faculty receive modest salaries, limited start up funds, and limited sustained financial support. It is no wonder that only about half the faculty that start in tenure-track positions are successful (Brown and Nielsen 2000). Different, but similar, demands await those that seek a career in management, as they have to " ... survive in a world governed by economics, sociology, and politics ... " (Brown and Nielsen 2000:496). It is not surprising that there is limited time to better mesh research and management.

CLOSING THE GAP BETWEEN RESEARCH AND MANAGEMENT

The scientific management of wildlife and fisheries resources demands that research-ers and managers work together so the discrepancies that divide them are minimized. Some recommendations for managers and researchers to use in closing the gap were provided by Knight et al. (2008).

1. *Recognize that this is a real gap.* Once both groups realize there is a gap, then we can take action to minimize the hiatus when possible. It is impor-tant that research is useful and used in the management of the nation's resources. This is just as important now as it has ever been as wildlifers are more conscious of how their actions influence other species. This approach, preached by Leopold, has been " ... reawakened with the resurgence of ecosystem management" (Krausman 2000:491).

2. *Research and management should work together to formulate research questions and management scenarios that will benefit wildlife and soci-ety.* Furthermore, a joint approach would improve management of nongame wildlife and enhance the emphasis that is being placed on biodiversity. One of the best and most efficient models for combining wildlife research and management is the U.S. Geological Survey Cooperative Research Unit Program. The Unit Program operates throughout the United States and cooperates with land-grant universities, state fish and game or conserva-tion agencies, the U.S. Geological Survey, and the Wildlife Management Institute to " ... facilitate cooperation between the Federal Government, colleges and universities, and private organizations for cooperative unit programs of research and education relating to fish and wildlife and for other purposes" (Public Law 86-686). One of the main benefits of the pro-gram is the connection of wildlife management agencies to the universities that may not otherwise have been available (Bissonette et al. 2000). "Units are conduits to state and federal funding for research projects conducted by university faculty and students. The ... program is well positioned to edu-cate a multitalented, ethnically diverse cadre of graduate students who will be prepared not only for their first professional job but also for their career by having been instilled with a desire for life-long professional accomplish-ment" (Bissonette et al. 2000:534). By instilling the importance of coop-eration between researchers and managers in the educational process, the distinctions between research and management naturally diminish. Much of the research accomplished by students and faculty in wildlife programs is orientated to assist management, and it is beneficial to have managers from the state agency serve on graduate committees when possible. In a similar vein, faculty and agencies should encourage students to attend state wildlife commission meetings to appreciate issues facing managers as they deal with wildlife, habitat, and multiple stakeholders.

   Other potential avenues of cooperation between researchers and state wildlife and fisheries are open. For example, state agencies could house a wildlife biologist within the university and have that person work on a graduate degree, give guest lectures, interact with students and student organizations, and serve as another link between the agency and university. Once the individual graduates, he or she would be replaced with another agency person. Researchers would also benefit from attending the planning meetings of the state fish and wildlife agency to find out what researchable issues the agency is facing, and together figure out collaboration and funding possibilities.

In a similar, albeit smaller fashion of the Cooperative Research Unit Program, the Boone and Crockett Club (the oldest conservation organization in the United States and one of the oldest in the world) initiated a Boone and Crockett Professor in Wildlife Conservation at the University of Montana in 1993. The program is based on an endowment created by the Boone and Crockett Club and the University of Montana; the professor contributes to teaching, research, and service to the wildlife profession. Because of the success of the program, other universities were interested and the program has expanded to Oregon State University, Texas A&M University, and in 2010 to Michigan State University. There are also considerations for Boone and Crockett Programs in Wildlife Conservation to expand to University of Wisconsin-Stevens Point, Oklahoma State University, University of Alaska-Fairbanks, Colorado State University, and other prominent institutions. As more Boone and Crockett professors are added across the country, there will be additional and increased emphasis on science, management, and policy.

3. *Conduct research and management on a landscape scale instead of local scales for maximum returns.* The challenges and competition that we face with land use on public lands is more representative of all Americans in the twenty-first century and not just trappers, hunters, and stockmen as in the past. The result will be a broader, and necessary, interest in landscape scale issues related to wildlife management.

4. *Incorporate the public into the beginning of the planning process for management activities.* We have to consider broader views and representation from all who are interested in wildlife and fisheries. Sheffer (1976:54) stated that the field of wildlife management has been weakened by inbreeding and needs a " ... whole-earth, web-of-life, one-people brand of philosophy."

5. *Use material from any discipline that will enhance management efforts.* This will force researchers and managers to stretch the comfortable bounds of their standard operating procedures into new and productive arenas. Since the mid-1980s with the inception of the "new" field of conservation biology, there have been discussions and debates about the differences between conservation biology, wildlife biology, and ecology (Jensen and Krausman 1993; Thomas and Pletscher 2000). Teams involved with solving natural resource problems will be enhanced by using talents from all disciplines. As we move away from single species management and work for biodiversity and some variation of ecosystem management, scientists realized that "Ecosystem management of natural resources will require a quantum increase in the use of interactive science and adaptive management. This

increase will be possible only in a 'science friendly' policy environment that will arise only if the scientific and policy communities establish a culture, protocols, and institutional arrangements that allow them to work closely and continuously together without weakening the basic purpose of either ...." (Gordon 1999: 54).

There is less discussion today about the difference between the disciplines because related disciplines have converged. "The three streams [conservation biology, wildlife biology, and ecology], to the extent that they were ever significantly different, are being forced into a single broad channel by rapidly evolving political circumstances and the need to respond to crisis" (Thomas and Pletscher 2000).

6. *Provide career incentives for academics to become involved in management.* The pitfalls of trying to meld management with academics were outlined by Brown and Nielson (2000) and unless faculty are compensated for management activities (e.g., the Cooperative Fish and Wildlife Unit program, Boone and Crockett Program in Wildlife Conservation) the gap between research and management will grow. The important aspect of management has to be recognized by university administrators.

7. *Educate students with the necessary skills to effectively manage and study natural resources.* Graduates of wildlife and fisheries programs should be effective in using science in the decision-making process that involves all wildlife and the habitats they depend on. " ... [W]e need to focus students on becoming problem solvers and advocates for wildlife in a manner that strikes a balance between the needs of those wildlife populations and the needs of people. Our goal is to produce not only clear thinkers but also decisive problem solvers and effective leaders" (Porter and Baldassarre 2000).

All of these issues add up to a different way of approaching wildlife, habitat, and other natural resource problems. Wildlife professionals need to continue to work to develop the relationship where they convey knowledge effectively to the public and policymakers as is being done by The Wildlife Society. This will require two-way communication between science and advocacy groups in a way that can be understood by the users (Schlesinger 2010).

Wildlife management needs to emulate translational medicine. The medical field significantly expanded when medical doctors realized that the results from basic research were not progressing fast enough for implementation and incorporated translational medicine. "Translational ecology should similarly connect the end-users of environmental science with the major funders of environmental research" (Schlesinger 2010). If the wildlife profession is not generating research that can be used to manage and better understand viable populations of wildlife, we will not pass the test. None of the hurdles that have created a disconnect between research and management are insurmountable; wildlife professionals generally do a good job of incorporating both in their activities. The key ingredient to solve the disconnect issue is to have more scientific management with the emphasis on management. There are no natural resources in the United States that are not managed. Even the decision not to manage something is a management decision. Wildlifers, therefore, have to ensure the management of our natural resources and the science that goes with it. Our profession and the wildlife we serve depend on it. The rewards are the resources.

ACKNOWLEDGMENTS

I appreciate the invitation from J. P. Sands, S. J. DeMaso, M. J. Schnupp, and L. A. Brennan to contribute to this book. Previous drafts were reviewed by W. B. Ballard and R. D. Brown. Time to prepare the document was provided by the Department of Ecosystem and Conservation Sciences, the Wildlife Biology Program, and the Boone and Crockett Program in Wildlife Conservation at the University of Montana, Missoula.

REFERENCES

Bhat, M., and A. Stamatiades. 2003. Institutions, incentives, and resource use conflicts: the case of Biscayne Bay, Florida. *Population and Environment* 24:485–509.

Bissonette, J. A., C. S. Loftin, D. M. Leslie, Jr., L. A. Nordstrom, and W. J. Fleming. 2000. The Cooperative Research Unit Program and wildlife education: historic development, future challenges. *Wildlife Society Bulletin* 28:534–541.

Bleich, V. C., and M. W. Oehler. 2000. Wildlife education in the United States: thoughts from agency biologists. *Wildlife Society Bulletin* 28:542–545.

Bonar, S. A. 2007. *The conservation professional's guide to working with people.* Island Press, New York.

Brown, R. D., and L. A. Nielson. 2000. Leading wildlife academic programs into the new millennium. *Wildlife Society Bulletin* 28:495–502.

Butcher, J. A., J. E. Groce, C. M. Lituma, M. C. Cocimano, Y. Sánchez-Johnson, A. J. Campomizzi, T. L. Pope, K. S. Reyna, and A. C. S. Knipps. 2007. Persistent controversy in statistical approaches in wildlife sciences: a perspective of students. *Journal of Wildlife Management* 71:2,142–2,144.

Caughley, G. 1977. *Analysis of vertebrate populations.* John Wiley & Sons, New York.

Czech, B., D. L. Trauger, J. Farley, R. Costanza, H. E. Daly, C. S. Hall, R. F. Noss, L. Krall, and P. R. Krausman. 2005. Establishing indicators for biodiversity. *Science* 308:791–792.

Davis, E. J. 2009. The rise and fall of a model forest. *BC Studies* 161:35–57.

Decker, D. J., T. L. Brown, and W. F. Siemer, Eds. 2001. *Human dimensions of wildlife management in North America.* The Wildlife Society, Bethesda, MD.

Ehrenfeld, D. 2000. War and peace and conservation biology. *Conservation Biology* 14:105–112.

Finewood, M. H., and D. E. Porter. 2010. Theorizing an alternative understanding of 'disconnects' between science and management. *Southeastern Geographer* 50:130–146.

Giles, R. H., Jr. 1978. *Wildlife management.* W. H. Freeman and Company, San Francisco, CA.

Gordon, J. C. 1999. History and assessments: punctuated nonequilibrium. In: *Bioregional assessments: science at the crossroads of management and policy.* K. N. Johnson, F. J. Swanson, J. Herring, and S. Greene, Eds. Island Press, Washington, D.C., pp. 43–54.

James, A. N., K. J. Gaston, and A. Balmford. 1999. Balancing the Earth's accounts. *Nature* 401:323–324.

Jensen, M., and P. R. Krausman. 1993. Conservation biology's literature: new wine or just a new bottle? *Wildlife Society Bulletin* 21:199–203.

Knight, A. T., R. M. Cowling, and B. M. Campbell. 2006. An operational model for implementing conservation action. *Conservation Biology* 20:408–419.

Knight, A. T., R. M. Cowling, M. Rouget, A. Balmford, A. T. Lombard, and B. M. Campbell. 2008. Knowing but not doing: selecting priority conservation areas and the research-implementation gap. *Conservation Biology* 22:610–617.

Krausman, P. R. 2000. Wildlife management in the twenty-first century: educated predictions. *Wildlife Society Bulletin* 28:490–495.

Krausman, P. R. 2002. *Introduction to wildlife management: the basics.* Prentice Hall, Upper Saddle River, NJ.

Leopold, A. 1933. *Game management.* Charles Scribner's Sons, New York.

Matter, W. J., and R. J. Steidl. 2000. University undergraduate curricula in wildlife: beyond 2000. *Wildlife Society Bulletin* 28:503–507.

McCleery, R. A., R. B. Ditton, J. Sell, and R. R. Lopez. 2006. Understanding and improving attitudinal research in wildlife sciences. *Wildlife Society Bulletin* 34:537–541.

Moody, G. L., and P. G. Hartel. 2007. Evaluating an environmental literacy requirement chosen as a method to produce environmentally literate university students. *International Journal of Sustainability in Higher Education* 8:355–370.

National Wildlife Federation. 2001. State of the campus environment—a national report card on environmental performance and sustainability in higher education. National Wildlife Federation, Reston, VA.

Porter, W. F., and G. A. Baldassarre. 2000. Future directions for the graduate curriculum in wildlife biology: building on our strengths. *Wildlife Society Bulletin* 28:508–513.

Rios, D. R., B. Thompson, and M. W. Hellickson. 2007. Application of the balanced scorecard to realize strategic management of wildlife resources. *Rangelands* 29:18–21.

Scalet, C. G. 2007. Dinosaur ramblings. *Journal of Wildlife Management* 71:1,749–1,752.

Scheffer, V. B. 1976. The future of wildlife management. *Wildlife Society Bulletin* 4:51–54.

Schlesinger, W. H. 2010. Translational ecology. *Science* 327: 609.

Thomas, J. W., and D. H. Pletscher. 2000. The convergence of ecology, conservation biology, and wildlife biology: necessary or redundant? *Wildlife Society Bulletin* 28:546–549.

Vitousek, P. M., H. A. Mooney, V. Lubchenco, and J. M. Melillo. 1997. Human domination of Earth's ecosystems. *Science* 277:494–499.

Section II

Research and Management Entities

3 Universities and the Disconnect between Wildlife Research and Management

Leonard A. Brennan
Caesar Kleberg Wildlife Research Institute
Texas A&M University-Kingsville
Kingsville, Texas

CONTENTS

> Universities pretend to measure their success by the light they throw on the world's problems, but the yardstick actually used is the number of graduates that they can get jobs for.
>
> **—Aldo Leopold (1937)**

ABSTRACT

Universities play a central role in education, research, and outreach for the wildlife profession. With respect to wildlife research, some have argued that what university faculty members do should be more practical and less academic. However, despite this opinion, the fact remains wildlife research is a signature product from research universities that have wildlife programs. There will always be a stream of new research that managers—here defined as wildlife professionals who have direct oversight of the activities on public or private properties that are intended to influence wildlife populations and their habitat—will be challenged to access, understand, and implement. The gap between research and application is not unique to the wildlife profession; the medical profession also struggles with this same disconnect. Medical researchers have identified that the characteristics of the organizations in which clinicians work are responsible for most of the barriers that prevent these practitioners from applying new research in their work. While the essential research and surveys have yet to be conducted, it is probably safe to assume that a similar situation exists in the wildlife profession; the agencies and organizations where many managers work may be creating similar barriers as those identified by our colleagues in medicine. Additionally, the research culture in university wildlife programs also contributes to this disconnect because: (1) the reward system at research universities focuses on production of research products, not necessarily the application of those products, and (2) wildlife researchers seem to get caught in analytical cultural confines that emphasize reportage of obscure statistical jargon and minutiae rather than practical results in a biological or ecological context. While universities with wildlife research programs can play a significant role in providing increased continuing education opportunities for managers, it is ultimately up to the organizations where these managers work to allow them to broaden their professional activities and perhaps encourage them to collaborate with wildlife research projects. Both researchers and managers have an obligation to stay current and active in their professional development activities and organizations as well as maintain open lines of communication with each other.

INTRODUCTION

By nearly any measure, a university education is essential for a career as a professional in wildlife science. The contemporary world of wildlife science, conservation, and management is diverse, complex, and challenging on many levels. Today's wildlife science professionals not only need to have an understanding of biology, ecology, and natural resource management, but also they must have excellent oral and written communication skills, as well as the ability to analyze and interpret complex data sets. Additionally, a high level of personal organization, the ability to manage projects, and develop and adhere to budgets, as well as skills in psychology and people management, while not typically taught in a university wildlife curriculum, can be helpful if not essential for a successful career in wildlife science.

Wildlife science, as a profession, is relatively new. Modern wildlife science, it can be argued, grew out of classical zoology and vertebrate biology programs in the early part of the twentieth century (Grinnell, Bryant, and Storer 1918; Miller 1943). We can identify the years of 1933–1937 as ground zero for the wildlife profession (see DeMaso, Chapter 1, this volume). Leopold's *Game Management* appeared in 1933 (Leopold 1933). The University of Wisconsin-Madison founded one of the first academic programs designed to educate and train wildlife professionals in 1933, and the *Journal of Wildlife Management* was established by The Wildlife Society in 1937. Thus, our profession is only about three quarters of a century young in North America, though older in Europe and Asia, at least from the perspective of applications. For example, Genghis Kahn established food plots for wildlife (Leopold 1933).

By the mid-1970s, Kadlec (1978) noted there were seventy-five wildlife programs in the United States and Canada. Today, The Wildlife Society website lists 120 university wildlife programs, an increase of 60% in a little more than three decades. This is an impressive increase by nearly any standard. Of these 120 university wildlife programs, 67 or 56% grant doctoral degrees. This is an indication that in North America, wildlife science has matured to the point where research, and training people to do research, has become a major component of our profession.

ROLE OF UNIVERSITIES IN WILDLIFE SCIENCE

The 120 university wildlife programs in North America range from institutions that offer only selected courses in aspects of wildlife ecology to universities that grant doctorate degrees. In perusing the list of programs, I see two broad types of higher education institutions that offer wildlife programs: (1) land grant universities, and (2) state and private universities and colleges that have broad offerings in liberal arts and the sciences. The historical distinctions between these two types of universities influence their current culture toward wildlife research and management.

Land grant universities were founded under the Morrill Act in 1862, with the mission to provide comprehensive programs in teaching, research, and extension, typically with an emphasis on agriculture and engineering. Although some of these programs now incorporate what some scientists might consider elements of basic research, applied research remains a key priority of the mission of these institutions, as it has been from the day they were founded. Typically, but not exclusively, land grant universities by far offer most of the doctoral programs in wildlife, and thus probably conduct most of the research.

Non-land grant state and private universities and colleges that offer wildlife programs are typically institutions that originally started as teachers' colleges, and then eventually grew to become relatively large universities. Two examples of such places that started as teachers' colleges and now have relatively large and well-established wildlife programs are Texas A&M University-Kingsville, which started as the South Texas Normal School (to train elementary and high school teachers), and Humboldt State University, which was founded as Humboldt State Normal School (again a teacher's college). At Texas A&M-Kingsville, we have been conducting wildlife research since 1982, about sixty years after it was founded in 1925. At Humboldt State, which was founded in 1913, the incorporation of wildlife research developed on a similar timeline, essentially beginning in or around the mid-1960s, a little more than half a century after it was founded. In situations like these, the result is an academic culture of research that has a much lower profile than at the land grant institutions. Nevertheless, today, both groups of universities produce high-quality wildlife research published in leading peer-reviewed journals.

RESEARCH DISCONNECT

Almost *de facto*, there is bound to be a disconnect between research and management in wildlife science. Research, when done correctly, not only will generate new knowledge and theory, it also will challenge and refute existing knowledge and theories. Thus, research from university wildlife programs is, almost by definition, likely to be a step ahead of the knowledge base of managers who are not involved with research. Given the dominant role universities have in conducting wildlife research, and the premium they put on making sure faculty develop and maintain productive research programs (getting tenure in a leading university wildlife program without doing research is virtually impossible), universities can't help but contribute to the disconnect between research and management. However, even if they can't help but contribute to such a disconnect, what can they do, if anything, to mitigate the perceived problem?

One of the criticisms leveled at universities with wildlife research programs is that students do not get adequate training for practical management and problem solving because faculty have little practical training or exposure outside of academia (Noss 1997). However, Porter and Baldassarre (2000:511) countered this argument by stating, " ... one of the primary functions of research universities (where most of the wildlife programs reside) is to conduct research. Arising from research are the innovative ideas ...." Although I will address the points of "practicality" versus what kinds of research products are generated by university wildlife programs later in this chapter, a key question is how can we get these "innovative ideas" from research to be embraced and used by managers? Clearly, there are barriers preventing this from happening, not the least of which is a time lag between when research results appear and when managers incorporate them, if at all. Both groups—researchers and managers—see themselves as leaders in wildlife conservation. However, despite the opinion of a trusted colleague who claims these problems can all be solved with a sufficient amount of money, power, and time, researchers and managers also need to work together closely, and in coordinated collaboration, to solve many of the seemingly intractable conservation problems that confront us today.

RESEARCH APPLICATION

INSTITUTIONAL BARRIERS

There is a body of literature in the field of medicine that indicates their profession is clearly struggling with what they call "getting research findings into practice" much like we are in wildlife science. About ten years ago, the *British Medical Journal* published a series of eight articles that analyzed the gap between research and practice (see Bero et al. 1998 for an example). Obviously, the wildlife profession is not alone in this dilemma. What the medical and nursing professions have done—and we in wildlife have not—is identified the barriers that are preventing practitioners from using new knowledge from research. By identifying these barriers, our colleagues in medicine also were able to identify ways to facilitate use of new research by medical practitioners (Table 3.1). This kind of background and prior work represents a huge conceptual opportunity for wildlife researchers who specialize in human dimensions.

In Table 3.1, points 1 and 2 are related to what Funk, Tornquist, and Champagne (1995) called "characteristics of the adopter" that indicated personal values, skills, and awareness of the practitioners. Points 3 through 10, which clearly were the

TABLE 3.1

Factors Responsible for the Disconnect between Clinical Research and Application in Practice in the Medical Profession

Funk et al. (1995) compiled a list of that which operated as barriers and prevented medical clinicians from using new information from research. Depending on the particular 11 topics listed below, somewhere between two-thirds ($n = 590$) and three-fourths ($n = 675$) of more than 900 medical practitioners holding clinical positions surveyed indicated the following factors were major barriers that prevented them from incorporating new information from research into their practice of medicine:

1. Lack of awareness of the research
2. Isolation from knowledgeable colleagues with whom to discuss the research
3. Insufficient authority to change procedures
4. Insufficient time to implement new ideas
5. Supervisors not cooperating with implementation
6. Administrators not allowing implementation
7. Other staff not supportive of implementation
8. Research results not generalizable to their own setting
9. Inadequate facilities for implementation
10. Insufficient time to read research
11. Statistical analyses were not understandable

It is my view that these 11 factors identified by Funk et al. (1995) are responsible for the disconnect between clinical research and practice in medicine and are also the primary factors that weaken the connection between research and management in wildlife science.

majority of the barriers, were related to "characteristics of the organization where the practitioners worked." The final barrier, point 11, indicated a "communication characteristic." At least it is satisfying, on some level, to know that the wildlife profession is not alone when it comes to issues related to statistics obfuscating research results. I will touch on this important point in more detail later in this chapter.

Assuming there might be parallels between wildlife and medicine when it comes to implementing new information from research into management and practice, does this mean that the organizations in which wildlife managers work—the vast majority who have a university education—are the primary barriers that cause the disconnect? This question certainly deserves further examination. Although the essential comparable research and surveys have yet to be conducted for the wildlife profession, it is not a stretch to imagine wildlife biologists who work for a state or federal agency, or in a private lands context for a non-governmental organization, are faced with the eleven (and perhaps even more?) barriers noted above. Palmer (Chapter 6, this volume) notes the importance of some of these cultural differences in the context of privately supported wildlife research institutes.

Additionally, the organizational culture of universities provides considerable barriers to the application of research results. The promotion and tenure system strongly influences what professors do, and do not do, when it comes to research. University tenure is an ancient tradition that rewards professors with lifetime job security after they have documented some kind of record of excellence with respect to research and teaching. It is designed to protect the academic freedom of university faculty who might, during the course of a research project, obtain data or develop a theory that runs counter to the prevailing values of a culture and its moral and financial leaders. This is the positive part of the tenure equation. The equivocal part of the tenure process is that it encourages young faculty to play relatively safe research bets and follow the grant and contract money, which are usually tied to resource agency objectives and agendas, if they are to maximize their chances of getting results published in the peer-reviewed literature. Having funding and research products linked to specific agency and donor objectives in many ways can be a good thing. After all, what better way to bridge the disconnect between research and application than to have management agencies offering research, and research funding, opportunities? I would argue, however, that barriers between research results and management applications occur when it comes time to publish results in scientific journals. This is where things often veer from being equivocal to unequivocal with respect to the disconnect between wildlife management and research. This is also where the problem of "statistical results not being understandable" comes into play. Ironically, this lack of understanding also affects the authors of papers, who are following statistical ritual rather than trying to pass along information from researchers to stakeholders.

During the 1960s, wildlife scientists became enamored with null hypothesis statistical testing (NHST). Whereas the virtues and shortcomings of NHST have been debated in huge depth and detail elsewhere (see Johnson 1999 and citations therein), what happened was that wildlife researchers became so caught up in NHST that statistics, rather than biology, became the foci of thousands of research publications. The problem was not with NHST per se, but with the way that wildlife researchers used it in a rote and unthinking manner to test trivial or even silly hypotheses that were obvious to anyone with management experience. Herman (2002:933) even went so far as to accuse wildlife scientists as having a "lust for statistics."

The challenge to NHST in wildlife science, which was instigated by Johnson (1999), did not solve this problem, however. A new statistical approach, using multimodel inference based on an Information Theoretical approach and Akiake Information Coefficients (IT-AIC), has now, for the most part, replaced NHST in a huge proportion of papers published in outlets such as the *Journal of Wildlife Management* (Guthery, Brennan, Peterson, and Lusk 2005). What it has not done, however, is enhance the alacrity with which wildlife research results are communicated to managers. This is because few researchers—or managers—really seem to understand the models generated and analyzed by the IT model selection process (cf. Reed, Bidlak, Hurt, and Getz 2011).

However, this chapter is not the appropriate forum to discuss the merits and limitations of IT-AIC in wildlife research. This debate rages on elsewhere (Stephens, Buskirk, and del Rio 2007; Arnold 2010). What has happened, whether proponents of IT-AIC want to admit it or not, is that the Ps, Fs, ts, and dfs of NHST have been replaced with AIC weights, corrected AIC values, and other coefficients that, in far too many cases, obscure the real and meaningful relationships of a data set collected for a particular research question. Today, unfortunately, it is not uncommon to see tables of basic descriptive statistics such as means, 95% confidence intervals, etc. eliminated from published papers, and replaced with obscure and sometimes dimensionless IT-AIC components. Wildlife researchers have purposefully created a communication barrier that has made it virtually impossible for managers to understand their results.

This issue also folds back to the problem between publication and tenure noted previously. I have personally heard of numerous instances where journal reviewers and editors have demanded that authors use IT-AIC analyses in order to get revised versions of their papers accepted for publication. Thus, the pressure on a tenure-track assistant or associate professor to revise a paper accordingly, or just use IT-AIC in the first place, must be enormous. A full professor with tenure can tell these editorial scofflaws to take a hike. An untenured assistant professor, or an associate professor still climbing the promotion ladder, can't. Therefore, researchers often put their results in print in a way that further exacerbates the disconnect between science and management. It will take nothing less than a sea change in wildlife science publishing to solve this problem.

The challenges and opportunities here are that the organizations and cultures in which researchers and managers work seem to be the primary culprits when it comes to forming the barriers that prevent implementation of research results into management practices. How can wildlife professionals, both at universities and in resource agencies and other organizations, overcome this problem and facilitate the use of new knowledge, should any of it appear, gained from research to improve management? Furthermore, how can universities play a role to help break down these barriers and facilitate a new direction that will more effectively allow managers to incorporate research findings in their work?

Facilitating Application

Universities with wildlife research programs can play a major, but not unilateral, role in facilitating the application of research by managers. After all, the Cooperative Extension Service branch of the land grant university system was designed to achieve this particular goal, which it does, but could probably do better.

With undergraduate students, universities need to instill, to the extent that they can, a motivation for students to become lifelong learners. Students, for whatever reason, are far too focused on "learning it for the test" rather than "learning it for life." Graduate students often suffer from this problem as well as a similar syndrome, which is something I call "thesis or dissertation myopia," or only learning or doing something because it is related to their thesis or dissertation, rather than for general intellectual curiosity.

For better or worse, some of the particular issues of various approaches to organizing university wildlife curricula have largely been laid to rest by the certification program developed by The Wildlife Society. The program requires specific hours of various coursework and classes by an individual if he or she is to be "certified" as a Wildlife Biologist by The Wildlife Society. In order for a wildlife professional to remain certified, continuing education is required at five-year intervals after certification. In general, this is probably a good thing, but for some reason—probably because certification is considered optional rather than mandatory by most employers—many practicing wildlife professionals, including myself, are not certified. I considered getting a Ph.D. adequate "certification" as a wildlife scientist, although I am sure more than a few of my respected and esteemed colleagues would disagree. Moreover, the issue of whether "certification" as a professional wildlife biologist should be based on some kind of standardized exam, rather than just a list of coursework, is hardly even considered a debatable topic in the wildlife profession these days.

So then, what this leaves us with is a situation where many practicing wildlife managers, whoever they are, must be entirely self-motivated to become lifelong learners. Or, continuing education issues related to certification notwithstanding, I would argue that it is incumbent on the various agencies and organizations where professional wildlife managers work to provide an environment where continued education, attending conferences and workshops, and briefing co-workers and supervisors on the merits and limitations of new findings from research might be a way to help close the gap between research and management.

MANAGERS INVOLVED WITH RESEARCH

In Texas, we have an enviable situation where a number of wildlife managers are directly involved as collaborators on research projects. Although Texas is largely a private lands state, there are two large state public wildlife management areas where involvement with research is a major part of their operating philosophy. Additionally, as researchers, we enjoy the collaboration with managers on many of the large, private ranches located throughout South Texas. Our university wildlife program, even though it does not have an explicit extension mission, regularly hosts workshops, short courses, and round table discussions for brainstorming research planning, all with and for managers. We also produce a steady stream of extension-promotional materials in both print and electronic format that are aimed directly at practicing wildlife managers. Most of the managers who participate in these events or subscribe to our products work for organizations that have either eliminated or drastically reduced the institutional barriers that prevent research from being applied as noted previously. This is similar to the situation described by Palmer in Chapter 6 (this volume).

The disconnect between wildlife management and research is clearly a phenomenon that has invaded the organizations and professional cultures in which we work. Unfortunately, the culture of organizations and agencies involved with wildlife research and management may be responsible for maintaining and perhaps even exacerbating barriers that foster this disconnect. Fortunately, the culture of organizations and agencies can change, and I hope will change in a way that helps facilitate closing the gap between research and management rather than perpetuating it. The first step toward solving a problem is admitting that it exists. Discussing the factors that contribute to and maintain this disconnect should be a prominent component of undergraduate and graduate level classes in all university wildlife programs.

ACKNOWLEDGMENTS

This chapter benefited greatly from review and editorial comments by B. M. Ballard, S. J. DeMaso, F. S. Guthery, W. P. Kuvlesky, Jr., J. P. Sands, M. J. Schnupp, and C. K. Williams. However, any errors in accuracy, interpretation, or logic are strictly my own. This is Caesar Kleberg Wildlife Research Institute Publication 11-128, with support provided by the Richard M. Kleberg, Jr. Center for Quail Research and the C. C. Winn Endowed Chair.

REFERENCES

Arnold, T. E. 2010. Uninformative parameters and model selection using Akaike's Information Criterion. *Journal of Wildlife Management* 1174–1178.

Bero, L. A., R. Grilli, J. M. Grinshaw, E. Harvey, A. D. Oxman, and M. A. Thompson. 1998. Closing the gap between research and practice: an overview of systematic reviews of interventions to promote the implementation of research findings. *British Medical Journal* 317:465–468.

Funk, S. D., E. M. Tornquist, and M. T. Champagne. 1995. Barriers and facilitators of research utilization: an integrative review. *Nursing Clinics of North America* 30:395–407.

Grinnell, J., H. Bryant, and T. I. Storer. 1918. *The game birds of California*. University of California Press, Berkeley, CA.

Guthery, F. S., L. A. Brennan, M. J. Peterson, and J. J. Lusk. 2005. Information theory and wildlife science: critique and viewpoint. *Journal of Wildlife Management* 69:457–465.

Herman, S. G. 2002. Wildlife biology and natural history: time for a reunion. *Journal of Wildlife Management* 66:933–946.

Johnson, D. H. 1999. The insignificance of statistical significance testing. *Journal of Wildlife Management* 63:763–722.

Kadlec, J. A. 1978. Wildlife training and research. In: H. P. Brokaw, Ed. *Wildlife in America*. Council on Environmental Quality, Washington, D.C., pp. 485–497.

Leopold, A. 1933. *Game management*. Charles Scribner's Sons, New York.

Leopold, A. 1937. The research program. *Transactions of the North American Wildlife Conference* 2:104–107.

Miller, A. H. 1943. *Joseph Grinnell's philosophy of nature*. University of California Press, Berkeley, CA.

Noss, R. 1997. The failure of universities to produce conservation biologists. *Conservation Biology* 22:267–1269.

Porter, W. F., and G. A. Baldassarre. 2000. Future directions for the graduate curriculum in wildlife biology: building on our strengths. *Wildlife Society Bulletin* 28:508–513.

Reed, S. E., A. L. Bidlak, A. Hurt, and W. M. Getz. 2011. Detection distance and environmental factors in conservation detection dog surveys. *Journal of Wildlife Management* 75:243–251.

Stephens, P. A., S. W. Buskirk, and C. M. del Rio. 2007. Inference in ecology and evolution. *Trends in Ecology and Evolution* 22:192–197.

4 Southeast State Wildlife Agencies' Research Priorities and Constraints

John Bowers
Georgia Wildlife Resources Division
Social Circle, Georgia

Chris Baumann
Georgia Wildlife Resources Division
Fitzgerald, Georgia

Reggie Thackston
Georgia Wildlife Resources Division
Forsyth, Georgia

CONTENTS

Actual cases of mis-appraisal of conservation questions … can be found by the dozen in any state, together with the historical evidence to prove them. More difficult is the planting of a conviction that research can gradually dissolve this ignorance, and that local institutions can and should undertake such research.

—Aldo Leopold (1933:386)

ABSTRACT

Wildlife management and the wildlife profession are based in scientific research. Development of the Cooperative Wildlife Research Unit program in 1935 and passage of the Pittman-Robertson Act in 1937 enabled state wildlife agencies to formalize science-based management programs that have resulted in numerous wildlife restoration and management successes. However, many wildlife species and ecosystems are in need of efficacious management. Increasing demands on state wildlife agencies for public services, coupled with budget and labor reductions, may reduce capacity and priority for research-based management. To evaluate this issue, we electronically surveyed the wildlife section chiefs of sixteen member states of the Southeastern Association of Fish and Wildlife Agencies relative to the following topics: (1) perceived importance and priority of research in achieving their management goals and objectives; (2) history, current status, and anticipated future organizational structure and allocation of labor and funding for research; and (3) perception of the role and effectiveness of universities, Cooperative Wildlife Research Units, and non-governmental organizations in meeting their agency's research needs. The results and implications of this survey are presented and discussed.

INTRODUCTION

Scientific research as a basis for wildlife management and the wildlife profession began developing in the early 1900s. Aldo Leopold, considered by many to be the "Father of Wildlife Management," chaired the committee that developed the first American Game Policy in 1930, and in 1933 at the University of Wisconsin was appointed Chair of one of the nation's first university wildlife programs (Meine 1988). In that same year his seminal book, *Game Management* (Leopold 1933), was published. Leopold recognized and championed the need to move beyond the regulatory protection of wildlife to the science-based management of wildlife species, habitats, and landscapes. He recognized scientific research as a prerequisite for effective conservation and intimated throughout his career that management must address not only individual species but also whole systems and landscapes, and should be on a continuum to account for changing environments. Leopold's efforts, in combination with those of many others, ultimately gave rise to the formation of the Cooperative Fish & Wildlife Research Unit program in 1935, and the development and passage of the Pittman–Robertson Act and formation of The Wildlife Society (TWS) in 1937. These were key developments that enabled the fledging state wildlife agencies to ultimately develop a science-based approach to wildlife restoration and management, which became the "backbone" of the agencies' management (U.S. Department of Interior Fish and Wildlife Service 1987). This research-based approach to wildlife management has resulted in many wildlife restoration successes. For example, southeastern state wildlife agencies have worked with private landowners, state, federal, universities, and non-governmental conservation partners to restore populations of eastern wild turkey (*Meleagris gallapovo silvestris*), white-tailed deer (*Odocoileus virginianus*), wood ducks (*Aix sponsa*), black bear (*Urus americanus*), bald eagles (*Haliaeetus leucocephalus*), and others. However, as grand as these successes have been, there are still many species and even entire ecosystems in peril. For example, the northern bobwhite (*Colinus virginianus*) and a number of other grassland obligate wildlife species are in severe decline and are in desperate need of effective conservation (Brennan and Kuvlesky 2005).

A cursory look at southeastern state agency wildlife research reveals that up through the 1970s efforts focused in great part on game species, while from the 1980s to the present emphasis shifted more toward nongame species, landscapes, and ecosystems (Figure 4.1). Jim Miller, TWS 2007 Aldo Leopold Award Winner, succinctly summarized the growth and development of wildlife science and the wildlife profession (The Wildlife Society 2007). During the last fifty years, the southeastern human population has increased by about 80% (U.S. Census Bureau 2010) while southeastern state wildlife agency budgets and labor have either decreased or at best remained stable. For example, Georgia's human population was estimated at 7.5 million people in 1997 (Population Reference Bureau 2011). At that time, Georgia's wildlife agency employed professional wildlife staff consisting of 345 personnel (i.e., wildlife biologists, wildlife technicians, conservation rangers). These personnel were responsible for ensuring scientific conservation and management across eighty-two wildlife management areas including 202,915 acres of state-owned lands, and ensuring compliance with the Game and Fish laws and regulations. By 2010, the state's

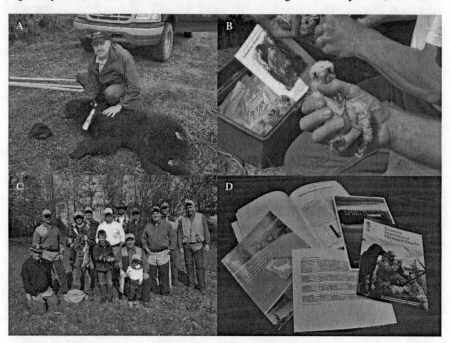

FIGURE 4.1 Historically, southeastern state wildlife agencies conducted research primarily on game species (A). Through time, research emphasis shifted to nongame, rare, threatened, or endangered species (B), and southeastern state wildlife agencies anticipate increased research emphasis in the future on hunter recruitment and retention (C), and (D) public opinion surveys. Photos courtesy of the Georgia Wildlife Resources Division.

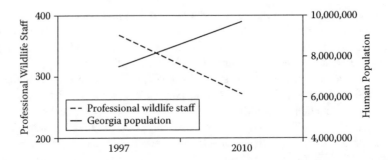

FIGURE 4.2 Comparative changes from 1997 to 2010 between Georgia's wildlife agency professional staff and human population.

human population was 9.7 million (U.S. Census 2010) and the number of wildlife management areas had increased to 100 including 287,521 acres of state-owned land. Conversely, due to budget reductions Georgia's wildlife agency's professional wildlife staff had declined to 277 during this time (Figure 4.2).

The expanding human population has placed increased demands on wildlife agencies for public services (e.g., nuisance wildlife abatement), which may in turn reduce capacity for conducting adaptive research as a basis for effective management, or redirect limited resources away from agencies' core responsibilities and higher priority conservation needs. This problem is further exacerbated by factors like climate change, invasive species, increased human demands for food and fiber, urban sprawl, and habitat fragmentation, which affect wildlife species' response to traditional management practices. Additionally, declines in hunter numbers along with new programs and initiatives (e.g., hunter recruitment and retention) may further reduce research funding and limit agency outsourcing of research.

As wildlife agencies struggle to meet ever increasing demands in the face of reduced budgets and labor, the question arises as to whether state wildlife agencies are sufficiently relying on the science-based approach that was vital to past conservation successes. To evaluate this question, we electronically surveyed the wildlife section chiefs of sixteen member states of the Southeastern Association of Fish and Wildlife Agencies (SEAFWA, hereafter Agencies, Figure 4.3.) relative to the following topics: (1) perceived importance and priority of research in achieving their management goals and objectives; (2) history, current status, and anticipated future organizational structure and allocation of manpower and funding for research; and (3) perception of the role and effectiveness of universities, Cooperative Wildlife Research Units, and non-governmental organizations in meeting their agency's research needs. The remainder of this chapter focuses on the results and implications of this survey.

A SURVEY OF SOUTHEASTERN WILDLIFE AGENCY SECTION CHIEFS

The survey was electronically distributed via "QuestionPro" (2011) to the wildlife section chiefs of the respective state agencies during June 2011. In responding to the

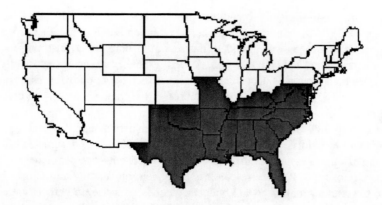

FIGURE 4.3 Highlighted states represent Southeastern Association of Fish and Wildlife Agencies member states.

survey questions or statements, agencies were instructed to interpret "wildlife" as referring only to vertebrate land species and not to include fish. Additionally, agencies were instructed to consider "research" as excluding annual or recurrent surveys. Unless otherwise annotated, all value ratings were determined using a Likert scale (e.g., importance, likelihood, satisfaction) that ranged from 1 (i.e., very important) to 5 (i.e., very unimportant).

A total of nine (56%) agencies responded to all, or a portion, of the survey questions. All respondents rated wildlife research as either very important (60%) or important (40%) as a foundation for scientific wildlife management. However, 60% indicated that wildlife research was not an explicit agency function within their mission statement.

Seven (78%) agencies reported allocating 6 to 10% of their annual operating budget on wildlife research. Two (22%) agencies indicated that funding exceeded this level and reported research expenditures in the 11 to 20% range. About two-thirds of research expenditures (Figure 4.4) were for game, nongame, and rare, threatened, and endangered (RTE) species, with an approximately equal amount spent on each.

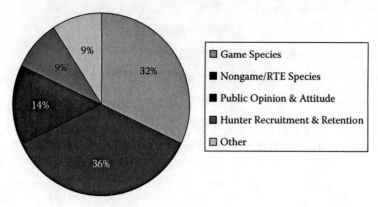

FIGURE 4.4 Southeastern State Wildlife Agency expenditures by research category.

The remaining one-third of expenditures was directed at research on public attitudes/opinions (PAO), hunter recruitment and retention (HRR), and other topics.

A series of questions was provided to assess and compare the past, present, and future importance of wildlife research. A total of 56% of agencies historically had wildlife research sections, but currently wildlife research sections are present in only 44%. However, 55% of the respondents reported that they were likely to create or maintain a research section in the future.

Historically, 78% of agencies had personnel primarily assigned (spending >50% of their time) to wildlife research but currently only 56% have designated wildlife research personnel. Agencies currently have four to twenty-one personnel assigned to research positions. Agency research is dependent on funding and is prioritized relative to statutory mandates and budgetary needs. Sixty-seven percent anticipate future funding for research to remain stable, 22% expect funding to increase, and 11% anticipate funding to decrease.

Agencies were asked to rate the importance of wildlife research for informing decision-making processes in the past, present, and future, relative to several categories. These categories were: development of hunting regulations; meeting management objectives for game, nongame, and RTE species; providing private landowner technical assistance; guiding public lands management decisions; PAO on wildlife management issues, and HRR. Wildlife research ratings were similar, ranging from very important to important for most categories across the past, current, and future periods (Figure 4.5). Exceptions were PAO and HRR, which were rated as historically being neither important nor unimportant but were viewed to be of increased importance currently and in the future.

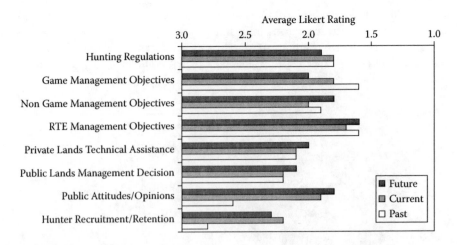

FIGURE 4.5 Relative importance of past, current, and future research in informing southeastern state wildlife agency decision-making processes for eight categories of decision making.

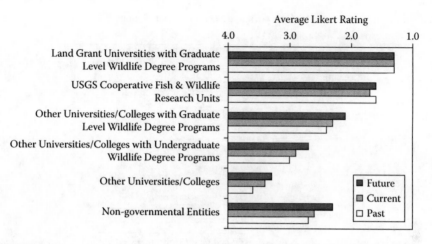

FIGURE 4.6 Importance of external research institutions in meeting southeastern state agencies past, current, and future research needs.

Agency research was conducted internally by agency personnel or externally through various research institutions. Agencies were asked to rate the importance of past, current, and future agency-conducted research in meeting management goals and objectives. Across all time categories, research conducted by state agency staff was rated between important to very important.

Agencies were asked to rate the importance of external research through the following entities to meet their past, current, and future research needs: land grant universities with graduate wildlife degree programs (LGU); United States Geological Service Cooperative Fish and Wildlife Research Units (CFWRU); other universities/colleges with graduate wildlife degree programs (OGWD); universities/colleges with undergraduate wildlife degree programs (UWD); other universities/colleges (OUC); and non-governmental entities (e.g., conservation groups, public/private foundations, NGE, Figure 4.6). The LGU and CFWRU received the highest importance ratings, respectively, across all periods. The OGWD were rated important in meeting past agency research needs and were rated of increased importance in meeting future research needs. The UWD, OGWD, and OUC were rated as neither important nor unimportant in meeting past, current, and future agency research needs. The OUC were rated as unimportant in the past but were viewed as increasing in importance into the future. The NGE rating was neither important nor unimportant in meeting past agency research needs but was rated as increasing in importance in meeting current and future agency research needs.

Agencies were asked to rate their level of satisfaction with each of the external research institutions in meeting the state agency's wildlife research needs. Overall, respondents indicated they were very satisfied that LGU and CFWRU were meeting their wildlife research needs. Satisfaction levels for OGWD and NGE overall were between moderately satisfied to very satisfied. The UWD and OUC had the lowest ratings with agencies' responses ranging from slightly to moderately satisfied.

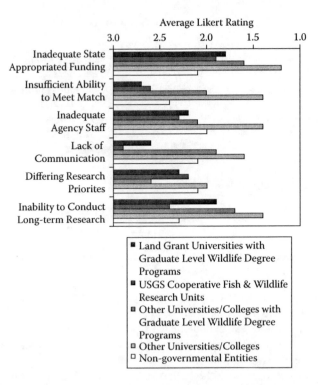

FIGURE 4.7 Relative importance of certain factors in limiting collaboration between state wildlife agencies and research institutions (*n* = 9; importance rating increases as numerical value decreases).

Agencies were asked to rate the importance of certain factors in limiting collaboration with research institutions. These factors included lack of state appropriated funding, lack of agency funding to meet funding match requirements, lack of agency staff, lack of communication between the agency and the research institution, differing research priorities; inability of institutions to conduct long-term research, and other factors (Figure 4.7). Generally, agencies indicated that inadequate funding, lack of staff, and the inability of research institutions to conduct long-term research were important to very important factors in limiting collaboration with all research institutions. The inability of agencies to meet match requirements and lack of communication were important to very important factors limiting collaboration with OGWD, UWD, OUC, and NGE; and differing research priorities was an important to very important factor with all institutions except LGU (Figure 4.7).

Additional respondent comments identified the following factors that limited collaboration with research institutions: (1) reduced in-state research as a result of increased emphasis on international relationships and associated projects, (2) increased governmental paperwork requirements and reporting processes, (3) inordinate focus by external institutions on theoretical problems as opposed to

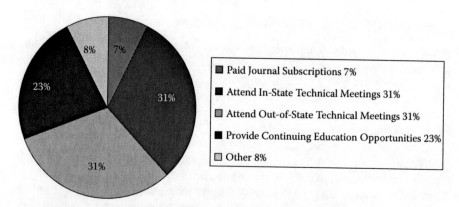

FIGURE 4.8   Actions employed by southeastern state wildlife agencies to ensure full-time wildlife biologists and managers stay knowledgeable of current wildlife research.

management applicable research, and (4) high overhead fees and match requirements by external institutions.

Agencies were asked to rate the importance of ensuring that full-time agency wildlife biologists and managers stay knowledgeable of current wildlife research. All respondents indicated that this was important with 56% indicating this was very important. Respondents also were asked to rate their agencies' effectiveness at ensuring wildlife biologists and managers stay knowledgeable of current wildlife research. Overall, agencies effectiveness rated between slightly effective (44%) to moderately effective (56%). Agencies were presented with a series of options that may be used to facilitate continuing education for wildlife biologists and managers relative to current research findings. These options included paid subscriptions to scientific journals, providing opportunities for biological staff to attend technical meetings in-state, providing opportunities for biological staff to attend technical meetings out-of-state, providing continuing education coursework for staff, and other (Figure 4.8). The two most selected options were providing opportunities for biological staff to attend technical meetings in-state and out-of-state. However, additional respondent comments revealed that such opportunities are too limited and constrained, and attributed some of this to budgetary restrictions, although these present great opportunities to keep staff informed and knowledgeable. The least selected option was paid subscriptions to scientific journals.

Finally, agencies were asked to list their five most important wildlife research needs in priority order. These were aggregated into broad categories and then each category was ranked based on the number of needs for each category. Overall, the five most important research needs in descending order of priority were: (1) applied research on game species, (2) applied research on nongame species, (3) research assessing public attitudes and opinions on wildlife management issues, (4) research assessing applied and cost-effective management strategies to address anticipated impacts of climate change on wildlife, and (5) applied research on the ecology and management of RTE and invasive species.

CONCLUSIONS AND RECOMMENDATIONS

Science-based wildlife management and the wildlife profession are based on a foundation of scientific research. This has served the public, wildlife resources, and the wildlife profession well as evidenced by numerous wildlife restoration successes. However, many wildlife species, ecosystems, and species guilds are in serious peril and in critical need of science-based restoration and management (e.g., native grassland systems and associated wildlife species).

Southeastern state agencies believe that research was and is important for sound wildlife resource decision making. However, most agencies do not include it in their mission statement and give it low priority in funding and labor. Perhaps related to this reduced emphasis, they are only slightly to moderately effective at providing continuing education to their biologists and managers so as to maintain awareness of current research findings. In part, the decreased emphasis on research has resulted from increased public demand for agencies' services coupled with budget and staff reductions. This has occurred at a time when research needs are expanding due to unprecedented increases in demand for and utilization of all natural resources coupled with changes in climate, human demographics, landscapes, and ecological conditions. Agencies largely depend on LGU and CFWRU to provide external assistance in meeting research needs. They are less engaged and communicative with OGWD, UWD, OUC, and NGE. Agencies reported inadequate funding, lack of agency staff, and the inability of research institutions to conduct long-term research as important factors in limiting agencies' collaboration with research institutions. Additionally, lack of communication and differing research priorities were also issues of concern. On a positive note, agencies are generally satisfied with the research results provided by all the research institutions.

We recommend that state wildlife agencies consider the following actions to ensure adequate emphasis on wildlife research and thereby improve the efficacy of resource management decisions:

1. Research should be included in agencies' mission statements.
2. Internally formalize the periodic assessment of research needs relative to ongoing and anticipated resource management decisions.
3. Conduct annual or periodic research meetings with external research institutions to open lines of communication and work collaboratively to set and address research priorities.
4. Collaborate with external research institutions to develop innovative ways to minimize fiscal impediments to agencies created by institutional overhead fees (e.g., indirect and administrative costs).
5. Dedicate adequate funding and labor to meet research needs as determined through the internal and external assessments.
6. Use social media and other electronic communications to share research findings with professional staff.
7. Commit to requiring and providing continuing education to biological staff through attendance at scientific meetings, membership in professional societies, and subscriptions to scientific and professional journals.

ACKNOWLEDGMENTS

We would like to thank our wives for their support, patience, and understanding through the additional time this effort required. An important contribution to this effort is the information provided by SEAFWA states; we greatly appreciate staff from the participatory SEAFWA states who took the time from their busy schedules to complete the survey. This chapter would not have been possible without their participation. We also appreciate the support of the Georgia Wildlife Resources Division. Finally, we thank the editors for their support throughout this effort, their beneficial comments to improve the final draft, and their faith in us to complete this important chapter.

REFERENCES

Brennan, L.A. and W.P. Kuvlesky, Jr. 2005. North American grassland birds: an unfolding conservation crisis. *Journal of Wildlife Management* 69:1–13.

Leopold, A. 1933. *Game Management*. New York: Charles Scribner's Sons.

Meine, C. 1988. *Aldo Leopold: His Life and Work*. Madison: The University of Wisconsin Press.

Population Reference Bureau. 2011. Population Reference Bureau. www.prb.org Accessed October 21, 2011.

Question Pro. 2011. Online research software. questionpro.com Accessed October 21, 2011.

The Wildlife Society. 2007. Full speech delivered by Aldo Leopold Award winner Jim Miller. http://joomla.wildlife.org/index.php?option=com_content&task=view&id=292 &Itemid=180 Accessed October 21, 2011.

U.S. Census Bureau. 2010. Resident population data. http://2010.census.gov/2010census/data/ apportionment-pop-text.php Accessed October 21, 2011.

U.S. Department of the Interior Fish and Wildlife Service. 1987. *Restoring America's Wildlife*. Washington D.C.: U.S. Government Printing Office.

5 Putting Science into Action on Forest Service Lands

William M. Block
U.S. Forest Service, Rocky Mountain Research Station
Flagstaff, Arizona

Victoria A. Saab
U.S. Forest Service, Rocky Mountain Research Station
Montana State University Campus
Bozeman, Montana

Leonard Ruggiero
U.S. Forest Service
Rocky Mountain Research Station
Missoula, Montana

CONTENTS

It can be safely said when it comes to actual work on the ground, the objects of conservation are never axiomatic or obvious, but always complex and usually conflicting.

—Aldo Leopold (in Flader and Callicott [1991:83])

ABSTRACT

The U.S. Forest Service includes three main branches: National Forest Systems, Research and Development, and State and Private Forestry. Herein, we focus on National Forest Systems and Research and Development. National Forest Systems is the management branch of the agency, and its charge is to administer national forests and grasslands throughout the United States. A number of laws, statutes, and policies guide and direct how forests should be managed, including provisions for considering and applying the best available science when planning and executing management actions. The Research and Development branch of the Forest Service is separate from the Management branch and this separation is purposeful. Its charge is to design and conduct research to provide the scientific basis for natural resource management on Forest Service lands. Even though these two branches of the Forest Service are separate, they are linked by virtue of being within the same agency. This linkage reflects the intent for management of Forest Service lands to be guided by the best available science, a concept deeply embedded within the National Forest Management Act (NFMA) and the enabling regulatory language. Despite the intent of Congress, application of science to management has been variable and has changed with each revision of the implementing language (i.e., planning rule) for NFMA. Our objectives here are to review the historical and current roles of science in guiding management of wildlife on national Forest Service lands. We will draw on case studies (Mexican spotted owl [*Strix occidentalis lucida*] and sensitive north-western woodpeckers) to illustrate situations where science was heeded and where it was not for managing these species, and discuss the ramifications of doing so.

INTRODUCTION

The U.S. Forest Service (hereafter Forest Service) is within the federal government's Department of Agriculture. The Forest Service was established in 1905 and is the largest natural resource organization in the world. The overall mission of the Forest Service is to "Sustain the health, diversity, and productivity of the Nation's forests and grasslands to meet the needs of present and future generations" (U.S. Forest Service 2012). It has management responsibility for 78 million hectares of forest, woodland, grassland, and desert, and partners with other organizations in the man-agement of an additional 200 million hectares of rural and urban lands.

The Forest Service is partitioned into five primary mission areas: Research and Development, National Forest Systems, State and Private Forestry, International Programs, and Business Operations. The focus of this chapter is the relationship of Research and Development to National Forest Systems. Although mission areas of both are somewhat intertwined, they are distinct in that they are not joined until the Office of the Chief. This is a key point because it conveys independence between science and management (Ruggiero 2009).

A number of laws, statutes, and policies guide and direct how forests should be managed, including provisions for considering and applying the best available science when planning and executing management actions. The Research and Development branch of the Forest Service is separate from the management branch and this separation is purposeful. Its charge is to design and conduct research to provide the scientific basis for natural resource management on Forest Service lands. Scientists require autonomy to pursue avenues of research unencumbered by the political reality of managers. This independence is critical to ensuring that research results are considered unbiased and not unduly influenced by managers' needs. Lacking this independence, credibility of scientific research could be questioned and management based on it subjected to challenge (Ruggiero 2010).

Even though these two branches of the Forest Service are separate, they are linked by virtue of being within the same agency. This linkage reflects the intent for management of Forest Service lands to be guided by the best available science, a concept deeply embedded within the National Forest Management Act (NFMA) and the enabling regulatory language. The act specifically states that " ... the new knowledge derived from coordinated public and private research programs will promote a sound technical and ecological base for effective management, use, and protection of the Nation's renewable resources." Despite the intent of Congress, application of science to management has been variable and has changed with each revision of the implementing language (i.e., planning rule) for NFMA.

Our objectives here are to review the historical and current roles of science in guiding management of wildlife on National Forest Service lands. We will draw on case studies (Mexican spotted owl [*Strix occidentalis lucida*], and sensitive northwestern woodpeckers) to illustrate situations where science was heeded and where it was disregarded for managing these species, and discuss the ramifications of doing so.

LAWS AND REGULATIONS ESTABLISHING THE ROLE OF RESEARCH IN THE MANAGEMENT OF NATIONAL FOREST SERVICE LANDS

A number of laws and regulations underlie establishment of the Forest Service and management of national Forest Service lands. We review them briefly here.

FOREST SERVICE ORGANIC ADMINISTRATION ACT (1897)

Governs the administration of national forest lands, outlines the establishment of forest reserves, and provides for their protection and management.

McSweeney–McNary Forest Research Act (1928)

Enables the Forest Service to conduct scientific research for a wide range of information users.

National Environmental Policy Act (1970)

Requires federal agencies to consider environmental impacts in their decision-making processes when considering their proposed actions and reasonable alternatives to those actions. During the scoping process, the action agency must consider all relevant information in developing a proposed action and alternatives. Information provided through research often provides the basis for that information.

Endangered Species Act (1973)

Specifies procedures for designating threatened and endangered species, recovery planning, and interagency consultation to evaluate effects of management that could affect species or their habitats. Listing species as threatened or endangered, developing recovery plans, interagency consultations, and decisions on whether to delist depend heavily on the best available scientific information.

National Forest Management Act (1976) or NFMA

Requires the Secretary of Agriculture to assess forestlands, develop a management program based on multiple-use, sustained-yield principles, and implement a resource management plan (also known as a forest plan) for each unit of the National Forest System. The act is implemented through planning regulations. These planning regulations, or planning rule, have been in flux since first developed in 1979. Individual forests and grasslands follow the direction of the planning rule to develop a land management plan specific to their unit. Presently, most forests and grasslands are operating under the 1982 planning rule or an interim planning rule that relies heavily on science consistency to ensure that forest plans and decisions consider the best available science. Through the 1982 regulations, each forest must provide habitat capable of maintaining viable populations of selected species, and are directed to select Management Indicator Species (MIS) to help ensure species viability. MIS are defined as "plant and animal species, communities, or special habitats selected for emphasis in planning," which are *monitored* during forest plan implementation to assess effects of management activities on their populations and other species populations with similar habitat needs.

Forest and Rangeland Renewable Resources Research Act (1978)

Authorizes expanded research activities to obtain and disseminate scientific information about protecting, managing, and using renewable resources. The Act contains an extensive list of research activities, including those related to maintaining and improving wildlife and fish habitats, and protecting threatened and endangered flora and fauna. The Act also authorizes international research. This authorization distinguishes Forest Service Research from other federal natural resources research organizations.

FOREST SERVICE MANUAL (FSM 1999)—NATIONAL FOREST RESOURCE MANAGEMENT (LAST MODIFIED 1999)

Defines sensitive species as those plant and animal species identified by a Regional Forester for which population viability is a concern, as evidenced by significant current or predicted downward trends in populations or in habitat capability that would reduce a species' current distribution. Objectives for designated sensitive species include implementing management practices to ensure that species do not become threatened or endangered because of Forest Service actions.

FOREST SERVICE RESEARCH (FSR)

The overarching mission of FSR is "to serve and benefit society by developing and communicating the scientific information and technology needed to manage, protect, use, and sustain the natural resources of forests and range lands" (U.S. Forest Service 2012). Most research is organized around seven strategic program areas: (1) fire and fuels, (2) resource management and use, (3) wildlife and fish, (4) recreation, (5) water and air, (6) inventory, monitoring, and analysis, and (7) invasive species. Specific national focus areas are dynamic and often reflect current issues and public concerns. Examples of current focus areas include inventory and monitoring, climate change, ecological restoration, and urban natural resource stewardship. Much of this research tends to be applied, but some work is conducted to acquire basic knowledge of ecosystem processes and functions.

The McSweeney–McNary Forest Research Act of 1928 (supplanted by the Forest and Rangeland Renewable Resources Research Act of 1978) mandates the Forest Service to conduct scientific research to provide the basis for the management of natural resources. Policy set forth in the FSM describes the role of research and specifies that research should function as a separate entity with scientific freedom to define its research portfolio. That is, ultimate decisions on topics to address and studies to conduct are made independently by research without undue influence from the management branch. That is not to say that managers and other stakeholders are left out of the process. Development of focal areas is done with extensive input from stakeholders and partners. Perhaps the chief stakeholder is National Forest Systems; thus, much of the research is geared to focus on their needs.

The research and development (R&D) arm of the U.S. Department of Agriculture (USDA) Forest Service works at the forefront of science to improve the health and use of our nation's forests and grasslands. Research has been part of the Forest Service mission since the agency's inception in 1905. Today, some 500-plus Forest Service researchers work in a range of biological, physical, and social science fields to promote sustainable management of our nation's diverse forests and rangelands. Their research occurs in all fifty states, U.S. territories, and commonwealths. The work has a steady focus on informing policy and land management decisions, whether it addresses invasive insects, degraded river ecosystems, or sustainable ways to harvest forest products. Researchers work independently and with a range of partners, including other agencies, academia, nonprofit groups, and industry. The information and technology produced through basic and applied science programs is available to the public for its benefit and use.

FSR is broken into seven research stations. Five are geographically oriented whereby the research done is to address pressing needs within that area. Two stations are national in scope, one directed at forest products and the other, tropical forests. Focus areas become more refined at the station level and then within individual research units. Research units develop five- to ten-year charters and these charters establish the umbrella under which research will be done.

NATIONAL FOREST SYSTEMS (NFS)

National Forest Systems is the management branch of the U.S. Forest Service. As with FSR, NFS is partitioned geographically with nine broad regions and individual national forests reside within each region. Management within national forests is articulated with forest plans (Land and Resource Management Plans). Direction for developing forest plans originated in the National Forest Management Act (1976), and this direction is implemented through established planning regulations and policy direction in the FSM. These planning regulations are in continual flux, reflecting both new priorities and approaches for managing forests and grasslands. Presently, a revision of the planning regulations is undergoing public review and comment.

At the national level, seven major goals comprise the Forest Service Strategic Plan through 2012. These include: (1) restore, sustain, and enhance the nation's forests and grasslands, (2) provide and sustain benefits to the American people, (3) conserve open space, (4) sustain and enhance outdoor recreation opportunities, (5) maintain basic management capabilities of the Forest Service, (6) engage urban America with Forest Service programs, and (7) provide science-based applications and tools for sustainable natural resources management. Although all are germane to the research-management interface, the primary goal relevant to the research-management interface is goal 7.

THE INTERSECTION OF SCIENCE AND MANAGEMENT

Science is the process of acquiring new information following a systematic design (Hull 1988). Science can range from observational studies to true experiments, and objectives range from describing natural phenomena to understanding effects of management options on selected response variables. As such, research runs the gamut of traditional hypothesis testing to parameter estimation. Research can be applied or basic; it can be visionary or responsive to specific information needs. Regardless, it must be objective such that the design should not be to support a particular viewpoint, but to acquire reliable information to guide resource management decisions.

This independence is key to maintaining research credibility within the scientific community and with the public. The separation between the research and management branches of the Forest Service is reinforced by appropriations law, whereby funds provided directly to research and development by Congress can be applied only to research activities and not management. Funds provided to the management branch, however, can be applied to administrative studies and monitoring. Administrative studies are typically short-term in nature and used to address a specific issue at a given location. Monitoring is the responsibility of NFS and is broken into four primary types: implementation, effectiveness, validation, and compliance (Morrison and Marcot 1995). Implementation monitoring focuses on whether a proposed management action was implemented as described in a project plan. For example, if a treatment calls for reducing tree basal area by 50%, did the treatment meet that goal? Effectiveness monitoring evaluates whether the treatment met the stated goal. For example, if a treatment was done to increase northern goshawk (*Accipiter gentilis*) populations, did population numbers actually increase? Validation monitoring relates to the concordance between management activities and standards and guidelines established by policy (e.g., forest plans). Compliance monitoring tracks how well the Forest Service meets the intent of actions proscribed by statute. An example would be meeting take limitations for an endangered or threatened species under the Endangered Species Act. Often, NFS lacks the expertise to design and implement key administrative studies or monitoring. In some cases, they will approach Research for assistance in which case Research will consult with NFS and provide advice, or actually conduct the study.

Monitoring is also critically important to the adaptive management process (Moir and Block 2001). The Forest Service has long espoused adaptive management as a cornerstone of its management efforts. As management actions are applied, information that details the efficacy of those actions is critical to future efforts. If the actions meet stated objectives, they should continue. If not, perhaps they should be revised and different approaches are warranted. Three key questions should be addressed with respect to monitoring within an adaptive management framework. First, was monitoring designed to correctly assess effects of management? Included here are considerations of the selection of the appropriate response variables and evaluation of the adequacy of the sampling design. Second, were protocols followed and was the design implemented correctly? If these first two questions are addressed adequately, then how do monitoring results influence subsequent management direction? Research can help to address the first two aspects, specifically to ensure that managers have reliable knowledge to form the basis of future management decisions.

Science is not done to support or refute management. The goal of science is to provide the best available information for informed decisions. This raises a key point regarding the separation of science from policy. The wisdom here is that science should inform but not dictate policy (Ruggiero 2010). If scientists are cast in the policy arena, their credibility and objectivity can be called into question. Further, numerous factors, not just science, must be considered when formulating policy. These decisions are best left to those who must weigh often competing information when selecting among management options or establishing long-term policies.

CASE STUDIES

WHITE-HEADED WOODPECKERS

The white-headed woodpecker (*Picoides albolarvatus*) has been considered a Management Indicator Species (MIS) by several National Forests in the inland

Northwest since the 1990s and designated more recently as a Sensitive Species by the Intermountain and Pacific Northwest Regions of the NFS. Both regions have sought the assistance of FSR on monitoring habitat and populations of the white-headed woodpecker and other woodpecker species for making science-based decisions for the conservation of sensitive woodpecker species. We provide this example as a case study of the procedures taken by the NFS in conjunction with FSR for the goal of applying the best available science for management of a Sensitive Species on public lands.

The white-headed woodpecker is a regional endemic species of the inland Northwest and California, and is strongly associated with dry coniferous forests dominated by ponderosa pine (*Pinus ponderosa*) (Garrett, Raphael, and Dixon 1996; Wightman et al. 2010; Hollenbeck, Saab, and Frenzel 2011). They are dependent on the seeds of large-coned pines (e.g., ponderosa pine, sugar pine [*Pinus lambertiana*]) for a portion of their diet (Raphael and White 1984). This woodpecker typically nests in mature, open forests with large-diameter trees and a relatively sparse canopy (Hollenbeck et al. 2011), but also nests in recently burned forests (Wightman et al. 2010). The white-headed woodpecker may be particularly vulnerable to environmental change because it occupies a limited distribution and has narrow habitat requirements. The loss of large-diameter snags and the conversion of pine-dominated forests to other forest types have been implicated in the potential decline of their populations (Garrett et al. 1996), although documentation of population declines is sparse (Marshall, Hunter, and Contreras 2003).

National Forests in the Pacific Northwest Region use results of a large-scale environmental assessment for land-use planning throughout the interior Columbia Basin ecosystem (Wisdom et al. 2000). This assessment reported that the white-headed woodpecker was one of only eight of the ninety-seven species analyzed that showed strong declines in habitat (>60% decline from historical conditions) (Wisdom et al. 2000). Presumably because of habitat decline, population declines and range retractions have occurred. In a Central Oregon study, reproductive success of white-headed woodpeckers appears too low to offset adult mortality (Hollenbeck et al. 2011). Survey efforts during the early 2000s yielded no white-headed woodpeckers at some locations in Oregon where they were once considered common in the late 1970s (Altman 2002). Managers are concerned about the status of the white-headed woodpecker not only because of declines in habitat and because of potential declines in populations, but also because dry coniferous forests are the focus of restoration activities in the inland Northwest (Marshall et al. 2003). Consequently, managers identified the need to predict potential wildlife habitat in landscapes affected by restoration activities to help with timely decisions regarding treatment options.

Beginning in 2002, the Pacific Northwest Region (R6) sought assistance on population and habitat monitoring of woodpeckers from Rocky Mountain Research Station (RMRS) researchers because they were considered experts on the ecology of cavity-nesting birds (e.g., Saab, Dudley, and Thompson 2004). With funding support by the NFS (R6) and FSR, RMRS researchers developed models of habitat suitability and nest survival of white-headed woodpeckers in burned and unburned forests (Wightman et al. 2010, Hollenbeck 2011). In 2009, the Pacific Northwest Region assembled a team of biologists to develop a conservation assessment and strategy for the white-headed woodpecker (Mellen-McLean et al. 2010). The team leader is the regional wildlife ecologist, co-lead is a research wildlife biologist, and other team members include NFS biologists, the Bureau of Land Management (BLM) biologists, and FS researchers.

The conservation assessment and monitoring strategy is a key element of a six-year commitment by R6 on "Dry Forest Habitat Condition and Trend Monitoring." The monitoring strategy was designed under an adaptive management framework. The intended outcome is to provide biologists and land managers with guidance on the locations, priorities, and types of landscape- and stand-scale prescriptions that can be used to maintain or increase areas of suitable habitat for the white-headed woodpecker and increase forest resiliency under current and future climate scenarios.

The conservation strategy has three components:

1. Broad-scale occupancy and vegetation monitoring—designed to provide reliable, standardized data on the distribution, site occupancy, and population trends for white-headed woodpeckers across their range in Oregon and Washington. A pilot of this monitoring occurred in 2010, providing the basis for sample sizes needed to estimate selected values of occupancy adjusted for detection. In addition to the occupancy monitoring, vegetation data are collected regionally for input into a fire-climate model to predict future conditions (Keane, Loehman, and Holsinger 2011). Beginning in 2011, the monitoring is expected to be funded by the NFS for six years with additional support by the BLM of Oregon and Washington for the fuels data collection.
2. Validation monitoring—designed to field test and refine models of nesting habitat suitability for current conditions (Wightman et al. 2010, Hollenbeck et al. 2011). Field data collection and model development for habitat suitability took place from 1997 to 2009 and was funded primarily by Oregon Department of Fish and Wildlife and NFS, with additional support by the Oregon chapter of The Nature Conservancy, and FSR. Starting in 2010, models were field tested in locations of model origin. Field-testing will be followed by model refinement, and finally models will be applied to new locations in the region. In addition to refinement of the habitat suitability models that describe current conditions for woodpeckers, a simulation-modeling component is being used to identify habitat areas that are vulnerable to effects of climate change, and to identify management strategies that may promote landscape resilience (Keane et al. 2011). This effort is primarily funded by FSR with additional funding and logistical support by the NFS.
3. Treatment effectiveness monitoring-adaptive management—designed to evaluate the effectiveness of silvicultural treatments that incorporate information from the validated habitat suitability models in reducing fuels, restoring dry forests, and creating-maintaining habitat for white-headed woodpecker across Oregon and Washington. This monitoring began in 2011 at one site in central Oregon. The plan is to add a network of sites over a six-year period. Currently, this effort is funded primarily by FSR with support from the NFS.

Time has not allowed for implementing adaptive management related to the R6 white-headed woodpecker conservation and monitoring strategy. However, the time and money (roughly $50,000 to $160,000 annually from 1997 to 2011) committed by the NFS indicate that adaptive management has a good chance of success. The procedures for the population and habitat monitoring are scientifically rigorous. The population monitoring will provide accurate and precise estimates of occupancy rates, and the vegetation monitoring will provide trends in habitat capability that could influence the current or future distribution of the white-headed woodpecker. Such information is necessary to help prevent the listing of Sensitive Species as a candidate for a threatened or endangered species.

Mexican Spotted Owl

The Mexican spotted owl was listed as a threatened species under the Endangered Species Act in 1993. A recovery team was assembled and a recovery plan was completed and approved in 1995 (USDI Fish and Wildlife Service 1995). The recovery team had strong representation from FSR. This was largely because Forest Service researchers either conducted or collaborated on most of the research done and were regarded by many as experts on the species (cf., Ganey, Ward, and Willey 2011). We present this example because it serves as a good case study of the application of science to management on National Forest Systems lands.

Although the Mexican spotted owl occupies a broad geographic range extending from Utah and Colorado south to central Mexico, it occurs in disjunctive locations corresponding to isolated mountain and canyon systems. The current distribution mimics its historical extent, with the exception of its presumed extirpation from some historically occupied locations in Mexico and riparian ecosystems in Arizona and New Mexico. Of the areas occupied, the densest populations of owls are found in mixed-conifer forests, with lower numbers occupying pine-oak forests, encinal woodlands, rocky canyons, and other habitats. Habitat-use patterns vary throughout the range of the owl and with respect to owl activity. Much of the geographic variation in habitat use corresponds to differences in regional patterns of vegetation and prey availability. Forests used for roosting and nesting often exhibit mature or old-growth structure; they are uneven-aged, multi-storied, of high canopy closure, and have large trees and snags. Little is known about foraging habitat, although it appears that large trees and decadence in the form of logs and snags are consistent components of forested foraging habitat. The quantity and distribution of owl habitat, as well as of areas that can be expected to support the necessary habitat correlates in the future, are poorly understood.

The recovery team assembled and reviewed all existing information on the ecology of the Mexican spotted owl, existing forest conditions and trends, and potential threats to the owl (Ganey and Dick 1995; USDI Fish and Wildlife Service 1995). This information provided the scientific basis for developing management recommendations, which were variously applied on the ground. The recovery plan was a combination of prescriptive site-specific guidance and descriptive desired conditions to strive for on the landscape. The underlying philosophy of the recovery team was that the plan should emphasize adaptive management, whereby recommendations would be adjusted as information was acquired to evaluate their effectiveness. As such, the plan was cast as a three-legged stool with management recommendations, habitat monitoring, and population monitoring representing the legs of the stool. The analogy to a stool means that if any one of the legs was removed, the recovery plan could fail. This concept was reinforced by the delisting criteria, which required strong evidence for stable or increasing habitat and populations.

The recovery plan designated three primary management areas: protected areas, restricted areas, and other forest and woodland types. The rationale was to protect areas currently occupied by owls and those with a high likelihood of occupancy (protected areas), manage for replacement nest and roost habitat (restricted areas), ensure the presence of key habitat elements for what the owl required for foraging and dispersal (restricted areas), and maintain connectivity (other forest and woodland types).

Management was most prescriptive within protected areas. Much of this entailed establishing a 243-ha (600-acre) protected activity center (PAC) around owl nest/roost areas within which little active management should occur. The recovery team recognized the need to reduce fuels within selected PACs. The failure to do so could render these PACs vulnerable to stand-replacement fire and loss of the habitat altogether. Treatments within PACs should be within an adaptive management framework, whereby effects of treatments on owls and their habitat would be monitored and assessed. The recovery team recommended a staged approach where up to 10% of the PACs could be treated by thinning trees <22 cm diameter breast height (DBH) and effects of those treatments would be assessed to identify the next course of action. Depending on the outcome of these assessments, treatments could continue, discontinue, or be adjusted. Despite this opportunity, an assessment never took place.

Management within restricted areas focused on creating replacement nest-roost habitat and managing to meet other land-use objectives. Restricted areas included mixed-conifer and pine-oak forests. Agencies were required to conduct landscape analyses to identify 10 to 25% of the restricted area and to manage those areas to become replacement nest-roost habitat. This percentage was based on modeling to identify that which could be sustained in this condition given ecology of these forest types. The remaining restricted area had few constraints other than to retain key habitat components such as large trees, snags, logs, and hardwoods. The habitat components were identified through a series of research studies that demonstrated strong correlations between them, the owl, and their prey (Ward 2002, Block et al. 2005).

As noted previously, monitoring both populations and habitat were critical to assessing how well the recovery plan was working at recovering the owl and leading to its delisting. With the exception of a few owl demographic studies and some general inventories, little is known about the population status of the Mexican spotted owl. Even these studies were limited in scope and objectives, and are unsuitable for projecting range-wide population trends. Consequently, the recovery team considered and re-analyzed existing data (White, Franklin, and Ward 1995) and used that as a basis for developing a rigorous approach for population monitoring. As conceived, the population-monitoring program in the Mexican spotted owl recovery plan was thought to be scientifically defensible and would provide accurate and precise estimates of population trend. However, a program that is conceived must be tested through a pilot study prior to implementation. The Southwestern Region of the Forest Service provided funding in 1999 to conduct a pilot study. The approach had merit, but logistical considerations and other factors led Ganey et al. (2004) to conclude that it was infeasible. No progress has been made to revise or implement population monitoring since then.

Habitat monitoring was proposed at two spatial scales—macro- and microhabitat. Macrohabitat monitoring was to evaluate changes in habitat quantity at large landscape or regional scales. It entailed conducting change-detection analysis based on remote-sensing data. Change detection analysis was conducted once and not continued. Microhabitat monitoring focused on evaluating trajectories of key habitat components (large trees, snags, logs) following habitat altering activities such as prescribed fire or mechanical treatments. A design for this was developed by an interagency team and is presented in Morrison et al. (2008). National Forest personnel collected these data, but have yet to conduct the analysis.

Gathering defensible scientific information is necessary for delisting the Mexican spotted owl. Data that clearly show that the population is stable or increasing and that adequate habitat is projected in the future would provide U.S. Fish and Wildlife Service with the basis for considering that the owl no longer requires protection under the Endangered Species Act. The recommendations for management and monitoring were based on the best available science, much resulting from studies conducted by FSR. Regardless, progress on implementing them has been delayed for various reasons. Although NFS requires high-quality information, its budgets and expertise are not keeping up with progress in wildlife science, namely the use of modern methods to monitor populations, and the need for experimental studies to realize cause and effect relationships.

LESSONS LEARNED

Wildlife science has grown exponentially over the past fifty years, especially since passage of the ESA (1973) and the National Forest Management Act (1976). Great strides have been made in the study of wildlife populations (Williams, Williams, and Conroy 2002), habitats (Morrison, Marcot, and Mannan 2006), and the design and implementation of research studies (Block et al. 2001, Morrison et al. 2008). The information garnered from these studies provides a defensible basis for managing species and their habitats, and for monitoring changes in populations in response to management practices. Are there obstacles in applying science to management? If so, are they real or perceived? Clearly, recent revisions to development of forest plans emphasizes the importance of basing management of National Forests on the best available science. Whether that occurs requires both commitment and practice. Perhaps the disconnect between the two is deeply embedded within the culture of the Forest Service. Brown and Squirell (2010) concluded that the Forest Service lacks a strong learning environment. Indeed, such a culture is required to identify, embrace, and apply new information as it becomes available. Our case studies suggest that the Forest Service is capable of applying science to management, but it is practiced unevenly. Wildlife science will continue to grow and improve. Wildlife management on Forest Service lands must keep pace with these changes and use that information to ensure conservation of species.

ACKNOWLEDGMENTS

We greatly value and appreciate the opportunity to work with managers from NFS. This partnership adds relevance to the science that we conduct and promotes wise management of wildlife resources. Comments from K. Mellen-McLean and M. Morrison greatly improved this paper.

REFERENCES

Altman, B. 2002. Conservation strategy for landbirds in the northern Rocky Mountains of eastern Oregon and Washington. Version 1.0. Oregon and Washington Partners in Flight. Unpublished report. http://www.orwapif.org/pdf/northern_rockies.pdf (accessed October 22, 2011).

Block, W. M., A. B. Franklin, J. P. Ward, Jr., J. L. Ganey, and G. C. White. 2001. Design and implementation of monitoring studies to evaluate the success of ecological restoration on wildlife. *Restoration Ecology* 9:293–303.

Block, W. M., J. L. Ganey, P. E. Scott, and R. King. 2005. Prey ecology of Mexican spotted owls in pine-oak forests of northern Arizona. *Journal of Wildlife Management* 69:618–629.

Brown, G. G., and T. Squirell. 2010. Organizational learning and the fate of adaptive management in the US Forest Service. *Journal of Forestry* 108:379–388.

Flader, S. L., and J. B. Callicott, Eds. 1991. *The River of the Mother of God and Other Essays by Aldo Leopold.* Madison: University of Wisconsin Press.

Forest Service Manual 1999. http://www.fs.fed.us/im/directives/dughtml/fsm_2000.html (accessed October 22, 2011).

Ganey, J. L., and J. L. Dick, Jr. 1995. Habitat relationships. In: *USDI Fish and Wildlife Service. Recovery Plan for the Mexican Spotted Owl,* Vol. 2. Albuquerque, NM: U.S. Fish and Wildlife Service, chap. 4.

Ganey, J. L., J. P. Ward, Jr., and D. W. Willey. 2011 Status and ecology of Mexican spotted owls in the Upper Gila Mountains recovery unit, Arizona and New Mexico. U.S. Department of Agriculture, Forest Service, Rocky Mountain Research Station. General Technical Report RMRS-GTR-256WWW. Fort Collins, CO.

Ganey, J. L., G. C. White, D. C. Bowden, and A. B. Franklin. 2004. Evaluating methods for monitoring populations of Mexican spotted owls: a case study. In: *Sampling Rare and Elusive Species: Concepts, Designs, and Techniques for Estimating Population Parameters,* W. L. Thompson, Ed. Washington, D.C.: Island Press, pp. 337–385.

Garrett, K. L., M. G. Raphael, and R. D. Dixon. 1996. White-headed woodpecker: *Picoides albolarvatus.* In: *The Birds of North America,* A. Poole, Ed. Ithaca, NY: Cornell Laboratory of Ornithology.

Hollenbeck, J. P., V. A. Saab, and R. Frenzel. 2011. Habitat suitability and survival of nesting white-headed woodpeckers in unburned forests of central Oregon. *Journal of Wildlife Management* 75:1061–1071.

Hull, D. L. 1988. *Science as a Process.* Chicago, IL: University of Chicago Press..

Keane, R. E., R. A. Loehman, and L. M. Holsinger. 2011. The FireBGCv2 landscape fire and succession model: a research simulation platform for exploring fire and vegetation dynamics. U.S. Department of Agriculture, Forest Service, Rocky Mountain Research Station. General Technical Report RMRS-GTR-255. Fort Collins,CO.

Marshall, D. B., M. G. Hunter, and A. L. Contreras, Eds. 2003. *Birds of Oregon: A General Reference*. Corvallis, OR: Oregon State University Press.

Mellen-McLean, K., V. Saab, B. Bresson, B. Wales, A. Markus, and K. VanNorman. 2010. White-headed woodpecker monitoring strategy and protocols for the Pacific Northwest Region. Unpublished report. U.S. Forest Service, Pacific Northwest Region, Portland, OR.

Moir, W. H., and W. M. Block. 2001. Adaptive management on public lands: commitment or rhetoric? *Environmental Management* 28:141–148.

Morrison, M. L., and B. G. Marcot. 1995. An evaluation of resource inventory and monitoring programs used in National Forest planning. *Environmental Management* 19:147–156.

Morrison, M. L., B. G. Marcot, and R. W. Mannan. 2006. *Wildlife-Habitat Relationships: Concepts and Applications*, 3rd ed. Washington, D.C.: Island Press.

Morrison, M. L., W. M. Block, M. D. Strickland, B. A. Collier, and M. J. Peterson. 2008. *Wildlife Study Design*. New York: Springer.

Raphael, M. G., and M. White. 1984. Use of snags by cavity-nesting birds in the Sierra Nevada. *Wildlife Monographs* 86:1–66.

Ruggiero, L. F. 2009. The value of opinion in science and the Forest Service research organization. *Journal of Wildlife Management* 73:811–813.

Ruggiero, L. F. 2010. Scientific independence and credibility in the sociopolitical process. *Journal of Wildlife Management* 74:1179–1182.

Saab, V. A., J. G. Dudley, and W. L. Thompson. 2004. Factors influencing occupancy of nest cavities in recently burned forests. *Condor* 106:20–36.

U.S. Forest Service. 2012. The U.S. Forest Service—An Overview. http://www.fs.fed.us/documents/USFS_An_Overview_0106MJS.pdf.

USDI Fish and Wildlife Service. 1995. *Recovery Plan for the Mexican Spotted Owl*, Vol. I. Albuquerque, NM: U.S. Fish and Wildlife Service.

Ward Jr., J. P. 2002. Ecological responses by Mexican spotted owls to environmental variation in the Sacramento Mountains, New Mexico. Dissertation, Colorado State University, Fort Collins.

White, G. C., A. B. Franklin, and J. P. Ward Jr. 1995. Population biology. In: *Recovery Plan for the Mexican Spotted Owl*, Vol. II. Albuquerque, NM: U.S. Fish and Wildlife Service, chap. 2.

Wightman, C. S., V. Saab, C. Forristal, K. Mellen-McLean, and A. Markus. 2010. White-headed woodpecker nesting ecology after wildfire. *Journal of Wildlife Management* 74:1098–1106.

Williams, B. K., J. D. Williams, and M. J. Conroy. 2002. *Analysis and Management of Animal Populations*. San Diego, CA: Academic Press.

Wisdom, M. J., R. S. Holthausen, B. C. Wales, C. D. Hargis, V. A. Saab, D. C. Lee, W. J. Hann, T. D. Rich, M. M. Rowland, W. J. Murphy, and M. R. Eames. 2000. Source habitats for terrestrial vertebrates of focus in the Interior Columbia Basin: broad-scale trends and management implications. U.S. Department of Agriculture, Forest Service, Rocky Mountain Research Station. General Technical Report PNW-GTR-485, Portland, OR.

6 Closing the Gap between Private Land Managers and Wildlife Scientists

The Importance of Private Wildlife Research Institutions

William E. Palmer
Tall Timbers Research Station and Land Conservancy
Tallahassee, Florida

CONTENTS

... the general effect of long-continued, or irregular but frequent, burning upon the vegetation of an area, and its indirect effect on animal life, present a complex problem, one that would require years of careful research on the part of a well-equipped experiment station to work out.

—Herbert Stoddard (1931:492)

ABSTRACT

Privately owned wildlands have significant conservation value, and wildlife managers influence habitat on large acreages of such properties. Private research institutes are suited for developing long-term collaborative relationships between wildlife scientists and managers because their mission is to conduct research over longer periods of time. Universities have access to the latest technologies and other knowledge in scientific and statistical methods that complement resources available at private research institutions. Thus, collaboration among private research institutes, universities, and land managers is a natural link to closing the gap between managers and researchers in wildlife science. Managers tend to have an inherent distrust of agendas, dogma, and short-term research results, as do many researchers. However, through adaptive management researchers can help managers answer pragmatic questions while testing ecological theory and benefiting conservation. To build these relationships requires researchers to strive to understand and respect a manager's objectives and remain open-minded and humbled by contextual influence and emergent patterns in complex ecosystems on management and research results. Managers need to be progressive, be willing to accept uncertainty, and be willing to apply changes to their management strategies and tactics in an adaptive framework. Private research institutes often foster independent thought that permits testing potentially controversial ideas. The benefit from working with private landowners for researchers is potentially large-scale, replicated, long-term research projects that benefit constituents (i.e., managers), as well as provide reliable inferences. The benefit to land managers is a structured learning process that provides improved and more efficient management and success toward attaining management objectives. Most importantly, such a process results in improved conservation of natural resources.

INTRODUCTION

Closing the gap between scientists and natural resource managers has been a topic of concern in the ecological sciences for over forty years (Howard 1968). The National Science Foundation among other major science institutions and natural resource agencies (McPherson and DeStefano 2003) has recognized this chasm as a significant issue. In wildlife science, a field in which the goal is to provide useful information to solve management questions, there has been growing concern among managers and researchers that this gap exists and is perhaps widening. Evidence that the gap existed was highlighted when The Wildlife Society ended publication of the *Wildlife Society Bulletin*, a manager-oriented journal that was more approachable to managers than the *Journal of Wildlife Management*. The recent revival of the *Bulletin*, along with the publication *The Wildlife Professional* indicates a sincere attempt to produce more accessible and useful information for managers. To a degree, the burgeoning literature on adaptive resource management is another indication of how efforts are being made to link research and management more closely (Grumbine 1994; Walters 1997; Mason and Murphy 2002).

Examples of successful collaborations between wildlife managers and researchers are relatively rare (Grumbine 1991; Lee 1993; McPherson and DeStefano 2003), but they do happen. For example, over the past twenty years, researchers and managers have collaborated on increasing northern bobwhite populations on private lands in the southeastern United States. Aspects of this successful collaboration include not only meeting management objectives, but also increasing acreages of habitat for a suite of declining species adapted to fire-maintained ecosystems, as well as protecting significant areas with conservation easements. The purpose of this chapter is to dissect aspects of how scientists and managers worked together to achieve these goals and helped close the gap. Much, but not all, of this work was conducted through Tall Timbers Research Station, a private nonprofit research, conservation, and education organization that works closely with private landowners. That said, much of the research conducted at Tall Timbers has been in collaboration with graduate students and universities; however, I focus the chapter on unique aspects of private research and experiment stations that work to help close the gap between researchers and managers. This focus should not be taken as diminishing the immense value partnering with university researchers brings to private research institutions, which includes many facets, but is outside the scope of this chapter (but see Chapter 3, this volume).

THE IMPORTANCE OF COLLABORATION

Tens of millions of acres of private lands are dedicated to wildlife management in the United States. At times, these lands encompass remnant ecosystems and have extensive management programs that benefit multiple species (Engstrom and Palmer 2005). Hunting estates often harbor tremendous biodiversity given they were protected from exploitation and managed over long time frames for exclusive use of people with high political or economic status (Engstrom and Palmer 2005). Today in the southeastern United States, over 1 million acres are managed for wild northern bobwhite populations of which over 400,000 acres are located in the Red Hills of Northern Florida and Southwest Georgia (Figure 6.1). There are another 300,000 acres near Albany, Georgia, and the number of acres has been increasing recently as new properties are established. These landscapes include properties that have been managed with frequent fire for over 100 years (Brennan, Lee, and Fuller 2000).

In contrast to what has happened on the private lands mentioned previously, the fact that the overall annual harvest of wild bobwhites has dropped from over 7 million birds in Georgia and Florida (circa 1960–1970) to less than 100,000 is an indication of the failure of management and research to tackle the large-scale conservation issue of fire-maintained successional habitats for birds (Brennan 2002; Chapter 9, this volume). In contrast, private lands dedicated to bobwhite management have sustained, and in many cases dramatically increased, densities of bobwhites. This is a conservation success story for fire-dependent species. Their success is not an accident but a result of collaboration among managers, property owners, and researchers. The long-term use of prescribed fire and compatible timber management practices has sustained pine savannas that also harbor significant populations of declining, threatened, and endangered fauna and flora adapted to fire-maintained ecosystems

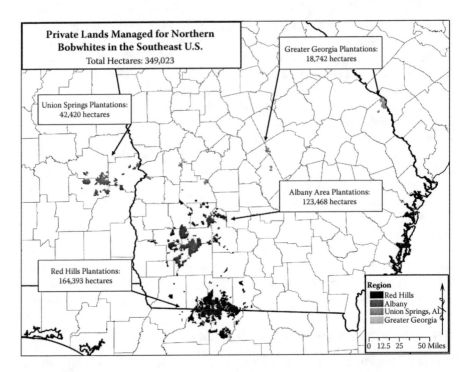

FIGURE 6.1 Distribution of private lands managed for northern bobwhite in the southeastern United States.

(Brennan et al. 1998; Engstrom and Palmer 2005). Adaptive research on bobwhite habitat and population ecology and reliance on feedback and collaboration with progressive managers has been a gateway to exploring the conservation value of these lands and changing management paradigms. While not the intention of the research specifically, the relationships developed through mutual interest in the land and landowner objectives provide opportunities to change management philosophies toward those that benefit multiple species while meeting management objectives and also changing managers' behaviors through collaboration.

The importance of closing the gap between researchers and managers is obvious when considering the conservation value of privately managed lands. Private land managers influence management on a lot of land, as noted previously, and the owners of these properties often are willing to fund research if such research is aligned with their goals. In the Red Hills region between Tallahassee, Florida and Thomasville, Georgia, there has been a long history of property owners supporting wildlife research to improve management of their properties. Long-term support for research is not unique to Tall Timbers; other examples include Archbold Biological Station, Welder Wildlife Foundation, and Delta Waterfowl Research Station, among others, but overall private long-term funding is rare in wildlife research. Landowner support for research in the Red Hills began when landowners funded Herbert Stoddard's classic studies on bobwhite in the 1920s. Ultimately, this support resulted in Stoddard and colleagues founding the research station in 1959, the genesis of which is noted in the epigraph to this chapter. During the early years of Tall Timbers, much of the research focus was on the effects of fire in southern pine ecosystems; only a modicum of quail research addressing management was conducted. In the 1980s, bobwhite populations declined on Tall Timbers and most of the private lands in the Red Hills and Albany area. Other regions of the Southeast were also experiencing severe declines in bobwhite populations, along with an entire suite of grassland bird species, as land use changes resulted in loss of habitat (Brennan 2002; Brennan and Kuvlesky 2005). Brennan's (1991) prediction that northern bobwhites would be extirpated across much of their geographic range was coming true, and it was happening sooner than he originally predicted. In many areas of the Southeast, large landowners gave up on wild bobwhite management and turned to releasing pen-reared bobwhites for sport, with resulting changes in management associated with pen-reared release systems including annual burning to reduce cover and increased timber density that competes with grassland ground cover important to many species. The Southeast Quail Study Group was founded (now the National Bobwhite Technical Committee) to address regional declines and provide new focus for determining how to return bobwhites to private and public lands. Finally, the National Bobwhite Conservation Initiative was developed to deal with range-wide habitat loss (Palmer, Terhune, and McKenzie 2011). Overall, the mood was bleak for bobwhites and grassland birds in the Southeastern United States during the last decade of the twentieth and most of the first decade of the twenty-first centuries.

Declines of wild bobwhite populations in the 1980s on most properties in the Red Hills and Albany areas, and the quest to improve management, once again led landowners to support research, both through Tall Timbers Quail Research Initiative and through Auburn University's Albany Quail Project, which is now a part of the research program at Tall Timbers. Over the past twenty years, applied northern bobwhite research, with a management focus, has provided information to landowners, who have implemented findings and reported on results from implementing alternative management techniques. Private properties maintain consistent hunting success records, which include data on "catch" per unit effort (e.g., coveys seen per hour of hunting) and as such, provide thousands of observations each year from which local and in some cases such as Red Hills, regional population trajectories can be assessed and comparisons of different management paradigms conducted (Brennan et al. 1997; Stribling and Sisson 2009). Through collaboration and sharing of ideas, managers have a high level of knowledge of how bobwhite populations operate, which helps to prepare them to consider intelligent new ideas related to management. Often a property manager will instigate such an idea, and later researchers will study it independently and determine its efficacy. Other times, researchers test an idea and managers adopt it if it turns out to be pragmatic.

Research in the Red Hills and Albany areas has focused on demographics of populations to assess the biological and management value of a treatment. Over 20,000 bobwhites have been radiomarked during these studies to assess changes in demographics as part of management experiments or other non-experimental factors. For instance, long-term telemetry data indicate there has been a steady decrease in breeding season survival on properties in the Albany area, and this will dictate new management research to address this problem. Long-term data (such as the forty-one-year mark-recapture data set, and twenty years of year-round telemetry-derived demographic data) provided baseline data from which to assess if demographic changes were relevant to management. Research projects have ranged from demographic analyses of populations, estimating density of bobwhites, brood-habitat management, timber management, effect of the timing and duration of the prescribed fire season, supplemental feeding techniques, predator control, predator-prey relationships, non-target species effects, translocation, population genetics, and breeding behavior, among others.

The result of the continuous collaboration between managers and researchers has had significant effects on the success of management on private lands. At this time, bobwhite populations across the properties in the Red Hills and Albany areas are at historically high densities (Figure 6.2). Generally, populations have increased greater than twofold relative to 1980 densities, and many properties have sustained densities over two bobwhites per acre for a decade or longer. Moreover, management has been refined to meet individual landowner objectives, including balancing natural resources or maximizing bobwhite populations (Masters et al. 2004). Landowners feel more confident supporting management that may be costly or that may significantly change habitat on their property when research has provided evidence that it will help them achieve their goals. Further, because research has provided information not just on *how* to manage, but also *why* management works, owners and managers are better able to make sound decisions. Regional declines of bobwhite

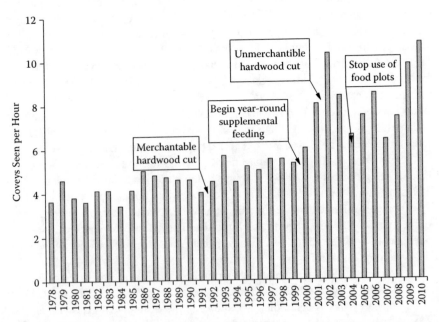

FIGURE 6.2 Long-term hunting records showing number of coveys of northern bobwhite seen per hour of hunting on a privately managed hunting estate in the Red Hills of northern Florida.

populations can be associated with many factors not directly related to management, such as below normal temperatures and above normal rainfall during the breeding season, which can lower chick survival rates and therefore reduce fall populations. If managers understand that the cause of a decline is not linked directly to limitations in habitat or other management, they can avoid pressures to make hasty decisions on what needs to be changed in their management system. That is, managers understand they could be doing everything right, and still see quail declines for a period of time. Long-term research also helps them know, however, that the decline may have been worse if it were not for sound management, which is about as positive as one can be when a target species of management declines. Successful collaboration has helped forge relationships that go beyond bobwhite management and include safe harbor protection of threatened red-cockaded woodpeckers (*Picoides borealis*) and over 100,000 acres protected under conservation easements, among other important conservation gains.

DOGMA IN SCIENCE AND MANAGEMENT

Research is a process in which knowledge is gained through application of scientific methods. The same overarching scientific principals are applied across disciplines in science, with variation of methods used largely dependent on the control researchers have over their experimental units and the potential to replicate them. The more complex the ecosystem under management, the more difficult it is to replicate experiments with clear and unambiguous results. As ecosystems are difficult to replicate in space and time, theory from ecological research often has lower external validity (Lee 1993), weakening strong inference. When sufficient information has been gathered through research, resulting or emergent paradigms tend to gather a sheep-like following in science and management from which dogma can be an eventuality. An example of dogma in science on bobwhites resulted in the concept of a "doomed surplus," which had huge implications on predation and harvest management for over fifty years (Errington 1967), and yet was never proven (Romesburg 1981; Roseberry 1993). The ability of a new theory or idea to gain wide acceptance by other scientists or managers is affected by more than just the idea and the science, but also by the timing of the idea relative to social views, differences in philosophies within the scientific community, the reputation of the researchers, and the strength of current competing theories in question, as well as the ability of the researchers to write well and speak well to engage an audience of peers (Peters 1991; Lee 1993; Mason and Murphy 2002). In wildlife science, the doomed-surplus concept stated that habitat could only support a fixed number of individuals through winter given limited space and protection from severe weather and, hence, predators only reduced the "doomed surplus" that would have died anyway due to weather. Thus, predators had no effect on the breeding population each spring and therefore carrying capacity of habitats. The timing of these results was perfect for gaining wide acceptance because they followed on the heels of mindless and large-scale destruction of predators across the country, which began to wear thin on biologists and the public which began to see predators for their cultural, economic, and ecological values. Paul Errington, the author of the doomed-surplus concept, was a skilled and prolific writer and he published several books and scientific papers espousing his ideas. As such, his ideas tended to gain wide acceptance. The doomed-surplus concept also supported the idea of compensatory mortality and that hunting replaced predators and skimmed the doomed surplus off the top of the population stock. Therefore, if hunting was compensated by reduced predation and did not lower the carrying capacity of a habitat, then hunting seasons and limits could be liberalized. For at least four decades, this dogma was used as a justification by state agencies to permit longer and more liberal bobwhite hunting seasons.

Recent research has shown that harvest tends to be additive (Williams, Lutz, and Applegate 2004) such that each bobwhite harvested is approximately one less potential breeder in the summer population. The additive harvest theory is a competing theory, and depending on which harvest management theory a manager followed, it had a major impact in the population dynamics of a bobwhite population. On most privately owned properties, a conservative approach to bobwhite harvest has been widely used for decades. Managers using common sense and experience knew the importance of having high breeding populations to maintain high fall populations, and that harvest should be adjusted seasonally and spatially. This contrasts with harvest management on public lands in which wildlife agencies often strive to maximize hunter opportunities on a management area, which can result in excessive harvest rates (Rolland et al. 2010). The point is that dogma misled management for decades while science, a much slower process, took a half-century to catch up to private land managers. Managers react to short-term changes and make decisions accordingly in order to maintain a long-term goal of population persistence, whereas science rightly operates on a much longer time scale. However, dogma can mislead because science is in the process of sorting out theories. Of course, dogma also occurs in management; for instance, the concept of food plots being important for wildlife is much overrated in management with no support in research (Guthery 1997). Managers and researchers should be wary of dogma in their fields, and utilize collaboration to test it whenever possible.

Context, or the time, place, and conditions under which an experiment occur, also can have tremendous effect on research results. In Errington's case, he formulated the doomed-surplus concept at the northern fringe of the bobwhites range where winter weather interacted with habitat on his study area to leave only those coveys that had access to good winter cover. He later expressed doubt about his theory when longer-term data he collected failed to fit his previous concepts. Scientists commonly recognize the time context issue in wildlife research; for instance, testing the effect of field borders along crop fields on wildlife populations may hinge upon suitable weather conditions for a study. Bobwhite research in the 1960s and 1970s occurred at a time when continental populations of their primary adult predator, the Cooper's hawk (*Accipiter cooperii*), were at their lowest versus today when populations have rebounded. This contextual difference resulted in harvest research showing that over-winter survival on managed lands was extremely high compared to bobwhite research today. Managers also recognize that context is important when they mention how each site has different qualities and requires slightly different management, or that no one "prescription" fits all sites. As such, research on wide-ranging species and across numerous contexts can result in differing results when researching management questions. Science is a pursuit of knowledge, and few researchers want to publish a study that suggests the results are only valid on the site on which they worked. Therefore, scientists have a tendency to attempt to extrapolate their results across space and time despite weak inference caused by complexity, emergent patterns, and context. This produces a fair amount of disagreement among scientists, and managers often have a hard time figuring out whom to believe, reducing their confidence in all research results. If managers are critical of the validity of research, they will not be likely to adopt the information and change their behavior, and rightly so; dogma is not always correct. For bobwhite management, this has resulted in policy that has misdirected management in more ways than just harvest, but also prescribed fire use, predation, and usefulness of translocation to establish populations. However, while context is important in complex ecosystems, I know of no bobwhite studies that tested a theory, or management practice, across range-wide study areas. For example, research has shown that response of bobwhite broods to the same management technique can be completely different on open frequently burned pine savanna only forty-five miles away from one another on areas (Palmer et al. In Press). Bobwhites with broods avoided burned pine forests and selected managed weed fields on sites with low soil fertility. Managing weedy fields is a common recommendation for bobwhites, but is expensive and lowers the conservation value of certain pine savannas. However, bobwhites with broods avoid managed weedy fields on sites with high soil fertility and select burned pine savanna instead.

In a related example, Wellendorf, Palmer, and Bromley (2004) found that density influenced bobwhite covey calling rate. In areas with high densities (>1 quail per acre), calling rates across the landscapes were high. However, on some agricultural landscapes with low densities of bobwhites, calling rates were also high because coveys moved from crop fields to limited amounts of winter cover where they existed locally in high densities. The landscape context influenced bobwhite behavior and therefore their calling rates.

Research on one study site, or in one region, over a fixed period of time, however, is often viewed by scientists as the best available knowledge, whereas managers may be more likely to lean on the experience of other managers that have similar experiences and challenges. Managers understand that properties are "different" and are wary of information derived from a different context. In closing the gap between managers and researchers, researchers need to acknowledge that global models and theories may have been developed under different contextual situations, and thus may have low relevance to a particular site. At the same time, managers must recognize that theories help provide a context for examining ideas relative to management.

To some degree, all hypotheses in science are value-laden to one extent or another (Peters 1991; Lee 1993). It is almost unavoidable not to have professional and personal philosophies embedded in research hypotheses, predictions, and data collection (Tauber 1999). This is not necessarily a problem in science as long as objectivity is maintained and intelligent criticisms of the results are openly allowed in the scientific community (Lee 1993). It can be a problem in management, however, because personal values usually influence how a property is managed. For example, a common philosophical conflict between ecologists and managers of private lands is the appropriateness of "single species management" versus "ecosystem management"

(Grumbine 1994). In the case of northern bobwhites, landowners across the Southeast spend significant resources to maintain bobwhites at densities around one quail per acre. The irony here is that these lands also harbor significant populations of species of high conservation value that are declining elsewhere because of the maintenance of high-quality early successional habitats that are managed at scales suitable for many wildlife species. Thus, gopher tortoise (*Gopherus polyphemus*), red cockaded woodpeckers, Bachman's (*Peucaea aestivalis*) and Henslow's (*Ammodramus henslowii*) sparrows, brown-headed nuthatches (*Sitta pusila*), pine snakes (*Pituophis melanoleucus*), Sherman's fox squirrels (*Sciurus niger*), and dozens of other species are doing quite well on these landscapes (Brennan et al. 1998). Perhaps the idea of managing for a single species conjures up images of cornfields or some other monoculture; however, single species management is a socio-cultural construct and not an ecological one. A landowner interested in a game species is by default managing for many species as the result of their efforts to sustain and elevate one species. This is no different from managing scrub habitats for Florida scrub jays (*Aphelocoma coerulescens*) or longleaf forests for red cockaded woodpeckers. What is important is how the scale and application of ecosystem processes influence the overall species composition. Only through working cooperatively can both landowner objectives and greater conservation objectives be optimized.

By recognizing that hypotheses are value-laden and that values among managers and researchers may be different, one can potentially avoid disagreement and ultimately inferences drawn from research that may seem at cross-purposes. The type of question asked and how it is asked can be shaped by values of the researcher or the manager, which is the point. Often, researchers have professional philosophies different from managers. A simple example is how bobwhite researchers deal with the issue of predation. Managers are quick to point out that predators are a controllable problem that can be actively managed, whereas researchers are more likely to doubt predators are a significant problem or are a natural "part" of the ecosystem. These differences in values can result in two researchers, depending on which belief they adhere to, designing two very different studies to "test" the same ideas on predators' impacts on bobwhite populations. When working with managers, care must be taken to understand their perspectives and goals to ensure research is designed such that it fairly tests pragmatic ideas, rather than fits a philosophy. In the end, the results of a well-designed study in which managers and researchers were involved are more likely to be adopted by managers than a study from which managers were excluded.

VALUE OF INDEPENDENCE AT PRIVATE RESEARCH INSTITUTES

During the 1920s, the U.S. Forest Service began a campaign against the use of fire to manage forests, which was a common practice in the southern United States to manage woodlands and fields. A group sponsored by the U.S. Forest Service called "The Dixie Crusaders" visited towns throughout the South with the objective of "educating" people on the negative consequences of burning on wildlife and forest. This highly successful public relations campaign helped to forever change the structure of southern forests and put in motion the long downward spiral of fire-dependent wildlife species. New northern property owners in the Red Hills Region of South Georgia and North Florida, having purchased large recreational tracts of land around the turn of the century, were aligned to some degree with the concept of stamping out fire. However, as time went by, the pine savannas and post-agriculture habitats became overgrown with brush and the primary quarry of interest to the landowners, northern bobwhites, quickly became scarce. Research now shows that in as little as two years, post-fire suitability of many forest stands is reduced for bobwhite and other species like Bachman's sparrows (Cox and Widener 2008).

At the request of the property owners, the United States Biological Survey agreed to conduct a study to determine the cause of the decline of bobwhite quail on the hunting preserves in the Thomasville-Tallahassee area, and landowners funded the research. Herbert Stoddard was hired to direct the studies over a five-year period, which resulted in the book, *The Bobwhite Quail: Its Habits, Preservation and Increase* (Stoddard 1931). He recognized that prescribed fire was essential to southern pine ecosystems. Stoddard called bobwhite the "fire bird" because they were so intricately linked with fire in these ecosystems. However, because Stoddard wrote of the benefits of fire for management of bobwhites, plants and other wildlife, and management of pine savannas, the Forest Service threatened to halt publication of his book if he did not water down the section on fire. Under threat of publication elsewhere, they agreed to publish his book with a much-diluted chapter on the use of fire. Stoddard later wrote in his memoirs,

> At one time I was classed by many as an enemy of these forests because of my written and spoken insistence that the pine forests not only *could* be burned over frequently enough to maintain their natural vegetation and associated wildlife but indeed *should* be burned, for the safety and the healthy development of the forests themselves. I did my part in bringing about "controlled burning," or "prescribed burning," as a routine practice in large acreages of pineland. (Stoddard 1969:195)

After his book was published, Stoddard consulted on wildlife and forestry management in the Southeast. Stoddard was a world-class ornithologist, a close friend of Aldo Leopold, and wrote one manuscript with Paul Errington. All three became giants in the nascent wildlife profession. Stoddard began what he called the Cooperative Quail Association, which relied on annual donations to support research on wildlife habitat, and produce a newsletter to the owners. Many of the privately owned shooting properties in the Southeast were designed by Stoddard, akin to Jack Nicklaus's fingerprint on golf courses. However, Stoddard saw the need for continued research on the role of fire in shaping plant communities. The negative press on prescribed fire, the industrial timber management paradigms (versus ecological forestry), and his experience with censorship by a governmental agency impressed on Stoddard and his colleagues the need for a research station independent of outside influence, as noted in the beginning of this chapter. As such, in 1958, Tall Timbers was created to study the effects of fire on plants, animals, and ecosystems in the Southeastern Coastal Plain, independent of the political and dogmatic forces against fire. Through its Fire Ecology Conference series, Tall Timbers had a lasting impact on the growth of fire, and it was at these conferences that "fire" and "ecology" were first linked. Stoddard and his protégés continue to be heralded today for their pioneering influence on fire ecology and management of natural ecosystems and most assuredly are responsible to a degree for Florida leading the country in the number of acres of wildlands prescribed burned each year (Karels, personal communication). Stoddard's lasting effect on management of pine savannas is a testament to how research provides information that influences management. The fact is that over the past 125 years, few places in the Red Hills area have not been without fire for more than three years. This has helped to sustain some of the rarest upland ecosystems in the Southeast.

Stoddard was not an academic and published few manuscripts in scientific journals during his career. Yet it could be argued that despite his lack of publishing in scientific outlets, he has had the greatest positive impact on management of upland pine forests in the South of any individual through his work in forestry and fire ecology and landowners' interest in single species management. A great example of this is the Wade Tract Preserve, an 80-ha, old-growth longleaf-pine wiregrass virgin forest with individual longleaf trees over 400 years of age. Stoddard recognized the ecological value of this tract, and worked diligently to educate the landowners on its value and protect the trees from over-harvest and the native ground cover from the harrow. Today it is protected by an easement with Tall Timbers and is used for ecological study as well as hunted for bobwhites.

A benefit of private research institutions is their orientation to allow researchers to think outside the box and test ideas that may not fit the recent trends in research or management. Based on my discussions with researchers at other private research stations, this is a notable constant appreciated by researchers. Perhaps it comes with a depth of understanding that occurs only after years of research in a topic on a specific area. The independence lends itself to working with managers because it helps to break down management dogma created by research, agencies, managers, or landowners. It also facilitates testing of and promoting ideas that are counter to recent trends in wildlife management.

MUTUAL RESPECT AMONG MANAGERS AND RESEARCHERS

The respect Stoddard had for people who managed the land came in part from a distrust of science driven by political or economic agendas rather than a depth of knowledge about the forest. This seems trite to say today, given all our advancements in research, modeling, and analyses; however, people who spend their days managing a system have something most researchers will never have—years of direct observation of the system they manage. The challenge for researchers is to harness that knowledge and work cooperatively to design research that challenges that information. For instance, managers of quail properties for years have surreptitiously conducted translocations of wild birds from one site to another with results that were encouraging. However, the research literature was neutral on the value of translocation for bobwhites and few states had policies that permitted the practice (Liu et al. 2000). An official research project was permitted in Georgia and the results indicated that translocated bobwhites had demographics equal to resident birds and densities of bobwhites quickly escalated because of translocation (Terhune et al. 2010). Now several states, including Florida, South Carolina, Alabama, and Georgia, have permitted translocations or have developed permitting systems for translocation of bobwhites to newly created habitats with low extant bobwhite populations. The new translocation policies have resulted in a resurgence of pine savanna restoration by landowners interested in having bobwhite populations. To date over 1,000 bobwhites have been moved from established populations on private lands to over 15,000 ha of new pine savanna habitats resulting in establishment of bobwhite populations >2.5 quail/ha, which now provides hunting recreation.

Managers also have a good sense of the importance of fire frequency on plant communities. Bobwhite require a mixture of grasses, forbs, and shrubs, and over many decades, managers have determined that burning every two years, or less, is essential for maintaining the open pine savannas with a diverse ground cover. If you read the literature, there are actually no research papers that "prove" that a two-year fire frequency is essential for bobwhite management in the pine systems of the South (although tremendous circumstantial and empirical evidence exists). Stoddard understood that fire shaped plant communities and set up a long-term experiment of different fire frequencies, and the results clearly indicate that one- to two-year fire frequency is essential for maintaining the grasses in the ecosystem. Further, managers understand that the extent of fire is an essential component of fire use. Large-sized fires (>200 acres) make no sense to a manager of a quail property, yet there are many examples of publicly managed lands where fires >1,000 acres are the norm and quail populations are notably low. The point is that private land managers often have significant knowledge that can help researchers better understand the system and design studies that provide useful information. In many cases, managers are already practicing what science has failed to refute. This is understandable because science is a much slower process than management. However, by respecting what managers do, and learning from them, it provides testable and realistic hypotheses that researchers can evaluate and will ultimately provide information that will improve management.

Managers were less clear on what effect different seasons of fire have on plant communities, bobwhite populations, and other wildlife (Cox and Widener 2008). Early researchers recommended dormant season fires be used to select for legumes over grasses, which should therefore increase food supplies and avoid destroying quail and turkey (*Meleagris gallopavo*) nesting habitat. Winter burning was also logistically easier for managers, fires were easier to control, and lower ambient temperatures reduced the potential to scorch valuable pine trees as well as making it easier on staff. Long-term research at Tall Timbers has shown that survival rates are more predictive of quail populations than nesting success and this has been supported by simulation modeling research (Sandercock et al. 2008). Burning later and spreading out fire use over a longer period of time benefits survival of bobwhites because it maintains higher amounts of cover at any one time, thus increasing useable space through time (Cox and Widener 2008). As such, fire season for bobwhite management has shifted with new research information; from December to February (1940s to 1960s) when winter burns were used to increase legumes, to March in the 1990s to burn soon after hunting season but before nesting, now most burning for quail habitat in the southeast occurs from March through May with benefits to bobwhite survival and conservation of a suite of fire-adapted species.

Other benefits include improving brood habitat over the summer (Carver et al. 2001), reducing hardwood encroachment, and increasing the flowering of a broader suite of plant species (Hiers, Wyatt, and Mitchell 2000) including viable wiregrass seeds, a cornerstone species in the longleaf pine-wiregrass system. These examples show that managers partnering with researchers to answer management questions can provide an avenue not only to improve management for stated objectives but also to learn why management works or fails. Managers willing to partner with researchers and change management in the face of uncertainty are managers who will be most likely to succeed in the long run.

LONG-TERM RESEARCH AT PRIVATE RESEARCH STATIONS

Long-term research is core to many private research institutions because it provides valuable management information in relation to ecological time frames. Long-term research provides a depth of knowledge about an ecosystem that a progressive manager develops over time through testing ideas on a property. Private research stations have the ability to conduct long-term research because the topic is often core to its mission. For Tall Timbers this mission is fire and wildlife ecology and management. Our research on fire frequency and its impact on fire began in the 1960s and the results of this research have significant implications. Research on fire frequency shows that if prescribed fire is not applied at least every two years in pine forest systems in the South, hardwood encroachment will change the ground cover composition from a grass-dominated system to one that is dominated by shrubs and vines, which has tremendous wildlife management implications. Longer rotations will result in a mixed pine-hardwood forest. Developing a long-term and productive relationship with constituents provides reliable funding mechanism for research, improves management, and ultimately conservation of natural resources. Progressive managers tend to distrust short-term research. Even for an r-selected species such as bobwhites, a two- to five-year study may be insufficient to understand how variation over longer time scales influences treatments tested during the duration of research. To understand the demographics of game bird populations, Dick Potts, a noted gray partridge biologist from England, remarked that at least thirty years of research are necessary, perhaps more (Potts 1997). For instance, shifts in weather patterns may have significant system-wide effects in most areas of the country. A short-term study may provide different answers as the varying impacts of El Nino and La Nina weather events come and go.

Some relief from the "publish or perish" science paradigm goes hand in hand with long-term research efforts at private research stations. Private landowners and land managers want the latest information coming from long-term research as it adapts and changes over time. As such, periodically providing results from these projects to constituents is important for influencing management actions. Ultimately, publishing in the scientific literature is critical for scientific credibility and influencing policy. However, managers value being kept in the loop and are supportive of long-term research, but they also don't want to wait for information from research projects until it is peer-reviewed and published. They view the value of information from either improving the efficiency of management or getting a desired outcome from implementing new management ideas. Therefore, when working with private land managers it is imperative that they be kept up-to-date on research results. This has direct benefits to the researchers as well because while controlled experiments or adaptive management provide information, they are typically too expensive to conduct on a large number of management areas. Having managers apply new management ideas while research is ongoing provides empirical results that support the research results, or not, on different sites. Having large numbers of managers apply new management ideas also provides empirical information on the likelihood of a practice working. In this way, managers provide a direct feedback loop at a broader scale than most researchers have time and resources to work on. Further, managers often refine practice implementation and therefore provide useful knowledge for outreach type publications.

ADAPTIVE MANAGEMENT

The long-term nature of research at private (and public) research and experiment stations lends itself to adaptive management. Research stations typically wrestle with the same issues as managers do because they have to manage the station's properties. At Tall Timbers during the early 1990s, the management staff attempted to use lightning-season fires to suppress hardwood encroachment. Tall Timbers is primarily composed of post-agriculture "old field" vegetation, which is distinctly different from the native longleaf-wiregrass, or shortleaf-bluestem habitats on sites without previous agricultural disturbance. Old-field plant communities tend to be less pyrogenic than native communities that were adapted to growing season fires. However, using "natural" lightning season fires was the ecological idea *de jour* in ecology at the time, but it failed to work on Tall Timbers because the old-field vegetation was not conducive to burning during the summer period and hardwoods continued to encroach. This management experience was instructive given that monitoring indicated that a season of fire was not a straightforward answer to hardwood encroachment and ecosystem restoration. Adaptive management recognizes that we cannot wait for perfect information to act and, therefore, must do some things to maintain the system. The mix of land management and research forces research stations to perceive the messiness and complexity of the ecosystems they manage just as a manager has to in order to be successful.

Active adaptive management, as opposed to trial and error, has two distinguishing characteristics. First, there is a direct feedback loop between science and management to allow management to change with the latest scientific information. Typically, monitoring a system response variable of importance to mutually agreed upon objectives is the link between science and management. Second, management is an experiment, in which its treatments and controls are experimentally embedded in management. In order to justify the effort to do active adaptive management, the outcome should be aligned with the goals of the management program. It is the combination of these two concepts that distinguishes adaptive management from learning by doing or traditional science (Walters and Holling 1990). Publication of results is important to keep from reinventing the wheel, but is not a necessary component of active adaptive management. This is sometimes confusing to administrators who want their scientists to abide by "publish or perish"; however, affecting change in management and building value with constituents may be as important to a private institution as publishing research results. Both have value, but when attempting to close the gap between science and management, one is clearly more important than the other to managers or scientists, and this can be a source of anxiety for both parties.

Managers and private landowners provided with an opportunity to participate in active adaptive management tend to appreciate its value over time. It requires a manager and scientist to agree that they must act given limited knowledge on what may happen and are willing to "risk" being wrong. There was a recent case in which a private landowner spent significant funds to test a technique that obviously failed to work. In this case, the test was of a new feeder designed to reduce the cost of supplemental feeding by reducing the amount that would be spilled or eaten by species other than bobwhites. As it turned out, video cameras documented that wild bobwhites did not use the feeders and began to show lower weights relative to bobwhites with supplemental feed spread along dedicated feed trails. As such, the landowner, well versed in active adaptive management on his property and a successful businessman, replied, "Well sometimes you have to spend money to find out you were doing things right all along, at least we know." He understood that uncertainty is a powerful force in business as it is in management and that adaptive management is a wise long-term approach.

Finally, adaptive management lends itself nicely to making relatively large shifts in management actions and testing dramatically different management philosophies. For instance, rather than "piddling" along, making small incremental changes in a more traditional management approach, once the objectives are clearly understood by all parties, then more than one management philosophy can be applied. In the case of pine savannah restoration, dramatic reduction in off-site hardwoods on upland forests is a prime example. On many lands with either post-agriculture or with a period of fire exclusion, off-site tree species encroach on the uplands, including species like live oak (*Quercus virginiana*), water oak (*Quercus nigra*), hickory (*Carya* sp.), and even loblolly pine (*Pinus taeda*). These changes occur gradually. The idea of conducting a major timber harvest to remove hardwoods and open the canopy of pines (to a degree) was counter to the concept of working with the forest you have. Essentially, the concern was that removing timber and destroying the ground cover could hurt the quail population versus renovating the pine ecosystem and creating a disturbance that shifted the predator-prey community, allowing bobwhites to respond to the improved habitat. Through active adaptive management, managers and researchers chose control and treatment areas and monitored the changes in bobwhite populations. The results were dramatic, as was the response of many other species of wildlife. Thus, hardwood "clean up" cuts have now occurred across most of the quail managed landscape of the Red Hills, perhaps over a half-million acres. While this is a significant result, it has not been published in the scientific journals. However, follow-up research is being published that documents what was already known from the adaptive management.

One of the greatest benefits of applying adaptive management on sites and monitoring results is that the degree of response obtained can be viewed across many different contexts. For instance, bobwhites' and other species' responses to hardwood reduction cuts can be monitored on old field lands, native ground cover areas, and xeric, mesic, and hydric soils. This provides value to outreach of information to other landowners because the likelihood of a response is addressed, rather than basing recommendations on one research project on a single site in the more classical scientific process. This has huge benefits on how private landowners view the credibility of the information being provided.

Adaptive management provides an excellent opportunity for collaboration among research scientists and graduate students at universities together with private research institutes and landowners. Students, who are provided with real world management experiences, can conduct individual research that is a component of long-term research supported by the private research institute. Private research stations benefit greatly from collaboration with universities, which typically have different resources, such as genetics laboratories, for instance, and professors tend to be knowledgeable on the latest analysis techniques. This collaboration can be highly productive for all parties involved.

PERCEPTIONS BY MANAGERS AND RESEARCHERS

Probably the number one concern the managers with whom I have interacted is that the goals of research, no matter how laudable, are not aligned with the goals for the property they manage. Landowners and managers can often have very specific management goals. Where there is mutual interest, there is ample opportunity to fund research. To engage and partner with managers requires understanding the questions they have regarding management. If research fails to meet the expectations of both parties, the partnership is doomed to failure. This occurs both on public and private lands.

Managers are leery of agendas, especially those driven by philosophical differences. This can be an obvious difference, such as having a philosophy of ecosystem management versus one of single-species management, but often such differences are much more subtle. The mismatch of agendas can result in research that is not as useful to management and can result in negative consequences. For instance, if a treatment is applied improperly, or not to the degree necessary to effect change in the ecosystem, then a potential management concept could be shown to have no effect. This provides the manager with false information and can lead to confusion. Researchers interested in testing the efficacy of a potential management technique should ensure the degree to which their treatments are applied cover the range of potential management needs.

Managers' questions may be perceived as mundane or naïve by researchers relative to testing ecological theory. For instance, managers may want to find ways to reduce costs, which is not nearly as interesting as testing ecological theory to most researchers. However, the questions are often timely, deal with conservation issues that affect people, and with appropriate attention may be quite complex to fully test and understand. For instance, what may appear to be a simple question such as, "Would it be cost-effective to reduce nest predators to increase bobwhites?" can easily become a lifetime of research, research that over time can test efficacy of trapping techniques, the role predators have on the predator-prey relationships, the effect of buffer prey species, cause-specific mortality, density dependency, compensation, and more. Alternatively, one can respond with dogma that nest predators are not an issue and recommend they ignore the issue. However, that invokes the agenda argument whereby the response is not from personal experience or research in the specific context, but through literature review based on a few studies, perhaps with ambiguous results. It can be perceived as arrogant rather than helpful. Indeed if we really consider the complexity of these systems, ecological theory can help guide research, but we must be humbled by the few studies that have been conducted and realize that given the permutations ecological systems can take, the effect of most treatments will be fuzzy and depend on context.

Managers consider the implications of an action today and in the future because their job to manage an ecosystem never ends. However, managers should not let fear of change, or testing new ideas, limit their ability to grow as managers. Good communication and collaboration in developing goals for a project and informing all partners helps to avoid the manager being in the "hot seat" if things don't work out well. Researchers often bring fresh ideas, new paradigms, and techniques to bear on management issues. With careful planning, managers will benefit greatly from these concepts.

GOING PUBLIC

The same concepts that have worked to close the gap among private land managers and researchers to increase, restore, and expand bobwhite populations are now being applied on public lands in Florida. The Upland Ecosystem Restoration Project is a joint venture among Tall Timbers, Florida Departments of Forestry, Environmental Protection, and Fish and Wildlife Conservation Commission to establish and maintain 100,000 acres of habitat for bobwhite and other species on public lands in Florida. The project, now in its sixth year, is working with agencies to choose suitable sites that need restoration and that have progressive managers willing to adopt new management paradigms, implement adaptive management, and risk failure.

The project coordinator works collaboratively with managers to develop management plans that fit the individual site, develops time frames that fit the budget and management logistics for a site, and establishes specific species target density goals. Agency leaders are involved as part of a Steering Committee and as such have buy in and provide support for the project. Everyone is aware that knowledge was incomplete and each area was very different and as such, the results would vary and active adaptive management was necessary.

Paradigm shifts were necessary for managers and agencies. Burning for acreage targets or fuel reduction had to be changed to burning for wildlife objectives. This required increasing fire frequency (among other management actions) and reducing the extent of individual fires such that a mosaic of habitats remained. Initial concerns that the amount of acres burned would be reduced because smaller-sized "patch" burns take more time were allayed as managers realized that burning an eighteen- to twenty-four-month "rough" required less personnel and equipment than burning areas with heavier fuel loads. Monitoring of plant and animal communities is ongoing and provides feedback to landowners (agency leaders) and managers, as well as information for researchers. Graduate students working with universities are assessing critical management questions that require additional information to better inform management, such as how extent of fire influences survival and productivity of target species. The point is that through mutual planning and setting objectives, avoiding pitfalls through communication, and implementing active adaptive management, the project has thus far been successful in improving habitat for fire-dependent wildlife and bobwhite populations have begun to respond to habitat improvements.

SUMMARY

Private lands have significant conservation value, and wildlife managers influence habitat on large acreages of such properties. Private research institutes are suited for developing long-term collaborative relationships between wildlife scientists and managers because their mission is to conduct research over longer time scales than typically used in graduate research at universities. Universities have access to the latest technologies and other knowledge in scientific and statistical methods that complement resources available at private research institutions. Thus, collaboration among private research institutes, universities, and land managers is a natural link to closing the gap between managers and researchers in wildlife science. Managers tend to have inherent distrust of agendas, dogma, and short-term research results, as do many researchers. However, through adaptive management researchers can help managers answer pragmatic questions while testing ecological theory and benefiting conservation. To build these relationships requires researchers to strive to understand and respect a manager's objectives and remain open-minded and humbled by contextual influence and emergent patterns in complex ecosystems on management and research results. Managers need to be progressive, be willing to accept uncertainty, and be willing to apply changes to their management strategies and tactics in an adaptive framework. The benefit to researchers is potentially large-scale, replicated, long-term research projects that benefit constituents (i.e., managers) as well as provide reliable inferences. The benefit to land managers is a structured learning process that provides improved and more efficient management and success toward attaining management objectives.

REFERENCES

Brennan, L. A. 1991. How can we reverse the northern bobwhite population decline? *Wildlife Society Bulletin* 19:544–555.

Brennan, L. A. 2002. A decade of promise, and decade of frustration. *National Quail Symposium Proceedings* 5:220–222.

Brennan, L. A., R. T. Engstrom, W. E. Palmer, S. M. Hermann, G. A. Hurst, L. W. Burger, Jr., and C. L. Hardy. 1998. Whither wildlife without fire? *Transactions of the North American Wildlife and Natural Resources Conference* 63:402–414.

Brennan, L. A., and W. P. Kuvlesky, Jr. 2005. North American grassland birds: an unfolding conservation crisis? *Journal of Wildlife Management* 69:1–13.

Brennan, L. A., J. M. Lee, and R. S. Fuller. 2000. Long-term trends of northern bobwhite populations and hunting success on private shooting plantations in northern Florida and southern Georgia. *National Quail Symposium Proceedings* 4:75–77.

Brennan, L. A., W. Rosene, B. D. Leopold, and G. A. Hurst. 1997. Northern Bobwhite Population Trends at Groton Plantation, South Carolina: 1957–1990. Miscellaneous Publication 10. Tall Timbers Research Station, Tallahassee, FL.

Carver, A. V., L. W. Burger, Jr., W. E. Palmer, and L. A. Brennan. 2001. Vegetation characteristics in seasonal-disked fields and at bobwhite brood locations. *Annual Conference Southeastern Association of Fish and Wildlife Agency Proceedings* 55:436–444.

Cox, J., and B. Widener. 2008. Lightning-Season Burning: Friend or Foe of Breeding Birds. Miscellaneous Publication 17, Tall Timbers Research Station, Tallahassee, FL.

Engstrom, T. E., and W. E. Palmer. 2005. Two species in one ecosystem. In: *Bird Conservation Implementation and Integration in the Americas: Proceedings of the Third International Partners in Flight Conference.* J. C. Ralph and T. D. Terrell, Eds. U.S. Forest Service General Technical Report PSW-GTR-191. Albany, CA.

Errington, P. L. 1967. *Of Predation and Life.* Ames, IA: Iowa State University Press.

Grumbine, R. E. 1991. Cooperation or conflict? Interagency relationships and the future of biodiversity for U.S. parks and forests. *Environmental Management* 15:27–37.

Grumbine, R. E. 1994. What is ecosystem management? *Conservation Biology* 8:27–38.

Guthery, F. S. 1997. A philosophy of habitat management for northern bobwhites. *Journal of Wildlife Management* 61:291–301.

Hiers, J. K., R. Wyatt, and R. J. Mitchell. 2000. The effects of fire regime on legume reproduction in longleaf pine savanna: is a season selective? *Oecologia* 125:521–530.

Howard, R. A. 1968. The practicality gap. *Management Science* 14:503–507.

Karels, J. Director, Department of Forestry, Tallahassee, Florida, personal communication.

Lee, K. N. 1993. *Compass and Gyroscope.* Washington, DC: Island Press.

Liu, X., R. M. Whiting, Jr., B. S. Mueller, D. S. Parsons, and D. R. Dietz. 2000. Survival and causes of mortality of relocated and resident northern bobwhites in east Texas. *National Quail Symposium* 4:119–124.

Mason, G., and S. Murphy. 2002. Redefining roles of science in planning and management: ecology as a planning and management tool. In: *Proceedings of the 2001 Northeastern Recreation Research Symposium.* S. Todd, Ed. U.S. Forest Service General Technical Report NE-289. Northeastern Research Station. Newtown Square, PA, pp. 239–245.

Masters, R., K. Robertson, W. Palmer, J. Cox, K. McGorty, L. Green, and C. Ambrose. 2004. Red Hills Forest Stewardship Guide. Miscellaneous Publication 12, Tall Timbers Research Station, Tallahassee, FL.

McPherson, G. R., and S. DeStefano. 2003. *Applied Ecology and Natural Resource Management.* Cambridge, UK: Cambridge University Press.

Palmer, W. E., D. C. Sisson, S. D. Wellendorf, T. M. Terhune, and T. L. Crouch. In press. Habitat selection by northern bobwhites in pine savanna ecosystems. *National Quail Symposium* 7.

Palmer, W. E., T. M. Terhune, and D. F. McKenzie. 2011. The national bobwhite conservation initiative: a range-wide plan for recovering bobwhites. National Bobwhite Technical Committee Technical Publication, ver. 2.0, Knoxville, TN. http://www.bring-backbobwhites.org/strategy/nbci-full-plan. Accessed October 7, 2011.

Peters, R. H. 1991. *A Critique for Ecology.* Cambridge, UK: Cambridge University Press.

Potts, G. R. 1997. Using the scientific method to improve game bird management and research: time. *National Quail Symposium Proceedings* 4:2–6.

Rolland, V. D., J. A. Hostetler, T. C. Hines, H. F. Percival, and M. K. Oli. 2010. Impact of harvest on survival of a heavily hunted game. *Wildlife Research* 37:392–400.

Romesburg, H. C. 1981. Wildlife science: gaining reliable knowledge. *Journal of Wildlife Management* 45:293–313.

Roseberry, J. L. 1993. Bobwhite and the new "biology." *National Quail Symposium Proceedings* 3:16–20.

Sandercock, B. K., W. E. Jensen, C. K. Williams, and R. D. Applegate. 2008. Demographic sensitivity of population change in northern bobwhite. *Journal of Wildlife Management* 72:970–982.

Stoddard, H. L. 1931. *The Bobwhite Quail: Its Habits, Preservation and Increase*. New York: Charles Scribner and Sons.

Stoddard, H. L. 1969. *Memoirs of a Naturalist*. Norman, OK: University of Oklahoma Press.

Stribling H. L., and D. C. Sisson. 2009. Hunting success on Albany, Georgia plantations: the Albany Quail Project's modern quail management strategy. *National Quail Symposium Proceedings* 6:338–347.

Tauber, A. I. 1999. Is biology a political science? *Bioscience* 49:479–486.

Terhune, T. M., D. C. Sisson, W. E. Palmer, B. C. Faircloth, H. L. Stribling, and J. P. Carroll. 2010. Translocation to a fragmented landscape: survival movements and site fidelity of northern bobwhites. *Ecological Applications* 20:1040–1052.

Walters, C. 1997. Challenges in adaptive management of riparian and coastal ecosystems. *Conservation Ecology* 1(2):1. http://www.ecologyandsociety.org/vol1/iss2/art1/ Accessed October 7, 2011.

Walters, C. J., and C. S. Holling. 1990. Large-scale management experiments and learning by doing. *Ecology* 11:2037–2068.

Wellendorf, S. D., W. E. Palmer, and P. T. Bromley. 2004. Estimating calling rates of northern bobwhite coveys and measuring abundance. *Journal of Wildlife Management* 68:672–682.

Williams, C. K., Lutz, R. S., and Applegate, R. D. 2004. Winter survival and additive harvest in northern bobwhite coveys in Kansas. *Journal of Wildlife Management* 68:94–100.

7 The Role of Professional Societies in Connecting Science and Management
The Wildlife Society as an Example

Laura Bies
The Wildlife Society
Bethesda, Maryland

Michael Hutchins
The Wildlife Society
Bethesda, Maryland

John Organ[*]
U.S. Fish and Wildlife Service
Hadley, Massachusetts

Stephen DeStefano
U.S. Geological Survey
Massachusetts Cooperative Fish and Wildlife Research Unit
University of Massachusetts
Amherst, Massachusetts

CONTENTS

---

[*] The views expressed in this paper are the author's and do not necessarily represent those of the U.S. Fish and Wildlife Service.

> We have the choice as a profession: We may be content to expertly tinker with the
> wildlife machine to keep it alive somehow; or we can give our profession the dignity
> and importance it deserves and help the public interpret the facts so as to contribute in
> [humanity's] struggle to find [itself].
>
> **—Olaus Murie (1954:293)**

ABSTRACT

Professional societies play a key role in connecting science and management. From
their roots in the early 1900s until now, such societies have provided professionals with opportunities to engage and network, with forums to share their research,
and with tools for information sharing. Through codes of ethics, certification
programs, and advocacy programs, professional societies ensure integrity within the
profession, as well as in the use of scientific information in the policy and decision-
making realms. Using The Wildlife Society as an example, we explore the valuable
role that professional societies play in connecting research and on-the-ground management, thereby advancing the field of wildlife management and conservation.

INTRODUCTION

Professional societies play a crucial role in the careers of many wildlife professionals. Offering networking opportunities, continuing education, certification programs,
political clout, and a community of peers, professional societies have existed as long
as the wildlife profession itself and have undergone similar growth and evolution.
Several professional societies serve the needs of those in the wildlife and fisheries field. These include The Wildlife Society (TWS), American Fisheries Society
(AFS), Society for Conservation Biology (SCB), and Ecological Society of America
(ESA), as well as smaller, more specialized societies like the American Society
of Mammalogists and the Society for the Study of Amphibians and Reptiles, and
consortiums of societies like the Ornithological Council or the Coalition of Natural
Resource Societies. Membership in professional societies provides an opportunity for
biologists and other professionals from different employers and specialties to interact,
network, and stay informed about the latest research and management developments.

Many professional societies were established in the late 1800s and early 1900s,
as natural resource management became a recognized profession in the United
States. These grew out of the desire of professionals to connect with each other, share
information, stay up-to-date in the latest developments in their fields, and have an

independent body responsible for establishing standards for the new profession. Today, many wildlife professionals belong to one or more relevant societies. A 2006 survey of employees of the U.S. Fish and Wildlife Service (USFWS) and the U.S. Geological Survey (USGS) found that about half of USFWS employees and over 90% of USGS employees surveyed were a member of at least one scientific society (Taylor and Lauber 2007). Staying informed of new research and management implications, strengthening scientific knowledge, networking, staying informed of natural resource policy, and sharing research results and management implications with other professionals were identified as reasons for joining and maintaining membership in a professional society such as TWS (Taylor and Lauber 2007). Many of the justifications for joining a professional society relate directly to the role that such societies play in integrating science and management.

There are many functions and services that professional societies can provide, such as promoting and maintaining serial scientific journals and other publication outlets, holding annual meetings, and providing guidance and advice on issues related to lobbying, advocacy, and ethics (e.g., professional conduct, using sound and objective scientific evidence for management decisions) are among the most important and obvious. We discuss these and other functions and services that professional societies provide to working professionals, using TWS as an example.

PUBLICATIONS

SCIENTIFIC PERIODICALS

One of the most important reasons for establishing TWS in 1937 was the collective desire for a journal of wildlife management so that professionals could communicate with each other (Swanson 1987). Volume 1 of *The Journal of Wildlife Management* was published in 1937, and in 1958, the *Wildlife Monographs* series was established to publish manuscripts too long for the *Journal* (over forty manuscript pages or over eighty pages overall). From the inception of *The Journal of Wildlife Management*, concerns were raised (particularly by its editors) that it did not contain enough content useful to wildlife managers, and that its target audience was primarily wildlife scientists. In May 1972, the *Wildlife Society Bulletin* was established as an outlet for management-related articles, which were geared even more toward field practitioners, such as state and federal agency biologists and managers, and included descriptions of techniques or methods that were directly applicable to management activities. The content of the *Bulletin* included opinion pieces as well, and offered news and information on Society issues, including obituaries. Additionally, the *Bulletin* published research papers on the human dimensions component of wildlife management. In 2007, the *Bulletin* was discontinued due to budget constraints and an effort was made to combine basic and applied scientific articles, including human dimensions, into one publication and to provide an opportunity for new publications that would appeal to a broader audience. However, in 2010 The Wildlife Society Council (the elected governing body of TWS) decided to re-establish the *Bulletin*, largely because practicing biologists and managers liked and missed the very practical and applied emphasis of the *Bulletin,* and because TWS's financial position had improved. During 2007, *The Wildlife Professional* emerged as a member magazine designed to provide current information, news, and analysis in a popular format for practicing wildlife professionals and people outside the profession. Because it appealed to non-academic practitioners, a group that makes up a significant portion of TWS's membership base, it is thought to be a primary factor in TWS's rapid growth from 2007 to 2010, during a time of economic recession.

Such a range of publications is common for professional societies in the natural resources fields. The ESA, which was founded in 1915, also has several publications for its members. The ESA's publications include four highly ranked, peer-reviewed journals (*Ecology, Ecological Monographs, Ecological Applications,* and *Frontiers in Ecology and the Environment*), as well as the online journal *Ecosphere,* the *Bulletin of the Ecological Society of America,* and a range of free, online publications that serve as outreach and teaching tools. Similarly, the AFS publishes a suite of academic journals for aquatic resource professionals. For example, *Transactions of the American Fisheries Society* features papers on basic fisheries science, the *North American Journal of Fisheries Management* covers management research and recommendations, the *North American Journal of Aquaculture* provides guidance for those who breed and raise aquatic animals, the *Journal of Aquatic Animal Health* concentrates on health maintenance and disease treatment, and *Marine and Coastal Fisheries* is an international, open-access online journal devoted to marine, coastal, and estuarine fisheries.

As more communication goes from printed pages to digital formats due to cost concerns and efforts to increase access, the importance of electronic forms of communication will continue to increase. Professional societies' websites already provide an important connection to the society as a whole and to individual members and their work, and TWS's continued widespread and broadening use of the Internet for gathering information, exchanging ideas, and communication is a reflection of this trend in society as a whole. Increasingly, members access journals online rather than subscribing to and receiving paper copies in the mail. In 2007, TWS digitized all of its legacy publications and made them available to subscribers online as an additional member benefit and to increase the impact factor of their publications. While providing quick and easy access to scholarly research anywhere there is Internet connectivity, the proliferation of online journals has also led to more integration, with societies able to imbed links that allow readers to explore subjects more deeply and research issues more thoroughly. Although online journals and other electronic publications offer easier access, as well as improved search capabilities, their cost of production is only slightly less than print publications. There are some savings in the cost of paper, printing, and postage, which can be passed along to the consumer. However, it is important to note that producing an online journal still requires personnel to handle submissions and communication with reviewers and authors, copy-editing, and formatting.

An increasing number of electronic journals have open access availability and many of these (e.g., The Public Library of Science or *PLoS*) are provided free of charge to allow unfettered access to information, and increase the impact factor of their content. It will be interesting to see how open access publishing models perform in the coming decades. The business model for such operations is untested and it is questionable whether such journals can be maintained over the long term. As all are subsidized by grants or donations, it remains to be seen if such publications can be sustained from a purely economic standpoint. However, many experts in technology transfer and other fields believe that open source information on the web is the wave of the future, and many such online sites have seen skyrocketing visitation rates and use. For online scientific journals, this can greatly increase the visibility and impact factor of the journal. Nevertheless, it is also important to realize that publications are a major source of revenue for professional and scientific societies. Such societies could not exist or maintain other programs (e.g., their policy activities) should their publication programs cease to exist. Recently, TWS has selected Wiley-Blackwell, one of the world's largest and most respected commercial publishers, to become its publishing partner for journals. It is hoped that the size and efficiency of a well-established global publisher will help to keep subscriber costs down, while at the same time maximizing TWS's revenue potential and global distribution and influence with their marketing and information transfer infrastructure. The more wildlife professionals and other interested individuals TWS can reach with its publications, the greater its ability to connect research and management and thereby advance wildlife management and conservation.

REPORTS, BOOKS, AND PROCEEDINGS

Many societies also publish other reports, books, and proceedings that serve to connect science and management. TWS publishes a technical review series on wildlife management or conservation issues of current concern. TWS Council President must approve a proposed topic before work can begin. Such reports serve to synthesize the existing knowledge on the issue at hand, identify data gaps, and make recommendations. These technical reviews provide a useful resource to managers looking for a compilation of all the relevant literature on a topic. Recent technical reviews by TWS have examined wind energy and its effects on wildlife and habitat, lead in ammunition and fishing tackle (co-published with the AFS), and the Public Trust Doctrine as it applies to wildlife management and conservation. Available to the public as free Portable Document Format (PDF) files, each completed report is distributed to relevant state and federal agencies upon its completion, and to relevant members of Congress. TWS often uses a completed technical review as the basis for a position statement, a topic discussed in more detail later. These technical reviews, while still considered scientific publications, serve a different role than scholarly journals. By consolidating and summarizing research, the reports make large amounts of scientific information more accessible and digestible to readers, especially managers and policy makers who may not have the time needed to perform the primary source research themselves or the expertise to synthesize the science.

The ESA has a similar series of reports, called policy papers, which capture current ecological knowledge about a particular topic and provide policy recommendations. The reports are peer-reviewed by outside reviewers and by ESA's Public Affairs Committee, and are then approved by the Governing Board. ESA's *Issues in Ecology* series uses commonly understood language to provide information on important environmental and ecological issues, such as biotic invasions, pollution, water issues, and climate change.

Professional societies also publish books, ranging from traditional textbooks to field manuals to conference proceedings. The AFS has a books department that serves as a full-service publisher, offering planning, peer review, manuscript development, editorial, production, distribution, and marketing assistance. In-depth studies on fisheries and interdisciplinary subjects, thematically related collections of papers, and general interest books are published by AFS, either alone or in conjunction with its subunits or other outside partners.

TWS has published a number of books internally, including six editions of a wildlife management techniques manual first published in 1960 that serves as a major textbook in university curricula (Braun 2005). A seventh edition will be published in 2012 (Silvy, In Press). TWS also publishes a textbook on the human dimensions of wildlife management (Decker, Brown, and Siemer 2001). Beginning in 2009, book publishing at TWS has been in cooperation with the Johns Hopkins University Press, one of the oldest and largest nonprofit university presses in the United States. As part of this agreement, TWS has also launched a new book series on wildlife management and conservation.

MEETINGS

Meetings of professional societies provide venues for members and others to come together face-to-face to network and share their knowledge. Such venues also provide attendees an opportunity to pursue continuing education and increase their skills through workshops and other events (Figure 7.1). Most professional societies have annual conferences, as do their subunits such as regional sections, state chapters, or working groups that focus on specific topics, such as endangered species, policy, biometrics, and international issues. Offering practical workshops, concurrent sessions featuring the latest research, and numerous social activities that provide opportunities for networking, mentoring, and job searching, conferences and meetings are a vital way for professional societies to help connect research with on-the-ground management.

Annual meetings of TWS were held in conjunction with the North American Wildlife and Natural Resources Conference from 1937 to 1993. This was a natural pairing, as TWS grew out of discussions among wildlife biologists at the first North American Conference in 1936. Beginning in September 1994, TWS has held its own annual conference sponsored each year by a state chapter, out of a desire to connect more with the state chapter and local members where the meeting was taking place. Conducting its own conference also gave TWS the opportunity to expand the scientific sessions and papers presented and emphasize student involvement at the conference. TWS Council continues to conduct its mid-year meetings annually in conjunction with the North American Wildlife and Natural Resources Conference.

Attendees at TWS's 2010 annual conference in Snowbird, Utah had the opportunity to choose from over 15 workshops and symposia, over 330 papers and 110 poster sessions presented concurrently, a dozen field trips, and numerous social events. One

FIGURE 7.1   Conference participants interact at the tradeshow at the TWS Annual Conference in Snowbird, Utah, October 2010.

of the panel discussions at the 2010 annual conference dealt specifically with the issue of integrating wildlife science and management. At the subunit level, many members and committees from different professional societies meet together, thus increasing the networking opportunities and the exchange of information. For example, TWS and AFS chapters often meet together; and in Minnesota, the local Society for Conservation Biology (SCB) chapter joins the state chapters of those two organizations for their annual meetings.

Other subunits, such as topical working groups, may also hold meetings for wildlife professionals. For example in 2009, the Urban Wildlife Working Group of TWS, together with the Massachusetts Division of Fisheries and Wildlife, Massachusetts Department of Conservation and Recreation, University of Massachusetts Amherst, and USGS Massachusetts Cooperative Research Unit, held an International Symposium on Urban Wildlife Ecology and Management at the University of Massachusetts campus in Amherst. Similarly, TWS's Wildlife Damage Management Working Group holds an annual conference that draws hundreds of specialists on human-wildlife conflicts and issues of wildlife population control.

PROFESSIONAL STANDARDS

Societies also play an important role in establishing standards for the profession, through codes of ethics, certification programs, and publication standards. Taken together, such standards provide guidelines for acceptable behavior within the wildlife professional community. TWS has developed a code of ethics, as have other societies such as AFS and ESA. Each of these societies also offers certification programs whereby members, and in some cases non-members, can become certified by the organization as biologists, fisheries professionals, or ecologists, respectively.

Codes of ethics and professional standards are means to facilitate the integration of science and management by virtue of establishing common standards for practitioners. What constitutes a professional and what governs his or her behavior are essential components for both development and application of science. Standards establish a foundation of scientific training, and codes of ethics typically mandate that practitioners base management actions on current science. Increasingly, individuals with little or no academic or professional training or non-science-based organizations with their own agendas are influencing the goals and direction of natural resource management and conservation, especially when their opinions are voiced through the Internet and social media. It is not that these individuals do not have a role to play or a voice in the debate—they certainly do—but scientific training and professional credentials are of critical importance and should be recognized as such.

TWS will have an increasingly critical role to play in establishing and documenting the background, training, and standards that individual professionals should maintain, and this information should be made known to policy and decision makers. Such standards are certainly the case for most professions, from law and medicine to physical therapy and plumbing. The need for recognized standards will be even more critical in the future of the natural resources professions. That being said, the profession also needs to embrace the growing diversity in the wildlife profession and the high degree of specialization that is occurring in this rapidly evolving field.

It is important to note, however, that the sources of information and their credibility have become more, not less, important in this situation. Violations of codes or standards, if proven after a thorough examination of specific cases, can be followed by sanctions, penalties, and removal from the professional society.

ADVOCACY AND SCIENTIFIC INTEGRITY

A key role for professional societies is serving as advocates for their members and enforcing scientific integrity. Most larger professional societies in the natural resources field employ professional staff charged with assisting the members and leadership in developing official positions on policy issues, representing their members' views, and advocating for the society's positions. Whether called public policy offices, government affairs programs, or legislative affairs, these staff members all exist to serve the important role of helping to translate the work of researchers, especially society members, into a language that policy and decision makers can understand, which helps to ensure that such science is considered when policies governing wildlife management and conservation are formed.

A primary role of these offices is helping to establish official society policy. For example, the TWS government affairs staff works with TWS leaders and members to develop position statements on important wildlife management and conservation issues. A "position statement" is a carefully prepared and concise exposition on a wildlife issue that defines the issue, contains factual background data, describes the most probable biological, social, and economic implications of alternative actions,

FIGURE 7.2   Meeting of TWS Council during the annual conference in Monterey, California, September 2009.

and may contain a recommended course of action. These statements are adopted by TWS Council (Figure 7.2), following a period of review and comment by the membership. TWS has over forty position statements. Once developed, these position statements allow TWS government affairs staff to express positions and educate key decision makers in government knowing that they have the full and unambiguous backing of the Society and its members. Of course, individual members may not agree with every science-based position taken, but they all have a chance to comment and leadership considers their comments during policy formation.

Other societies develop similar documents. The ESA has a dozen policy statements—two- to four-page statements of the Society's position on critical national or international issues, which provide policy recommendations, and are reviewed by ESA's Public Affairs Committee and approved by ESA's Governing Board. The AFS has thirty-two policy statements.

THE WILDLIFE SOCIETY AS AN EXAMPLE

TWS was founded in 1937 and is a nonprofit scientific and educational association of professional wildlife biologists and managers, dedicated to excellence in wildlife stewardship through science and education. The mission of TWS is to represent and serve the professional community of scientists, managers, educators, technicians, planners, and others who work actively to study, manage, and conserve wildlife and its habitats worldwide. As a professional society, TWS values its over 11,000 members and is committed to providing excellent and responsive membership services, including opportunities for professional development and recognition, information sharing, communication, and networking. TWS also values science as a necessary tool to understand the natural world and supports the use of science to develop rational and effective methods of wildlife and habitat management and conservation, and to inform policy decisions that may affect wildlife and wildlife habitats (The Wildlife Society 2010).

Throughout its history, TWS has worked to connect wildlife researchers and the science they create with the managers working directly with wildlife and habitats. As noted previously, publications and annual conferences have been one of their primary methods of accomplishing this goal. Recent changes to the TWS annual conference format have included adding more interaction among attendees, through more panel and round table discussions, with the goal of facilitating interchange about current and important topics.

TWS's publications are also changing to encourage this dialogue between research and management and strengthen the connection between these two aspects of the profession. As noted previously, in 2007 TWS's member magazine, *The Wildlife Professional*, was launched. Containing news and analysis, it is designed to keep today's wildlife professionals informed and up-to-date about critical advances in wildlife science, conservation, management, and policy. *The Wildlife Professional* features in-depth articles as well as brief summaries of relevant scientific articles and a profile of a modern professional wildlife manager. Additional columns cover topics such as health and disease, human-wildlife connections, and ethics in practice. By providing articles on the latest science and research development in a more readable format, *The Wildlife Professional* provides an outlet for on-the-ground managers to connect with emerging science.

Also noted previously is the re-launch of the *Wildlife Society Bulletin* in March 2011. In 2010, TWS Council decided to begin re-publishing the *Bulletin*, to provide a focus on papers dealing with applied wildlife management and field techniques, as opposed to more basic scientific inquiry. The target audience for the *Bulletin* is the field-based wildlife practitioner. The Council's action was in direct response to membership appeal to reinstate the *Bulletin*, as well as the financial ability to re-launch. The perception among many, in spite of the Council's mandate and the efforts of the *Journal* editors to solicit applied papers, is that content formerly published in the *Bulletin* was not being published in the *Journal*. The *Journal of Wildlife Management* (JWM) will continue to publish articles on basic wildlife biology and ecology. However, the editor of JWM will also continue to ask authors to provide brief summaries of the management implications of their work. In this way, both journals will place a premium on the integration of science and management, but their emphases will be different, with the *Bulletin* designed to be more accessible to field managers interested in submitting manuscripts.

The way TWS interacts with its members through its website and other online media is also changing. No longer a venue for one-way information exchange, the TWS website, and other professional society websites, is increasingly interactive, encouraging members to connect with TWS staff and each other. TWS has a Facebook page with almost 7,000 fans, a Twitter account with over 4,000 followers, and a LinkedIn account. The connections that these programs offer, between TWS and other members as well as among members, serve to strengthen the tie between science and management. TWS has also launched a new blog, Making Tracks, to provide another forum for discussion about wildlife management and conservation. TWS also offers various newsletters to its members, such as The Wildlifer (a monthly member newsletter), Wildlife Policy News (a monthly publication that updates members on wildlife policy issues), and a weekly Wildlife News Update (which provides a compilation of wildlife-related new stories from the preceding week).

Many examples of TWS's work in connecting science and policy come from the government affairs program. The TWS government affairs program covers a broad arena of policy issues related to wildlife research, management, and conservation. Communicating with Congress, the administration, and federal and state agencies, preparing position statements, and educating our members on policy issues are all part of the government affairs program. Drawing on the expertise of TWS Council and individual members, and the local connections and expertise of over 150 sub-units (chapters, student chapters, sections, and working groups), TWS staff work to ensure that management plans, regulations, and other wildlife policies are based on the best available science.

During 2007, TWS Council formed an ad hoc Science and Policy Committee to explore the growing concern among wildlife professionals about the misuse of science in formulating wildlife policy (Table 7.1). The committee developed a white-paper with recommendations, which was discussed during the plenary session at TWS's 2009 annual conference and featured in an article in *The Wildlife Professional* (Haufler et al. 2009). The committee's work also formed the basis of a new position statement by the Society, on the Use of Science in Policy and Management Decisions.

In its report, the committee noted that the role of science in policy and decision making is to inform the decision process, rather than to prescribe a particular outcome. Policy and decision makers may make determinations that do not always provide maximum benefits or minimize impacts to wildlife and their habitats. However, such determinations are appropriate if the best available science and likely

---

TABLE 7.1

Examples of the Misuse of Science in Formulating Wildlife Policy

Type of Abuse	Example (Haufler et al. 2009)
Changing research results or conclusions to support desired policies.	Assistant secretary changed findings and recommendations of scientists regarding endangered species.
Ignoring science that contradicts desired policy outcomes.	A 2009 report by the National Corn Growers Association claimed there are many misconceptions about the hypoxic zone in the Gulf of Mexico, due to a lack of data. No such lack of data exists.
Reporting as a universal finding the results of science that apply only to a subset of situations.	Both sides in the debate over forest management are guilty of this, claiming either that any management is good or that all management is bad.
Promoting alternative "hypotheses" that have no underlying scientific foundation in order to raise doubts.	A study by a group in Florida and financed by developers implying that wetlands "discharge more pollutants than they absorb" prompted EPA to issue a development permit. The study contradicted the current state of knowledge.

consequences from a range of management options have been openly acknowledged and considered (Hutchins 2010). The committee noted that there are several ways that science can be abused by decision makers in setting policies and making management decisions to the detriment of wildlife and their habitats, as seen in Table 7.1. The committee made several recommendations for ways that TWS can address and correct abuses of science in determining policy, including increasing awareness and encouraging discussion of the issue; identifying abuses; protecting members muzzled by their employers relative to the application of sound science; working with federal and state agencies to establish guidelines for the use of science in policy formation; and expanding its role in providing peer review of documents for agencies, companies, and organizations to ensure that information has been compiled and summarized appropriately (Haufler et al. 2009).

TWS has a long history of holding agencies and others accountable for the scientific integrity of their actions. In early 2009, TWS and six other professional societies wrote to President Obama thanking him for his memo on scientific integrity, which called for the development of scientific integrity polices for each agency. A report by the Inspector General, released on April 28, 2010, noted that the Department of the Interior did not yet have a scientific integrity policy. TWS and three other organizations wrote to the Secretary of Interior to encourage development of such a policy, pursuant to the President's 2009 memo. In late August 2010, the Department of the Interior released a scientific integrity policy, and TWS provided comments and feedback.

TWS has also been involved in the controversy over management of the northern spotted owl (*Strix varia*) for years. The goal has been to ensure that the recovery plan, which will direct future management actions, is based on science. In 2007, the USFWS released a draft of the Northern Spotted Owl Recovery Plan. The USFWS worked with a stakeholder committee, including federal and state agencies, the timber industry, and conservation groups, to draft the recovery plan. The team's plan was sent to the Department of the Interior, where it was subject to internal review that resulted in an overall weakening of the plan. Upon its public release, TWS asked experts in population dynamics, spotted owl ecology, forest ecology and management, and fire ecology to conduct an independent review of the draft recovery plan. This group included persons who have participated in spotted owl research, planning, and recovery for the past thirty years.

TWS's review team concluded that although the spotted owl is one of the most studied raptor species in the world, and there is no other species listed under the U.S. Endangered Species Act for which such extensive information is available upon which to build a scientifically credible recovery plan, the draft plan did not adequately avail itself of the depth and breadth of this information, resulting in a seriously flawed plan for recovery. The group concluded that neither option presented in the 2007 plan would lead to recovery of this species and, in fact, the plan would reverse much of the progress made over the past twenty years to protect this species and the habitat upon which it depends. The USFWS contracted the SCB (North American Section) and the American Ornithologists' Union to review the draft plan, and their reviews were similarly critical of the science in the plan. It later became clear that high-ranking officials within the Department of the Interior intervened in the development of the plan, likely compromising the science-based protections in order to reduce barriers to increased logging in old-growth forests.

When the final plan was released in 2008, TWS re-engaged the same experts to review and comment on the plan. The experts concluded that USFWS responded constructively to some of the comments and suggestions submitted by TWS in 2007, and there were a few areas of improvement, but the overall underlying strategy for recovery of northern spotted owls was not improved from the 2007 draft plan, with some components weakened in the final recovery plan. In June 2011, the USFWS released a revised plan which addressed some of the concerns of TWS's reviewers and those from other scientific societies. Regardless, TWS's comments on the various incarnations of the plan serve as an example of the value of a professional society both in working to ensure that the best available science is reflected in management plans and policy, and in enforcing scientific integrity.

Other work by TWS also illustrates how professional societies can promote scientific integrity. For instance, in 2008, TWS provided valuable, science-based comments to the National Park Service in support of their efforts to remove nonnative ungulates from Point Reyes National Seashore. Similarly, that same year TWS also supported the USFWS in its plan to remove feral cats from San Nicolas Island in the Channel Islands of California. Providing science-based expert review of management plans, environmental impact statements, and other documents is a key role for TWS and a clear example of how professional societies can work to integrate science and management, even when public opinion may not favor a science-based management action.

Professional societies also play a valuable role in ensuring that professionals, especially those who are federal government employees, are able to participate fully in the activities of such societies. Federal conflict of interest legislation and executive branch ethics rules that are inconsistent and overly broad limit the ability of federal employees in some agencies to serve in leadership positions within their professional societies. Several natural resource societies have been working for years with the Office of Government Ethics, the Department of Justice, the Office of Science and Technology Policy, and the Office of Personnel Management to work toward exempting professional societies from the limitations on federal employee board service, and ensuring consistent government policies on such service. Since many TWS leaders, both on Council and in our chapter and sections, are federal employees this is a key issue for TWS. Federal agencies are important places where the integration of science and management occurs, and it is detrimental to resource management and conservation when federal scientists and managers are only speaking to one another. This can create a narrow worldview and institutional culture that is insular and therefore out of touch with the broader society. To be effective, federal managers and scientists must be kept closely informed of the work going on outside their agencies and programs (e.g., in academia, state wildlife agencies, NGOs, and industry), and one of the best ways for this to occur is for them to participate directly in their relevant professional and scientific societies (Hutchins 2009). Networking and interaction of this kind also helps to maintain scientific credibility and to limit the often-crippling effect of politics on science-based decision making. It is critical that such decision making is transparent and that the basis for it and potential consequences are fully explained to the public (Hutchins 2010).

CONCLUSION

Since its establishment in 1937, TWS has served as a voice and meeting place for wildlife professionals. TWS has played an active role in connecting science and research with on-the-ground management and conservation. Professional societies like TWS provide the necessary network among scientists, researchers, managers, and administrators committed to wildlife management and conservation. This allows translation of research innovations to managers and conveyance of research needs to scientists, helping to ensure that wildlife policy is based on the best available science.

ACKNOWLEDGMENTS

The authors thank David Kittredge and Paul Fisette, both of the Department of Environmental Conservation at the University of Massachusetts-Amherst, for their thoughtful reviews of this chapter.

REFERENCES

Braun, C., Ed. 2005. *Techniques for Wildlife Investigations and Management*, 6th ed. Bethesda, MD: The Wildlife Society.

Decker, D. J., T. L. Brown, and W. F. Siemer, Eds. 2001. *Human Dimensions of Wildlife Management in North America*. Bethesda, MD: The Wildlife Society.

Haufler, J. et al. 2009. The abuse of science. *The Wildlife Professional* 3(3):44–45.

Hutchins, M. 2009. Partnering with government agencies. *The Wildlifer* 353. http://joomla. wildlife.org/index.php?option=com_content&task=view&id=503&Itemid=304#act

Hutchins, M. 2010. To advocate or not to advocate? *The Wildlifer* 363. http://joomla.wildlife. org/index.php?option=com_content&task=blogcategory&id=92&Itemid=312

Murie, O. J. 1954. Ethics in wildlife management. *Journal of Wildlife Management* 18:289–293.

Silvy, N. (Ed.). In Press. *The Wildlife Techniques Manual: Research and Management*, 7th ed. Baltimore, MD: Johns Hopkins University Press.

Swanson, G. A. 1987. Creation and early history. *Wildlife Society Bulletin* 15:9–14.

Taylor, E. J., and B. Lauber. 2007. Values and functions of scientific societies. *The Wildlife Professional* 1(2):28–31.

The Wildlife Society. Strategic Plan 2008–2013. http://joomla.wildlife.org/index.php?option= com_content&task=view&id=267&Itemid=273 (accessed August 31, 2010).

8 Funding Research as an Investment for Improving Management

Matthew J. Schnupp and David S. DeLaney
King Ranch, Inc.
Kingsville, Texas

CONTENTS

The big thing is that the King Ranch is one of the best jobs of wildlife restoration on the continent, and has almost unparalleled opportunities for both management and research. Still more important: it is a gem among natural areas, and must be kept intact.

—**Aldo Leopold (1947a)**

ABSTRACT

King Ranch and its South Texas Farm are located in Kingsville, Texas and have been in operation since 1853. Due to King Ranch's influential research and political efforts and the dedication of its family shareholders to wildlife conservation, it has established a reputation as a leader in both ranching and wildlife management. King Ranch strives to make conservation decisions based on sound research and has a rich history of research projects where management-based research was conducted to assist in decision-making processes. Research conducted on King Ranch can be broken into two categories: (1) management-based research, and (2) research-based management. In the context of this work, the nature of management-based research is analytical and didactic, and the results of such research function as a tool for determining appropriate management decisions. Research-based management is experimental and ascertainable, and results are applied to achieve specific management goals. Thus, by these definitions, research-based management is subject to consequences. Our objectives are to provide insight on how King Ranch has historically made wildlife, range, and other management decisions based on research and how King Ranch biologists Val W. Lehmann, William "Bill" H. Kiel, Jr., and Mickey W. Hellickson improved the welfare of King Ranch by conducting and applying sound management-based research. We also discuss the current King Ranch and how the family has always taken an innovative approach to wildlife and habitat management, and this commitment has remained steadfast as non-family professional management has taken a more active role in recent years in all aspects of ranch operations. In addition, we discuss how research-based management decisions were made and what the resulting unforeseen ramifications were. We will draw on the introduction of nonnative grass on King Ranch to illustrate this objective. Ultimately, we encourage landowners, managers, and biologists to make management decisions based on accurate and long-term research and consider all consequences before making decisions.

INTRODUCTION

King Ranch is a 333,865-ha (825,000-acre) ranch comprised of four divisions (Santa Gertrudis, Laureles, Encino, and Norias) that encompass a diversity of habitat types including cordgrass-bluestem, bluestem-prairie, mesquite-granjeno, mesquite-bluestem, oak-bluestem, and mesquite (*Prosopis glandulosa*) savanna. King Ranch is unique in that there is interdependency among wildlife and cattle, both of which are dependent upon range health. Range health is defined by the status of the rangeland ecosystems soil, ecological processes, and range condition, which is the site's vegetation occurring under climax condition. Ranching operations for King Ranch are entirely self-sufficient and the ranch's primary management goal is to generate a profit so as to sustain its stewardship of natural resources such as rangelands and wildlife populations. King Ranch strives to make conservation decisions based on sound research and the family shareholders' commitment to the protection of native flora and fauna for succeeding generations. Since the early 1900s, King Ranch has been a national leader in focusing on the economic and societal value of its natural resources.

MANAGEMENT-BASED RESEARCH

KING RANCH MANAGEMENT BEFORE BIOLOGISTS

Before King Ranch hired biologists, game management decisions were based mostly on field observations and intuition; however, it was well ahead of its time in almost every aspect of management. For example, from 1916 to 1918 a severe drought in South Texas contributed to the extirpation of the wild Rio Grande turkey (*Meleagris gallopavo intermedia*) population on the Laureles and Santa Gertrudis divisions of King Ranch (Lehmann 1948). Ten years later, approximately 300 turkeys were captured at the Norias division and relocated to the Laureles (100 turkeys) and Santa Gertrudis (200 turkeys) divisions.

In the 1930s, Caesar Kleberg, Norias division foreman, knew specific habitat requirements were important for bobwhite quail (*Colinus virginianus*). Thus, he experimented with brush shelters and fenced areas for bobwhite cover (Lehmann, unpublished data). With the help of current ranch management, Kleberg modified water wells in the pasture so they would overflow into earth tanks, which in many cases were fenced off from cattle, to provide drinking areas for wildlife (Lea 1957). The influence Caesar Kleberg had on the wildlife conservation went well beyond the borders of King Ranch. In 1917, he was appointed to the Texas Game, Fish, and Oyster Commission in which he served for two decades. During his tenure on the Commission, he was successful at petitioning in Austin for stronger state game laws, many of which were implemented unilaterally on King Ranch as "rules" prior to their legal status.

Harvest Management

In 1912, which was an era before statewide game laws and regulations in Texas, Caesar Kleberg suggested the first hunting codes (i.e., regulations) which were designed to increase sport and reduce crippling loss (Lehmann 1957:761). These codes were approved and put in place by King Ranch largely because, despite a perception to the contrary, King Ranch was never a "vast game preserve closed to all except a select few, who hunt largely without restraint" as most assumed at that time (Lehmann 1946). Therefore, King Ranch managers had the foresight to establish rules and regulations to conserve and increase wildlife populations on the Ranch (Forgason and Fulbright 2003).

In addition to following state and federal regulations, wildlife harvest on King Ranch was regulated by unwritten laws and traditions, which were even more restrictive than the government regulations (Lehmann 1946). For example, King Ranch bag limits were usually lower than were those mandated by government regulations. King Ranch harvest regulations for daily and seasonal limits on white-tailed deer (*Odocoileus virginianus*), turkey, and collared peccary (*Tayassu tajacu*) were one of each species per eligible hunter. In addition, hunters were allowed two geese (most likely lesser snow geese, *Chen caerulescens*) and eight bobwhites per person per day (Lehmann 1946). These limits were well below those allowed by the state at the time.

Additionally, hunting in areas where game were artificially concentrated was not considered ethical; therefore, dove (most likely mourning dove, *Zenaida macroura*) and waterfowl shooting was done prior to the opening of white-tailed deer season when water was normally abundant (Lehmann 1946), thus causing the game to be dispersed. In addition, areas that concentrated the most wildlife, like the Laguna Larga (one of the largest freshwater wetlands in Texas) were maintained as sanctuaries where hunting was prohibited (Lehmann 1946). King Ranch hunting seasons were shorter than those allowed by government regulations. White-tailed deer hunting season officially ended on King Ranch on December 15 when bucks were most vulnerable due to the rut (Lehmann 1946). Most quail hunting was ended by January 16 to preserve the remaining quail as brood stock for the next season. This was early recognition of the importance of winter survivors and the reality of compensatory versus additive harvest.

The method of hunting was also strictly regulated. It was felt that the customs of King Ranch concerning the kinds and sizes of guns would contribute to reducing crippling loss of wildlife (Lehmann 1946). For example, borrowed or defective guns with which the hunter was not familiar were not allowed. Additionally, to avoid crippling loss of turkey, shotguns were not allowed for turkey hunting. The original hunting regulations were based on foresight, biological intuition, or possibly unpublished data; thus, "hunting on the King Ranch … [was] different from hunting on most other ranches. The differences … [were] in the direction of efficient harvest and good sportsmanship" (Lehmann 1946).

KING RANCH MANAGEMENT AFTER BIOLOGISTS

By the early 1940s, Robert J. Kleberg, Jr., family member and acting manager of King Ranch, began to consider whether wildlife could be developed into a major auxiliary crop in south Texas (Lehmann 1957:762). In 1945, Kleberg hired Val W. Lehmann, who became the first known wildlife biologist to work for a private ranch (Forgason and Fulbright 2003). Lehmann's first undertaking was to develop an intensified research and management program (Lehmann 1957:762). The objectives of the program were: (1) to determine if game could be produced in significant quantities, and (2) to determine if game production was desirable from a ranching standpoint (Lehmann, unpublished data). Lehmann's program concentrated solely on objective 1 using bobwhites, white-tailed deer, and Rio Grande turkey as his research focus.

The wildlife management program Lehmann developed was extensive and investigated a number of different research areas. Some of his investigations included parasitology, water requirements, hunter harvest effects, habitat requirement, vitamin deficiency, range manipulation, and predator control. From 1940 to 1952, the predator control program was the most intensive ever applied in this hemisphere at that time, killing an estimated 14,086 large predators (mostly coyotes) and countless skunks (*Mephitis mephitis*), raccoon (*Procyon lotor*), opossum (*Didelphimorphia spp.*), ground squirrels (*Sciuridae spp.*), and rattlesnakes (*Crotalus atrox*) (Lehmann, unpublished data; Lehmann 1984). Results from this intensive predator removal showed that "where habitat during breeding season was poor, or when weather conditions were unfavorable, coyote control proved far short of compensating for these deficiencies" (Lehmann 1984). Lehmann was the first ever to conduct a comprehensive Texas study of white-tailed deer food habits and diets in addition to white-tailed deer and cattle range use relationships (Lehmann, not dated). His research

FIGURE 8.1   Val Lehmann (right) removing bobwhites from a walk-in trap in the Canelo Pasture on the Santa Gertrudis Division of King Ranch so they can be sexed, aged, and banded before they are released.

on bobwhites produced one of the cornerstone references of bobwhite management (Figure 8.1), *Bobwhites in the Rio Grande Plain of Texas* (Lehmann 1984). He also established the only significant turkey populations in predominantly mesquite (Santa Gertrudis Division) and scrub oak (Laureles) range in all of Texas at that time (Lehmann, not dated).

By the end of Lehmann's career, he was not only a well-respected scientist, but his accomplishments earned him the respect of Aldo Leopold, founder of the Department of Wildlife Management at the University of Wisconsin and a pillar of modern wildlife management. In January 1947, Robert J. Kleberg, Jr., King Ranch president, and Howard Dodgen, executive secretary of the Texas Game, Fish and Oyster Commission (today known as Texas Parks and Wildlife), invited Aldo Leopold and fifteen other leading conservationists of the nation to visit King Ranch on February 6 and 7, 1947, after attending the conference of the Wildlife Management Institute in San Antonio, Texas (Dodgen 1947). Their intention on inviting Leopold was to "gather from you valuable information on conservation problems that will be applicable to Texas" (Dodgen 1947). All invited conservationists accepted the invitation, including Leopold, and they spent two days touring King Ranch with Val Lehmann and Robert J. Kleberg, Jr.

Following Leopold's visit, Lehmann and Leopold corresponded through several (5) known letters. Leopold and Lehmann discussed various topics ranging from

Leopold's needs for ranch pictures (e.g., customized heavy equipment implements, wildlife, etc.) for a lecture he was hoping to formulate on the ecology and management of the King Ranch, the importance of goatweed as quail food, how "top-heavy" densities like quail and turkey are not affected by raptor control (Leopold also states that he cannot "speak too loudly" of the futility of raptor [i.e., predator] control on top-heavy densities "for I myself got high-pressured into assenting to a wolf bounty in Wisconsin on a range top-heavy[and then some] with deer. Worse still, the timber wolf is nearly extinct. I mention this only to show we are all in the same boat"), the relationship of jackrabbit populations with plant succession, and how Mr. Bob Kleberg instructed ranch employees to stop harvesting mountain lions, mostly found on the Norias division, based on a recommendation Leopold made during his King Ranch visit (Lehmann 1947a,b; Leopold 1947b,d). Ultimately, their letters illuminate the respect Leopold had for Lehmann, "Your opportunities for study are so extraordinary and your actual management work has been so good …" (Leopold 1947c), and Lehmann had for Leopold " … (I) hope (this letter) will serve as another step in cementing the closest possible relations and cooperation between your office and ours" (Lehmann 1947b).

In January 1962, William "Bill" H. Kiel, Jr., was hired as King Ranch chief wildlife biologist. While employed with King Ranch, Kiel was active with the white-tailed deer and turkey-trapping program in conjunction with Texas Parks and Wildlife Department (Forgason and Fulbright 2003). He assisted in the capture of somewhere between 12,000 and 14,000 white-tailed deer, which were moved to areas in Texas that had low white-tailed deer populations (Figure 8.2). He also aided in capturing between 3,000 and 4,000 wild turkeys, which were transplanted to other

FIGURE 8.2 Norias division cowboys releasing a young white-tailed deer buck from its transportation pen.

sites in Texas, as well as Nebraska and Hawaii (Forgason and Fulbright 2003). Many of the white-tailed deer, if not most, and Rio Grande turkey population in Texas today carry genetics that can be traced to the original King Ranch population.

From 1968 to 1974, Kiel banded 8,026 bobwhites on the Santa Gertrudis division of King Ranch (Kiel 1976). In addition, he collected 2,447 hunter-harvested bobwhites to determine band recovery rates and age ratios (Kiel 1976). From these data, Kiel was able to support the conclusion that bobwhite production is directly tied to rainfall (e.g., $r = 0.96$). Additionally, he determined that the rainfall received from May through July could be used as a predictor for the following Fall's juvenile to adult ratio. Kiel also found that female bobwhites are subjected to higher mortality rates than males (Kiel 1976). The band recovery data were also used to determine movement patterns in bobwhites. In addition, Kiel suggested that bobwhites are dependent on various habitat attributes. He suspected that nonnative grasses were not favorable for bobwhites and may become a future management problem because they inhibit weeds and insects vital to bobwhite survival (Kiel 1976). Flanders et al. (2006) confirmed this thirty years later.

In January 1999, King Ranch hired Mickey W. Hellickson as King Ranch chief wildlife biologist. When Hellickson arrived in 1999, he designed and implemented ranch-wide helicopter surveys to assess trends in wildlife populations (M. Hellickson, personal communication 2010). He conducted these surveys during September and October on every King Ranch pasture for a total of 4,000 miles of survey effort per season (M. Schnupp, unpublished data). These surveys became the cornerstone of wildlife harvest management recommendations and provided estimates of the wildlife populations found on King Ranch.

Hellickson was also instrumental in developing the white-tailed deer harvest quota system that produced some of the largest antlered white-tailed deer in the United States. From 1999 to 2009, Hellickson worked in conjunction with the Caesar Kleberg Wildlife Research Institute (CKWRI) at Texas A&M University–Kingsville to develop the South Texas Buck Capture Project and King Ranch Deer Project. This research resulted in several important discoveries regarding white-tailed deer breeding and reproduction, DNA, antler growth, home range, aging techniques, and movement patterns that have gained the researchers, and King Ranch, international recognition. Specifically, some noted discoveries include the following: antler and body size has little or no influence in deciding whether a buck is successful at siring fawns; the average home range of a breeding buck during breeding season is 2,950 acres (1193 ha) and has a high degree of variability; 35% of all fawns are sired by young bucks between 1.5 and 2.5 years old; females can be bred successfully by more than one buck; old mature bucks do not monopolize breeding; and translocating bucks to different areas negatively affects antler growth, and they may travel thirty to fifty miles to return to the original capture site. Many of the results are and will continue to be useful in directing the future of south Texas white-tailed deer management.

In 1998, Hellickson and other King Ranch biologists collaborated with CKWRI on the South Texas Quail Project. From 1998 to 2008, researchers captured and radiomarked more than 2,000 bobwhites and investigated over 300 bobwhite nests (F. Hernandez, unpublished data). This research was the first King Ranch project in which modern techniques like VHF radio collars were used. From this project, researchers obtained valuable data on survival, nesting success, causes of mortality, and hunter dynamics (Teinert 2009; Rader et al. 2007; Rader et al. 2011; Hardin et al. 2005). These findings provide more insight into the appropriate management practices necessary to increase bobwhite abundance through suitable habitat and harvest management.

CURRENT KING RANCH MANAGEMENT

King Ranch is no less dynamic today as it was when the first artesian water well was drilled a mile from the headquarters on June 6, 1899. The King Ranch family has always taken an innovative approach to wildlife and habitat management, and this commitment has remained steadfast as non-family professional management has taken a more active role in recent years in all aspects of ranch operations. Individuals with B.S., M.S., and Ph.D. degrees are now part of daily operational decisions. King Ranch management has developed and implemented one of the most efficient and dynamic private wildlife programs in North America, if not the world. The current ranching and wildlife operations have made numerous advancements for a vast array of interests (e.g., private landowners, cattle operations, government entities, etc.). However, for the sake of brevity, we will concentrate on the three that we deem the most important: (1) development of an efficient wildlife density estimation technique and resulting density-based harvest quotas, (2) development of a long-term data storage and analysis program, and (3) continued support for wildlife and ranching research.

Step one in managing any wildlife population that receives significant harvest pressure is to determine a baseline pre-season population estimate. As such, estimating population abundance and setting harvest quotas for game species, particularly quail and deer, on King Ranch are key components of our wildlife management program. From 1998 to 2009, the months of September through November were used to collect and analyze wildlife survey data conducted from a helicopter. On a yearly average, King Ranch biologists spent one to two months surveying 5,646 km (3,508 mi) of transects collecting data with a pencil and data sheets, two to three weeks entering data sheets into the computer, and two to three weeks organizing these data into harvest quota tables. Field data were stored in over 400 different Microsoft Excel files, 200 of which had ten different tabs resulting in approximately 2,400 Microsoft Excel sheets of data. Density was estimated, long-term trends were reviewed (e.g., buck:doe ratio, etc.), and harvest quotas were typically set based on a maximum harvest per acre basis. For example, the maximum quail harvest was set at one harvested quail per 6 ha (15 ac). Ultimately, harvest quotas were not based on a biologically justifiable harvest quota.

In October 2008, the first privately funded helicopter-based distance sampling quail survey was initiated on King Ranch (Figure 8.3). The quail surveys, which are conducted on 120,000 acres, utilized electronic survey equipment (e.g., rangefinders, field computer, etc.) and a CyberTracker survey database developed by ranch biologists. In December 2008, King Ranch biologists used results from the helicopter-based distance sampling surveys to determine a pasture density, prescribe a varying

FIGURE 8.3   Helicopter-based distance sampling has been determined to be a viable technique for estimating densities of bobwhite quail and plays a vital role in the King Ranch quail program.

percentage harvest (typically 10 to 20%), determine a per hunter harvest quota, and equitably distribute these quotas among hunters. King Ranch has continued to conduct the quail program in this manner and has found it to be a successful way to manage hunter satisfaction. Additionally, the quail program now determines a harvest quota that is based on the number of birds present at that time (i.e., density) so the harvest is more sustainable. The quail surveys and survey software have become an essential component to managing the King Ranch quail population. Given our success with the density-based harvest quotas for quail, King Ranch has also begun using the same type of approach with the deer population. Although we are still developing the program, basing harvest rates on the estimated densities of deer has shown to be a realistic option given that the survey data are precise and available in a timely manner.

In 2010, in an effort to provide precise and immediate wildlife survey results, the survey software developed for quail surveys was utilized on all wildlife surveys conducted on King Ranch. King Ranch and CKWRI have invested over $30,000 from 2008 to 2011 for advancements in the CyberTracker survey software. The software enables biologists to collect survey data with an electronic, easy-to-use, streamlined data collection system and software that minimizes errors and requires no post-survey data entry. The survey software has made the King Ranch wildlife surveys 80% to 90% more efficient (based on field data entry and data transcription) than the traditional pencil and paper technique. Additionally, results are nearly instantaneous and the data format is far more useful in statistical analysis programs. In efforts to make the software applicable in most common scenarios, modifications to the software have been made to enable users to collect a wide variety of survey data (e.g., spotlight counts, call counts, etc.). To ensure public access, King Ranch has made the survey software available at no cost to the user. Although we provide no technical support for the software, King Ranch collects over 25,000 lines of data each year with few complications.

In addition to the wildlife survey data, King Ranch collects thousands of lines of data on a weekly basis (e.g., game harvest, rainfall, brushwork, etc.). With over 825,000 acres and various operations (e.g., cattle, wildlife, range, etc.), there is a huge demand for a safe and secure storage and retrieval system for the various data. These data are not only necessary for daily operation, but they are essential to developing future strategic management plans. These data can be used to investigate research questions like "How does rainfall affect the number of harvested bucks with a Boone and Crocket score greater than 170?" In a hostile and erratic environment like south Texas, these questions can only be answered with a multifaceted and long-term database. In 2007, King Ranch began developing what has become the King Ranch Natural Resources Management System (KRNRMS). The original vision for this system was to provide an all encompassing storage and retrieval system that includes data like rainfall, game harvest, habitat treatment, stocking rates, forage and range condition, and wildlife survey data so any relationship and resulting correlation could be determined. After four years, five of the six systems have been built and the final component to the system, wildlife survey data, is projected to be completed by the end of 2012. To our knowledge, the KRNRMS is the only single database that includes most of the common factors that contribute to range, wildlife, and cattle production. The data are collected on 309,584 ha (765,000 ac) that encompass six habitat types. Over fifteen wildlife species are included, and the harvest management system alone includes 128,000 lines of source code. Moreover, this database already has more than ten years of data stored and has been designed to collect future data in the same manner. Therefore, the system can be used to analyze and store data for current and future management personnel and students.

Funding and providing data for researchers, biologists, and students has and will continue to be important to King Ranch and its family shareholders. In 2010 alone, there were more than seventy-five research projects being conducted on King Ranch. Projects ranged from quail, cattle, and bobcats to invasive grasses, brush control, and how large ranches measure and incentivize employee performance. In 2009, King Ranch made a four-year commitment to provide $20,000 per year for funding for a Ph.D. student to begin analyzing and publishing data that were stored in the KRNRMS. Additionally, it works with companies like DuPont to promote the development of new, inexpensive, and effective alternatives for treating problem brush like Huisache (*Acacia farnesiana*). In 2003, to commemorate the 150th anniversary of King Ranch, the King Ranch family shareholders and friends created the King Ranch Institute for Ranch Management (KRIRM). KRIRM is designed to train students in a multi-faceted systems approach to managing large, complex ranches and the biological and financial issues that confront them. Not only do they educate future managers, they also conduct a wide variety of research that serves the ranching industry and provides the students with tools to become capable and competent ranch managers. Given the sheer size and capabilities of King Ranch, ranch management and its family shareholders feel a certain obligation to be on the forefront of innovative technology and then to provide insight for others based on the research results. More importantly, ranch management strives to make management decisions based on accurate and long-term research so they can thoroughly understand and consider the ramifications before making any management-based decisions.

Management-based research should be an important component of contemporary wildlife science. From the research conducted on King Ranch, scientists have built models, tested hypotheses, and have ultimately advanced the field to what it has become today. King Ranch 2011 is similar to King Ranch 1900 in that it continues to apply management procedures based on sound management-based research. However, some historic decisions made by King Ranch have become controversial, as is the case of the King Ranch grass program. The King Ranch grass program could be viewed as research-based management that had unexpected consequences.

RESEARCH-BASED MANAGEMENT

THE KING RANCH GRASS PROGRAM

For more than a century, grazing livestock was the sole source of income for most Texas ranchers. The success of cattle in Texas is largely attributed to King Ranch. Prior to the turn of the century, Robert J. Kleberg, Sr., family member and acting manager of King Ranch, felt that the essence of cattle ranching was to produce high quality grass and to harvest this grass with livestock adapted to the climate, which would ultimately show the highest efficiency in rate of gain and quality of carcass (Kleberg and Díaz 1957:743). In addition, he felt that highly nutritive grasses that had the hardiness to spread and invade existing species under natural grazing conditions needed to be found or developed (Kleberg and Díaz 1957:743). By the 1900s, King Ranch was on the forefront of introducing nonnative grasses to increase the carrying capacity of livestock (Forgason and Fulbright 2003; Figure 8.4).

From 1900 to 1925, Robert J. Kleberg, Sr., began testing grasses with an introductory garden that was used to evaluate new forages (Kleberg and Díaz 1957:743), most of which were species native to Africa. Around 1914, King Ranch began experimenting with the plantings of Rhodesgrass (*Chloris gayana*). Rhodesgrass had a high grazing value and was very drought tolerant. The carrying capacity of Rhodesgrass (one animal to 6 acres [2 ha]) was nearly quadruple the carrying capacity of the native vegetation (one animal to 25 acres [10 ha]; Kleberg and Díaz 1957:744). By the height of the planting program, King Ranch had about 75,000 acres (30,351 ha) planted with Rhodesgrass (Kleberg and Díaz 1957:745). Rhodesgrass became one of the most popular grasses ever planted on south Texas ranges and came to be known as "the wonder grass" of south Texas (Kleberg and Díaz 1957:745).

On November 26, 1942, Nick Díaz, a Soil Conservation Service agronomist, who was later hired by King Ranch as their resident agronomist, found a minute parasitic insect was attacking the basal crown and nodes of Rhodesgrass (Kleberg and Díaz 1957:745). The parasite, which was suspected of arriving in south Texas via packing material used to protect chinaware during shipment from Japan, virtually

FIGURE 8.4 The seeds of target grass species (Rhodesgrass, King Ranch Bluestem, etc.) were harvested using relatively complex pieces of machinery for its time. The harvested seeds were distributed elsewhere on the ranch and across south Texas using various types of machines.

eliminated all Rhodesgrass on King Ranch in two years. Although the Rhodesgrass was decimated, King Ranch continued to break away from the traditional ranching paradigm of hoping for rain and letting Mother Nature take care of the range, and instead reestablished its nonnative grass program with other species (Kleberg and Díaz 1957:746).

During the grass improvement program, Nick Díaz and Robert J. Kleberg, Jr. worked closely with various projects associated with nonnative grasses and King Ranch donated grants to various institutions such as the University of Texas, the Texas Research Foundation at Renner, Texas A&M College (currently Texas A&M University–College Station), and Texas College of Arts and Industries (currently Texas A&M University–Kingsville). The King Ranch grass improvement program resulted in a monumental amount of knowledge and advancements in modern ranching. By May 1955, the Texas Research Foundation, which was supported financially by King Ranch and private donations, completed 4,498 studies to analyze soil, water, feed, and plants to further livestock and in some cases wildlife management (Kleberg and Díaz 1957:751). Most importantly, the policy of King Ranch was to release findings from projects as public information so all ranchers could benefit from the research (Kleberg and Díaz 1957:751). Ultimately, the goal of researchers, Robert J. Kleberg, Jr., and his staff was to provide an endless search for more practical and effective ranching methods (Kleberg and Díaz 1957:750). An important method they determined to be effective was to plant more productive nonnative grasses for livestock grazing.

In 1939, Nick Díaz found for the first time Kleberg bluestem (*Andropogon annulatus*) in Kleberg County growing along the side of a road near an abandoned grass nursery, and in 1944 he found King Ranch bluestem (*Andropogon isch-aemum*), which was found on the Santa Gertrudis division of King Ranch (Kleberg and Díaz 1957:747). Reports from ranch employees stated King Ranch bluestem had been on King Ranch for more than thirty-five years before being discovered by Díaz, and it possibly originated from an introduced grass plot planted near the ranch dairy barn (Kleberg and Díaz 1957:747). Both grasses proved themselves to be among the best-adapted forage grasses in south Texas (Kleberg and Díaz 1957:747).

By the 1950s, King Ranch was planting these grasses and others like Buffelgrass (*Pennisetum ciliare*) and Bermuda (*Cynodon spp.*) to invade native grass areas and increase carrying capacity by providing grass that produces large quantities of forage under harsh conditions. The general procedure of the King Ranch grass planting program was first to clear brush while plowing the ground and sowing these grasses (Kleberg and Díaz 1957:748; Figure 8.5). Ultimately, the value of grass was determined by a combination of its productivity and nutritional value for livestock (Kleberg and Díaz 1957:749). King Ranch had considered the effects of nonnative grasses on wildlife populations, specifically bobwhites, because both the brush clearing and grass program were closely coordinated with the game conservation

FIGURE 8.5   King Ranch and Holt Machinery Company developed the first Twin D8 funnel dozer (front implement) and root plow (rear implement) used to remove brush and sow introduced grasses. The funnel dozer could knock down 12-m tall mesquite trees and the root plow would remove the stumps and root buds up to 40 cm while dropping grass seeds.

and management program (Kleberg and Díaz 1957:749). However, it would take nearly twenty years (and the establishment of monocultures on large areas) before the effects of the nonnative grass program would become evident on non-target species such as bobwhites.

CONSEQUENCES OF THE KING RANCH GRASS PROGRAM

From 1945 through 1949, Lehmann determined that nonnative grass species were not abundant and did not have a negative impact on game populations (Lehmann, not dated). Twenty-five years later, the nonnative grass program began to reveal the negative impacts it had on the King Ranch wildlife population, specifically bobwhites. During 1976, King Ranch biologist Bill Kiel suspected that dense stands of nonnative grasses inhibited weeds, seeds, and insects that were vital to bobwhite survival. In addition, the densely grown grasses drastically reduced bare ground and obstructed the movement of bobwhites and thus areas dominated by nonnative grasses would hold few to no bobwhites (Kiel 1976). However, the economic incentive ($2.00 per acre [$0.80 per ha] for a quail hunting lease; Kiel 1976) for landowners to provide bobwhite habitat had not progressed to the point where landowners were willing to remove nonnative grass stands (and reduce grazing value) in efforts to increase bobwhite habitat.

By the 1960s and 1970s, revenues based solely on cattle profits were becoming difficult and most ranches were looking for ways to increase revenues (Genho et al. 2003). In the late 1980s, Stephen J. "Tio" Kleberg, family member and King Ranch manager, began to implement the first hunting leases on King Ranch (Genho et al. 2003). During the 1990s, King Ranch and various south Texas landowners began reporting income from wildlife that was equivalent to, or exceeded, profits made from livestock (Rhyne 1998). By 1996, Kleberg County, a county that contains a large portion of King Ranch, attributed 42% of the real estate market value to hunting and outdoor recreation (Baen 1997).

With the increase in financial incentives for landowners to provide hunting opportunities and the decrease in profitability of cattle, landowners' perspectives on the value of wildlife began to change. Because of this, research shifted to developing ways to improve wildlife habitat and increase the abundance of game species. By the early 1990s, it became evident to researchers that nonnative grasses were detrimental to wildlife populations. Pimm and Gilpin (1989) ranked the invasion of nonnative species as the second greatest threat to endangered species in the United States, second only to habit loss. In addition, it was noted by Wilcove (1998) that nonnative grass might be even more of a problem given that a nonnative plant invasion represents a form of habitat loss. However, by 2002, south Texas researchers could only speculate on the issue of the effect nonnative grasses had on wildlife populations because scientific facts were limited. Only about six studies were completed by 2001 and, unfortunately, most were conducted in the Southeastern and Midwestern United States (Kuvlesky et al. 2002). Research scientists and King Ranch management became frustrated because while they suspected there was a problem, they lacked empirical evidence on the subject due to the lack of research (Kuvlesky et al. 2002).

A few years later, information on the effects of nonnative grasses became abundant and ultimately the results were dismal. Flanders et al. (2006) determined, as expected, that not only was nonnative grass extremely invasive but it would outcompete and replace native grasses in sites where it was introduced. It also affected forb, insect, and grass species richness and diversity. In addition, total bird abundance was 32% greater on native grass sites than on nonnative grass sites and birds that foraged on the ground, including bobwhites, were nearly twice as abundant on native grass sites (Flanders et al. 2006). Flanders et al. (2006) also determined that spiders, beetles, and ants were 42% to 83% more abundant on native grass sites than on bufflegrass (a nonnative species) sites. Sands et al. (2009) determined that areas with bufflegrass had a 73% reduction in available forb canopy and a 64% reduction in forb species richness. With this research, it confirmed that nonnative grasses negatively affect wildlife species, especially bird species regarding their food-producing areas used for foraging. However, these areas can provide adequate nesting cover for ground-dwelling bird species.

Within the past decade, there has been interest and concern from landowners, ranchers, and biologists to fill the knowledge gap of how nonnative grasses affect wildlife and how the potential effects might be mitigated. In efforts to form a bridge in the knowledge gap, the South Texas Natives Program was initiated in conjunction with Caesar Kleberg Wildlife Research Institute (CKWRI) in January 2000. King Ranch family members were instrumental in funding this initial program. South Texas Natives provides economically viable sources of native plants and seeds to both the private and public sector in efforts to restore the native plant communities in south Texas. In addition, CKWRI has initiated scores of research projects to understand nonnative grasses and their effects. In 2009 alone, CKWRI initiated eleven different research projects investigating nonnative plants and fourteen projects examining native plant habitat enhancement and relationships.

Although the research is not complete, researchers suspect that effects of nonnative grass will reach far beyond ground-dwelling birds. In addition, it will be very expensive and difficult to return a nonnative grass stand back to native grass because the grass was specifically designed to make it resilient and enduring. As with the Rhodesgrass, time will reveal the fate of the south Texas nonnative grasses. However, we suspect that researchers should concentrate future research on controlling and managing nonnative grass stands instead of trying to completely eradicate them over large landmasses. More importantly, we urge researchers and managers to design management and research to work in conjunction with the environment, rather than attempt to change the species within an environment to fit the management practices. Although research-based management does have its place in wildlife science, the consequences, if any, are typically unexpected and long term.

RECOMMENDATIONS

King Ranch has always been and will continue to be an influential leader in both ranching and wildlife management. We believe that the primary reason is that King Ranch has successfully and profitably operated on a large-scale agricultural and natural resource basis while sustaining and enhancing the natural resource values. Additionally, King Ranch strives to make conservation decisions based on sound research. This is evident in the support given to CKWRI and KRIRM, and the scores of research projects that are conducted annually on the ranch. Although natural resource management is the primary goal of King Ranch, no program is perfect, as illustrated by the nonnative grass program. Thus, we recommend having long-term goals and a strategic plan when making important management decisions. Sound, long-term research should be the guiding light of the decision. The core long-term goal of our management strategy is to keep King Ranch intact and profitable. With the ever-increasing loss of wildlife habitat, it is vital that large areas remain available. Aldo Leopold stated it best: "I hope and pray that the King Ranch will never be broken up: it is a gem the value of which few can appreciate" (Leopold 1947a).

Wildlife research is far too variable, especially in the south Texas environment, to draw conclusions from studies that last only two to three years. As managers, we need to be patient, heed the science, and work with research institutes such as CKWRI on a long-term basis. Additionally, we need to develop a sense of responsibility to base management on current and sound science and be willing to provide input and guidance to researchers instead of becoming weary and pessimistic. As researchers, we need to provide firm results that encompass the appropriate amount of time and effort necessary to disclose reliable results and possible consequences and provide these results in a clear and available outlet.

The "publish or perish" mentality for researchers and the "perception is reality" mentality for mangers are short term and unreliable. Thus, we feel that these are the origin of the disconnect between managers and researchers. We understand researchers have to publish to remain competitive and, by nature, managers tend to draw conclusions based on observations and opinions, but ultimately the importance of both entities working together is essential for the advancement and conservation of wildlife and the wildlife management profession. We will close with a quote that we believe epitomizes the problems we face as wildlife scientists and managers, and we leave it up to you to overcome them.

> Too often, what are at first possibilities become probabilities and then facts, not so much on the basis of additional study and field test, but apparently largely in progress through a number of papers. Advanced thinking is wonderful, but I wonder if it may also be dangerous, particularly when the pillars of the structure vary in strength and in analysis are not always so evaluated. (Lehmann 1947a)

ACKNOWLEDGMENTS

This chapter benefited greatly from review and editorial comments by Joseph P. Sands, Leonard A. Brennan, Marc L. Bartoskewitz, and Jessica L. Ruiz. We would like to thank Lisa Neely, archivist for King Ranch, who located articles and provided input on the historical articles.

REFERENCES

Baen, J. S. 1997. The growing importance and value implications of recreational hunting leases to agricultural land investors. *Journal of Real Estate Research* 14:399–414.

Dodgen, H. D. 1947. Personal letter from Howard D. Dodgen to Aldo Leopold. January 16, 1947.

Flanders, A. A., W. P. Kuvlesky, Jr., D. C. Ruthven, III, R. E. Zaiglin, R. L. Bingham, T. E. Fulbright, F. Hernández, and L. A. Brennan. 2006. Effects of invasive exotic grasses on South Texas rangeland breeding birds. *The Auk* 123:171–182.

Forgason, C. A., and T. E. Fulbright. 2003. Cattle, wildlife and range management on King Ranch over the years. In: *Ranch Management; Integrating Cattle, Wildlife, and Range*, C. A. Forgason, F. C. Bryant, and P. C. Genho, Eds. Kingsville, TX: King Ranch, pp. 9–21.

Genho, P., J. Hunt, and M. Rhyne. 2003. Cattle, wildlife and range management on King Ranch over the years. In: *Ranch Management; Integrating Cattle, Wildlife, and Range*, C.A. Forgason, F. C. Bryant, and P. C. Genho, Eds. Kingsville, TX: King Ranch, pp. 81–107.

Hardin, J. B., L.A. Brennan, F. Hernandez, E. J. Redeker, and W. P. Kuvlesky, Jr. 2005. Empirical tests of Hunter-Covey interface models. *Journal of Wildlife Management* 69:498–514.

Kiel, W. H., Jr. 1976. Bobwhite quail population characteristics and management implications in South Texas. *Transactions of the North American Wildlife and Natural Resources Conferences* 41:407–420.

Kleberg, R. J., and N. Díaz. 1957. The grass program (as told to Francis L. Fugate). In: *The King Ranch*. Volumes 1 and 2, T. Lea, Ed. Boston, MA: Little, Brown and Company, pp. 743–760.

Kuvlesky, W. P., Jr., T. E. Fulbright, and R. Engel-Wilson. 2002. The impact of invasive exotic grasses on quail in the southwestern United States. In: *Quail V: The Fifth National Quail Symposium*, S. J. DeMaso, W. P. Kuvlesky, Jr., F. Hernández, and M. E. Berger, Eds. Austin, TX: Texas Parks and Wildlife Department, pp. 118–128.

Lea, T. 1957. *The King Ranch*. Volumes 1 and 2. Boston, MA: Little, Brown and Company.

Lehmann, V. W. not dated. Wildlife production studies on the King Ranch. Unpublished manuscripts. James C. Jernigan Library, Texas A&M University-Kingsville.

Lehmann, V. W. 1946. Hunting customs on the King Ranch. Unpublished manuscripts. King Ranch, Inc., Kingsville, TX.

Lehmann, V. W. 1947a. Personal letter from Val Lehmann to Aldo Leopold. February 24, 1947.

Lehmann, V. W. 1947b. Personal letter from Val Lehmann to Aldo Leopold. March 7, 1947.

Lehmann, V. W. 1948. Restocking on King Ranch. *Transactions of the North American Wildlife Conference* 13:236–242.

Lehmann, V. W. 1957. Game conservation and management. In: *The King Ranch*. Volumes 1 and 2. T. Lea, Ed. Boston, MA: Little, Brown and Company, pp. 761–766.

Lehmann, V. W. 1984. *Bobwhites in the Rio Grande Plain of Texas*. College Station, TX: Texas A&M University Press.

Leopold, A. 1947a. Personal letter from Aldo Leopold to Mr. Robert Kleberg. February 10, 1947.

Leopold, A. 1947b. Personal letter from Aldo Leopold to Val Lehmann. February 14, 1947.

Leopold, A. 1947c. Personal letter from Aldo Leopold to Val Lehmann. Date Unknown.

Leopold, A. 1947d. Personal letter from Aldo Leopold to Val Lehmann. March 12, 1947.

Pimm, S., and M. Gilpin. 1989. Theoretical issues in conservation biology. In: *Perspectives on Ecological Theory*, J. Roughgarden, R. May and S. Leven, Eds. Princeton, NJ: Princeton University Press, pp. 287–305.

Rader, M. J., L. A. Brennan, K. A. Brazil, F. Hernandez, and N. J. Silvy. 2011. Simulating northern bobwhite population responses to predation, nesting habitat and weather in South Texas. *Journal of Wildlife Management* 75:61–71.

Rader, M. J., T. W. Teinert, L. A. Brennan, F. Hernández, N. J. Silvy, and X. B. Wu. 2007. Nest-site selection and nest survival of northern bobwhite in Southern Texas. *Wilson Journal of Ornithology* 119: 392–399

Rhyne, M. Z. 1998. Optimization of wildlife and recreation earnings for private landowners. Thesis, Texas A&M University-Kingsville.

Sands, J. P., L. A. Brennan, F. Hernández, W. P. Kuvlesky, Jr., J. F. Gallagher, D. C. Ruthven, III, and J. E. Pittman, III. 2009. Impacts of Buffelgrass (*Pennisetum ciliare*) on a forb community in South Texas. *Invasive Plant Science and Management* 2:130–140.

Teinert, T. W. 2009. Overwinter survival of northern bobwhites in two ecoregions of Texas. Thesis, Texas A&M University–Kingsville.

Wilcove, D. S. 1998. Quantifying threats to imperiled species in the United States. *BioScience* 48:214–222.

Section III

Species Case Studies

9 The Disconnect between Quail Research and Quail Management

Leonard A. Brennan
Caesar Kleberg Wildlife Research Institute
Texas A&M University–Kingsville
Kingsville, Texas

CONTENTS

As to how research and management interact, well, I know how they are supposed to interact. Research is supposed to accumulate and synthesize knowledge about a particular subject, and management is supposed to apply this knowledge to achieve certain goals.... Sounds simple enough, but we all know it is not.

—John Roseberry (2000:243)

.... there is nothing new in quail research. We are just rediscovering, and taking a subject that we studied in the past a little farther or perhaps in a different direction. What we face today is not a dilemma with managers *or* researchers, but *between* managers and researchers.

—Hunter Drew (2000:247)

ABSTRACT

The disconnect between quail research and management is the result of cultural differences that have evolved between tribes in the natural resource management profession. These differences became evident at the dawn of modern quail research when the promotion of prescribed fire to sustain northern bobwhite (*Colinus virginianus*) habitats and populations in the southeastern states collided with the prevailing resource management policies of fire suppression. The disconnect between quail research and management further deepened as the widespread decline in bobwhite populations indicated that predominate land uses in forestry and agriculture could not sustain huntable numbers of quail in the absence of active management.

The irony of this situation was that, over numerous decades, research had developed many of the basic management concepts and details required to sustain and elevate bobwhite populations. However, the well-documented literature on bobwhite habitat and population management was largely ignored by most agency resource managers, and many other landowners and conservationists who claimed to be interested in quail, as the bobwhite decline attained a near-crisis status in contemporary wildlife science. Fortunately, today, managers, policy makers, and researchers seem to be bridging this disconnect via an unprecedented, emerging program, the National Bobwhite Conservation Initiative, version 2.0.

INTRODUCTION

John Roseberry and Hunter Drew addressed the banquet audience at the conclusion of the Fourth National Quail Symposium with their respective observations and opinions noted previously. As chair of the Quail IV program committee, I deliberately recruited John and Hunter to address that conference and share their respective views about quail management and research. Both men have reputations for excellence in what they do. Both also have long careers as wildlife professionals with keen interests in promoting quail conservation.

John Roseberry is the visionary scientist who produced, among other key publications in his career, the classic *Population Ecology of the Bobwhite* with his colleague Bill Klimstra (Roseberry and Klimstra 1984). At the time, Hunter Drew was the manager for Ocmulgee Properties who had developed a *ne plus ultra* reputation for producing abundant populations of wild quail for hunting in South Georgia. We (the Program Committee) needed Quail IV banquet speakers who could summarize, compare, and contrast the different worldviews of quail managers and quail researchers, and each did exactly what we asked them to do. The motive for orchestrating such a contrasting set of banquet speakers was that, during the previous decade, we were continually amazed at how two groups of people—managers and researchers—with a seemingly common interest—the stabilization and increase of quail populations—could have such different philosophies and attitudes toward quail conservation as well as toward each other. Although a few sparks flew during the course of their talks, everyone in the audience, which was composed of a near-equal mix of quail researchers and mangers, parted as friends or colleagues after the banquet. Nevertheless, I could not help but realize that I had orchestrated some kind of council meeting between two tribes. In retrospect, this is exactly what I had done.

My experience with the disconnect between quail management and research actually began in California during the early 1980s. After completing my M.S. on a project that modeled the common elements of mountain quail (*Oreortyx pictus*) habitat in northern California, I explored the interest of funding agencies in supporting a Ph.D. project that used radiomarked mountain quail to assess their habitat use and migratory paths from low elevation wintering areas to high elevation breeding areas in the Sierra Nevada mountains. As the only migratory species of quail in North America, I thought it would be useful to know what kinds of habitat mountain quail used, and when and where they used these habitats during the critical migration phases of their annual life history cycle. The response of personnel in the California Fish and Game Department to my queries about such a study was that " ... it would be a waste of time and money ... we already know everything that we need to know about how to manage quail in California." Needless to say, I forged onward by seeking other Ph.D. research opportunities.

Several years later, when I went to Mississippi State University to conduct research on factors responsible for the northern bobwhite decline, the disconnect between quail management and research hit me squarely in the face again. Even though administrators and program directors at the Mississippi Department of Wildlife, Fisheries and Parks (MDWFP) had budgeted a generous amount of funding (>$100,000/year at the time) for a research program to identify factors behind the quail decline in Mississippi, several area managers who worked for MDWFP made it clear that they thought such a program was a waste of time and money. Their opinions spanned the extremes between "we already know how to manage quail," and "the quail situation is hopeless and there is nothing we can do about it." Either way, these attitudes pointed to a position that did not consider research to have value. The fact that professional wildlife managers in a state agency could have such a position about quail research astonished me yet again.

BACK TO THE CLASSICS

I spent most of my first year in Mississippi traveling around the state and the Southeast in general. My strategy was to get an understanding of why bobwhites had declined from a situation where hunters bagged >2.2 million birds in Mississippi in 1980 to <500,000 by 1990. Nearly every place I went, I heard the refrain about "how many quail we had on grand-daddy's place back in the day" and how "now they are all gone." However, by comparing habitat features on places that still had bobwhites in relation to places that no longer had bobwhites, it was clear that habitat changes from forestry and agricultural land uses were the primary culprits behind the declines (Brennan 1991). It was quite clear. Bobwhites were gone because their habitat was gone.

One of the intimidating luxuries about conducting research on bobwhites is that this species is one of the most studied species of wild birds in the world. Herbert Stoddard's classic *The Bobwhite Quail: Its Habits, Preservation, and Increase* (Stoddard 1931) was the first modern, comprehensive study of a wild animal in relation to the habitat and other factors that influence their populations. Following Stoddard, there was a flurry of key bobwhite research resulting in scientific papers on nutrition, physiology, and population dynamics by people such as Errington, Nestler, Robel, and Roseberry, among others (Hernandez, Guthery, and Kuvlesky 2002). By the late 1960s, the monumental *Bobwhite Quail: Its Life and Management* by Walter Rosene (1969) was published. In 1984, a third bobwhite *magnum opus, Bobwhites in the Rio Grande Plain of Texas* by Val Lehmann (1984) did for bobwhites in Texas what the titles by Stoddard and Rosene did for these birds in the Southeast: it put, and kept, them in the forefront of modern wildlife science. Interestingly, Lehmann (1984) was followed quickly by a pithy but critically important book by Guthery (1986) that made an important set of links between bobwhite research and management. Scott (1985) documented that from 1822 to 1982, more than 2,700 titles about some aspect of bobwhite biology or management had been published, much of it in the scientific literature.

Frankly, this was a huge body of literature that documented when, where, and how to manage habitats to sustain and elevate bobwhite populations. Furthermore, I found it incredible that this literature was virtually ignored by people who claimed to have a keen interest in management for sustaining and elevating bobwhite populations. How could such a disconnect develop? Why does such a disconnect seem to persist? I strive to answer these two questions in this chapter.

DISCONNECT DEVELOPMENT

Steve DeMaso (Chapter 1, this volume) did an excellent job of identifying the historical roots of the disconnect between wildlife research and management. The bobwhite situation is certainly a microcosm in the broader wildlife context of how this disconnect developed. In any case, it can be argued that the disconnect between quail research and management has its roots in the cultural-tribal affinities of the respective groups. It is also important to consider the limitations of what each group can actually accomplish when it comes to their respective backgrounds and worldviews. Researchers can provide excellent and high quality information in the form of various publication formats (Figure 9.1). If managers do not know of, or read, this information, or if they consider it wrong, it might as well not exist. Conversely, if researchers submerge their results and findings in opaque jargon and obscure language, it will not be understandable to managers no matter how hard the managers strive to read and understand it.

Despite that we accept most of Stoddard's results today, his research findings that abundant bobwhite populations were closely tied to the regular and purposeful use of fire as a tool to control understory vegetation in the southern piney woods were heretical to the broader forest management community at that time. During the first half or more of the twentieth century, the nascent forestry profession dominated natural resource fire management policy, and by extension, many options for wildlife management in North American forests. Stoddard's conclusions were a problem; this was a major disconnect between research results and what many people considered to be palatable management options. Many people who could implement these research results to benefit quail were unwilling to do so because the purposeful use of fire was a cultural anathema to them.

FIGURE 9.1   Banding (note band on leg) and radiotelemetry (note radio and antenna) are two research tools that have been used to generate a huge body of research on northern bobwhite life history. Photo courtesy of Leonard A. Brennan collection.

The chapter on "The Use and Abuse of Fire on Southern Quail Preserves," which promoted the use of fire under specific conditions that we now call prescriptions, created a maelstrom of controversy during the development and production of Stoddard's book (Way 2006). The fact that Stoddard was forced to revise and tone down his recommendations for the use of fire in quail management is well documented in both Stoddard's autobiography (Stoddard 1969) and in correspondence archived at Tall Timbers Research Station in Tallahassee, Florida (Way 2006).

I postulate Stoddard most likely realized he had, in some ways, let the research genie escape from the management bottle when it came to fire. He understood that the use of fire in wildlife and forest management " … was a complex problem … that would require years of careful research on the part of a well-equipped experiment station to carry out. Such research is greatly needed and should be carried on, for fire may well be the most important single factor in determining what animal and vegetable life will thrive on many areas" (Stoddard 1931:402). As it turned out, the issues related to fire were only the first of what would be many disconnects between quail research and management over the years.

DISCONNECT PERSISTENCE

One of the factors that caused the disconnect between quail research and management to persist over the years is that the importance of land use was largely neglected, or perhaps overlooked, by many people who were in a situation to influence bobwhite management. By the late twentieth century, abundant, huntable bobwhite populations persisted only in relatively small areas where active, intensive management was being practiced, such as the Red Hills in South Georgia–North Florida or in regions such as South Texas where land uses remained favorable to these birds. During the 1800s, there was a quail boom across most of the central and southern states. The landscape disturbances from logging and primitive agriculture provided a landscape saturated with usable space for quail. As time progressed, however, small weedy crop fields became sterile large ones, and vast regions of open piney woods were replaced by closed-canopy plantations. In looking for scapegoats, people who were able to influence or implement quail management pined for the "good-old days when we used to have quail" and repeatedly blamed "those damn coyotes and fire ants" when coyotes and fire ants usually had little to do with the bobwhite decline. The real cause of the bobwhite decline was habitat loss from changing land use. One part of the problem was that far too many of these management stakeholders, for whatever reason, did not avail themselves of the available literature and related information from research on bobwhite habitat relationships. A notable exception to this dynamic seems to have played out in the Red Hills region of South Georgia and North Florida, where a strong connection between quail research and management has developed (Palmer, Chapter 6, this volume). However, the situation in the Red Hills, as well as the one in South Texas (Brennan, Chapter 3, this volume) are exceptions to the more general rule that a profound disconnect exists between quail management and research.

The irony of this situation, as noted previously, is that a vast library of information on how to manage, sustain, and perhaps even increase quail numbers was widely available to people who groused about the quail decline, yet bobwhite numbers have continued to decline over much of their geographic range. How could so many people with such a proclaimed and vocal interest in quail conservation be so unable to implement the management actions that could result in a "prescription for plenty" that were so clearly formulated by Rosene (1969:224)? How could bobwhite numbers continue to decline and local extinctions continue to happen when we have such a vast storehouse of information on how to conserve and sustain wild populations of this bird that many seem to adore? The answer, as we shall see, is that today's world of bobwhite management is upside down.

WHY QUAIL MANAGEMENT IS UPSIDE DOWN

In 2004, Chris Williams and colleagues got my attention with a paper on scaling bobwhite management for the twenty-first century (Williams et al. 2004). In this paper, Williams et al. argued that the fundamental approach to bobwhite management during the past half-century or more has been wrong. It has been wrong because our approach to bobwhite management is upside down. The ways that we have been approaching bobwhite habitat management and bobwhite population management are 180 degrees from where they should be directed. We typically manage bobwhite habitat on a local, intensive scale and we manage populations on an extensive, statewide scale. Truth be told, we should be doing exactly the opposite; we should be scaling habitat management up to the landscape level and we should be scaling population management down to the individual property level.

When bobwhite management is upside down, among other things it further exacerbates the disconnect between bobwhite research and bobwhite management. Many people tend to think they are conducting effective bobwhite management, which they might be doing on a small scale, while populations continue to decline on the landscape scale. Furthermore, state agencies like to think they are the "regulatory authority" for quail populations when in fact regulations have done nothing to restore, sustain, or elevate declining quail populations (Cooke 2007).

Managing Habitat

As noted previously, we have a large and diverse body of knowledge from which to draw when it comes to managing bobwhite habitat (Figure 9.2). In a synthesis of this vast literature, Guthery and Brennan (2007) noted that there are four pillars of knowledge (*r*-selection, successional affiliation, adaptive plasticity, and weather influences), and if mangers can understand these factors, this understanding can be used as a basis for effective bobwhite management. The problem, however, as noted by Williams et al. (2004) is that virtually all bobwhite habitat management or restoration efforts are conducted on a local or individual property, farm, plantation, or ranch scale, when they need to be conducted on a landscape scale that includes multiple properties, farms, plantations, or ranches. Until or unless what we know from research about how to manage bobwhites can be implemented across millions of acres instead of thousands of acres, these populations will continue to erode. This is why Farm Bill programs, at least until the past five years or so, have only held false hope for restoration of bobwhite populations at the landscape scale (Brennan 2002); although implementation of the recent CP-33 Bobwhite Buffers program holds promise (Evans et al. 2009), as does the longleaf option (CP-36) in CRP in the southeast.

FIGURE 9.2 Measuring components of northern bobwhite habitat in South Texas. Photo courtesy of Leonard A. Brennan collection.

MANAGING POPULATIONS

Although management of r-selected animals such as quail typically focuses on habitat, increasing or relaxing hunting pressure can clearly influence population abundance. This is because repeated, excess fall and winter harvest at local scales can reduce the density of spring breeding pairs and thus severely limit summer gain and ultimately the subsequent fall density of birds available for harvest. From the standpoint of both biology and logic, it seems odd that fixed liberal harvest policy and regulations persist at a time when bobwhite numbers are continuing to decline over most of their geographic range. In Texas, for example, an allowable harvest of fifteen birds per person per day, over a 120-day season, allows an individual quail hunter to potentially bag 1,800 quail in a season. Fortunately, no quail hunters do such a thing. However, in Mississippi, state agency managers found themselves in the unenviable position of having twenty public hunting areas where hunting pressure was significantly increasing, but populations were declining, and there seemed to be little that anyone was capable or willing to do about it (Brennan and Jacobson 1992).

Based on what we have learned about quail population ecology through research, which is considerable, we have the basic tools to develop specific, individual property-based harvest prescriptions that can be aimed at sustaining a predetermined crude density of spring breeding populations (Guthery 2002). For example, Brennan et al. (2008) examined three scenarios where a combination of variable (high, medium, and low) summer gain and assumed constant overwinter survival resulted in harvest prescriptions that allowed moderate (360 birds), light (170 birds), and no harvest (0 birds) under a range of these conditions for a 400-ha (1000 acre) pasture. While admittedly conservative because Brennan et al. (2008) assumed complete additive mortality, their scenarios had at least a modicum of biology and sound conservation theory behind them, especially compared to the present laws that would allow a hunter to legally bag many more birds from such an area in a single season.

BRIDGING THE GAP?

Based on the material discussed thus far, it is not hard to imagine that the disconnect between quail research and quail management seems doomed to persist. It has persisted for more than seventy years, the lifespan of the contemporary era of wildlife science. Why should this disconnect not then continue to persist? Most modern land uses in agriculture and forestry remain hostile to quail, and the inexorable toll of habitat fragmentation and other factors of our prodigal society remain (Leopold 1978). How can there be hope for bobwhites and other quails under such circumstances?

There is hope because I think a critical mass of interest in quail conservation and management is finally beginning to take hold across many states within the bobwhite geographic range. This hope is embodied by the recent expansion of the National Bobwhite Conservation Initiative (NBCI), which is now in its second iteration with respect to planning habitat restoration efforts, and it is doing this on a scale never before considered. The Joint Venture model of cooperative initiatives that was developed and refined as a method of program delivery for the North American Waterfowl

Management Plan is being expanded to focus on quail and grassland birds for the southeast and central plains ecological regions of the United States (see Chapter 19, this volume). The scope, breadth, and depth of the NBCI planning efforts, as my colleague Steve DeMaso put it, "Is unprecedented for a resident upland game bird since restoration of wild turkeys (*Meleagris gallopavo*)." I agree. The NBCI, if implemented, and if successful, will be a way to scale up habitat management for bobwhites to a level that may actually stabilize and increase populations where they are now declining. For the NBCI to be successful, it will have to use, in a management context, the vast body of information on bobwhite life history, habitats, and populations that we have gained from research. The NBCI has the potential to bridge the gap and connect the disconnect between quail research and management. The key to NBCI success will be if it can implement a comprehensive quail management initiative based on an impressive and ever-growing storehouse of knowledge based on research.

ACKNOWLEDGMENTS

This chapter benefitted greatly from review and editorial comments by B. M. Ballard, S. J. DeMaso, F. S. Guthery, J. P. Sands, W. P. Kuvlesky, Jr., M. J. Schnupp, and C. K. Williams. However, any errors in accuracy or logic are strictly my own. This is Caesar Kleberg Wildlife Research Institute Publication 11-127, with support provided by the Richard M. Kleberg, Jr. Center for Quail Research and the C. C. Winn Endowed Chair.

REFERENCES

Brennan, L. A. 1991. How can we reverse the northern bobwhite population decline? *Wildlife Society Bulletin* 19:544–555.

Brennan, L. A. 2002. A decade of progress, a decade of frustration. *National Quail Symposium Proceedings* 4:230–232.

Brennan, L. A., F. Hernandez, W. P. Kuvlesky, Jr., and F. S. Guthery. 2008. Upland game bird management: linking theory and practice in South Texas. In: *Wildlife Science: Linking Ecological Theory and Management Applications*. T. E. Fulbright and D. G. Hewitt, Eds. Boca Raton, FL: CRC Press, pp. 65–74.

Brennan, L. A., and H. A. Jacobson. 1992. Northern bobwhite hunter use of public wildlife areas: the need for proactive management. *Gibier Faune Sauvage* 9:831–836.

Cooke, J. L. 2007. Quail regulations and the rule-making process in Texas. In: *Texas Quails: Ecology and Management*. L. A. Brennan, Ed. College Station, TX: Texas A&M University Press, pp. 299–312.

Drew, H. 2000. Concluding remarks: the manager's perspective. *National Quail Symposium Proceedings* 4:246–247.

Evans, K. O., W. Burger, M. Smith, and S. Riffell. 2009. CP-33–Habitat buffers for upland birds. Bird monitoring and evaluation plan. 2007 Annual Report. Mississippi State University, Starkville.

Guthery, F. S. 1986. *Beef, Brush and Bobwhites: Quail Management in Cattle Country*. Kingsville, TX: Caesar Kleberg Wildlife Research Institute Press.

Guthery, F. S. 2002. *The Technology of Bobwhite Management*. Ames, IA: Iowa State Press, Blackwell.

Guthery, F. S., and L. A. Brennan. 2007. The science of quail management and the management of quail science. In: *Texas Quails: Ecology and Management*. L. A. Brennan, Ed. College Station, TX: Texas A&M University Press, pp. 407–421.

Hernandez, F., F. S. Guthery, and W. P. Kuvlesky, Jr. 2002. The legacy of bobwhite research in South Texas. *Journal of Wildlife Management* 66:1–18.

Lehmann, V. W. 1984. *Bobwhites in the Rio Grande Plain of Texas*. College Station, TX: Texas A&M University Press.

Leopold, A. S. 1978. Wildlife in a prodigal society. *Transactions of the North American Wildlife and Natural Resources Conference* 43:5–10.

Roseberry, J. L. 2000. Concluding remarks: the research perspective. *National Quail Symposium Proceedings* 4:243–245.

Roseberry, J. L., and W. D. Klimstra. 1984. *Population Ecology of the Bobwhite*. Carbondale, IL: Southern Illinois University Press.

Rosene, W. 1969. *The Bobwhite Quail: Its Life and Management*. New Brunswick, NJ: Rutgers University Press.

Scott, T. G. 1985. *Bobwhite Thesaurus*. Edgefield, SC: International Quail Foundation.

Stoddard, H. L. 1931. *The Bobwhite Quail: Its Habits, Preservation, and Increase*. New York: Scribner's.

Stoddard, H. L. 1969. *Memoirs of a Naturalist*. Norman, OK: University of Oklahoma Press.

Way, A. G. 2006. Burned to be wild: Herbert Stoddard and the roots of ecological conservation in the southern pine forest. *Environmental History* 11:500–526.

Williams, C. K., F. S. Guthery, R. D. Applegate, and M. J. Peterson. 2004. The northern bobwhite decline: scaling our management for the twenty-first century. *Wildlife Society Bulletin* 32:861–869.

10 Sage-Grouse and the Disconnect between Research and Management on Public Lands in the American West

Steven G. Herman
The Evergreen State College
Olympia, Washington

CONTENTS

.... it is difficult to be optimistic about the future of the sage-grouse in North America.

—Johnsgard (1973:159)

ABSTRACT

The greater sage-grouse (*Centrocercus urophasianus*), hereafter sage-grouse, is an icon of shrub-steppe habitats in the American West. During the past two centuries, anthropogenic activities related to land use, disturbance, and human settlement of this region have resulted in highly significant and widespread sage-grouse geographic range contractions and population declines. Livestock grazing, especially on public lands where the vast majority of sage-grouse habitat remains, is at least qualitatively, a major driver in the downturn of sage-grouse populations and habitat. Curiously, however, many research investigations, along with a major contemporary monograph on sage-grouse biology, ecology, and conservation, have failed to make a direct and quantitative link between livestock grazing on public lands and the decline in sage-grouse numbers. In this chapter, I explore some of the possible reasons for this disconnect that I base on more than fifty years of observing sage-grouse populations, as well as their leks, and their habitat.

INTRODUCTION

During mid-September of 1958, I was traveling in Eastern California with my friend Jerry Cook. We were scouting for a trip where we planned to capture living yellow-bellied marmots (*Marmota flaviventris*) for a hibernation researcher at the University of Chicago. Jerry had learned that a series of these marmots in the Museum of Vertebrate Zoology at the University of California–Berkeley had been collected near the ghost-mining town of Bodie, north of Mono Lake. In my field journal for September 12, 1958 I wrote, "We saw a flock of Sage Hens about 2 blocks east of the town." My next recorded connection with sage-grouse came in the early 1960s, not far from this same location. Back then, I was a technician with the Division of Biological Control on the Berkeley campus of the University of California, interested in the effects of insectivorous birds on a native insect, the Lodgepole Pine Needle Miner (*Coleotechnites milleri*).

Bordering the conifer forests in this largely ignored part of the world are vast expanses of shrub-steppe (a landscape of sagebrush and native grasses, at least aboriginally), and driving to and from my study sites, I was introduced to the magic world of sagebrush. Sage-grouse were, in those days, an easily encountered and

FIGURE 10.1 Female greater sage-grouse (*Centrocercus urophasianus*), Inyo National Forest, Inyo County, California. Photo courtesy of Ron Wolf, CalPhotos.

dazzling component of that world (Figure 10.1). While visiting a spring for water or walking behind a dog in the sagebrush, I was often startled by the rumbling rise of a dozen or more birds. I remember a place on California State Highway 120 between Benton and Benton Station where I could always find sage-grouse, and did. I also saw them again, in abundance, when I returned to the vicinity of Bodie.

SOUTHEASTERN OREGON

Eleven years and a long college education later, I began a forty-year career teaching at the newly opened Evergreen State College in Olympia, Washington. I have taught natural history, wildlife science, and related programs and courses at Evergreen since then, specializing in long field trips to wild and semi-wild places where we camp out and work hard. Most of the extended field trips I have led have been in the arid regions east of the Cascades, where shrub-steppe landscapes dominate. Most of those were in Oregon.

My love affair with shrub-steppe blossomed as I learned more about this diverse, rich landscape. Over many years, I taught bird census techniques in shrub-steppe, where the fact that I was taller than most of the vegetation was welcome; everything was much more easily quantified.

In those early days, I bivouacked with my students at the Malheur Field Station, formerly a Job Corps outpost and perfectly set up as a facility to host college students and faculty. Dr. Denzel Ferguson, a research scientist who specialized in pesticide-wildlife relationships, was director at the Station in those days. There were barracks, a mess hall, and a recreation facility called the Greasewood Room. The property belongs to Malheur National Wildlife Refuge. Malheur headquarters, an oasis of trees in an essentially treeless high desert, is ten minutes away from the Field Station, and every spring dozens of ace birders assemble there to seek rarities and enjoy the abundant migrant passerines drawn to the huge trees.

One day Denzel told me about a sage-grouse lek that was accessible from the Station. I had already seen a good number of sage-grouse, but the sketch map he drew led me and a batch of cold, sleepy students to my first lek, and an iconic display of a species that has fascinated me and been the focus of much of my attention in the decades since. With this lek experience, I was hooked, and I have returned to that lek many times over the years. My sage-grouse species account for May 2, 1978 includes a detailed description of this lek, which that morning held eighteen male sage-grouse and two females. A sketch map shows the positions of the males on the lek. This lek is still active and still much attended by the birds, and fortunately remains quite robust.

SAGE-GROUSE AND THE DISCONNECT BETWEEN MANAGEMENT AND RESEARCH

Since those times in Oregon and California, I have visited sage-grouse leks in Washington as well at sites where both of the struggling subpopulations in that state are still hanging on. All four sage-grouse populations with which I have had some familiarity have declined over the years. The Mono Lake population has declined by nearly 60% from the mid-1960s to 2007; the South Mono Lake population is estimated to have declined by 65%. The Washington populations are estimated to have declined from 73% to 79% from peak lek counts between the period 1975–1979 and 2000–2007.

The appropriateness of including sage-grouse in a volume that addresses the disconnect between research and management is emphatically validated by Schroeder, Young, and Braun (1999:20), who remark on sage-grouse in their section on Priorities for Future Research, "Unfortunately, previous research has not been able to reverse the declines. This failure is due, in part, to the relatively narrow focus of the research, and to the fact that many of the research results have not been directly applied to management objectives ...." This comment is furthermore supported by Crawford et al. (2004:14): "Such efforts must have active participation from both management and research entities; without management buy-in, significant amounts of time and energy can be wasted developing models that will never be used."

Many palpable and some impalpable actions and lack of actions have contributed to the decline of sage-grouse over the couple of centuries that its geographic range has been occupied, used, and abused by people. However, it is my contention, which is based on hundreds of my own observations dating back to 1958, that essentially unregulated grazing by domestic livestock over the eleven western states where sage-grouse are found has been and continues to be the prime suspect in the range-wide decline of this species.

The sage-grouse decline has not occurred in an arena short on research. Surely, sage-grouse have enjoyed a huge amount of attention from agency and academic researchers—certainly more than any other bird species in their range—perhaps more than all other shrub-steppe species combined. That there has been a disconnect between research and management is without question; I seek to explore here the extent of the research and the reasons for the disconnect.

LIVESTOCK GRAZING AND SAGE-GROUSE

A.C. Bent (1932:309) wrote, "The range of the sage-grouse has been greatly restricted through the development of the West and through grazing activities, particularly of sheep, which do much to extirpate the birds over wide areas." Six years later, Rasmussen and Griner (1938) observed that grazing has a detrimental effect on sage-grouse reproduction (citing "reduced vegetation height"). The body of literature now supporting that observation is—to say the least—substantial. The science that has tested these early anecdotal observations concerning the negative effects of livestock on sage-grouse grazing was decades coming, but is now rich with confirmation.

The earliest recent work was done by Gregg, Crawford, Drut, and DeLong (1994), who found that at two Oregon study sites, "cover of tall grasses was greater at non-predated nests than at predated nests or random locations." They also documented the importance of shrub cover to nest success. An experimental test of this nest cover-predation hypothesis was conducted by DeLong, Crawford, and DeLong (1995), who used 330 artificial sage-grouse nests. The fate of these nests turned on grass cover height and medium-height shrub cover. Summarizing, they stated, "Greater amounts of both tall grass cover and medium-height shrub cover were associated collectively with a lower probability of nest predation." Additionally, DeLong et al. (1995) cautiously advised, "Some rangelands may need rest from grazing to increase herbaceous cover and height to desired levels." Elevated grass height has repeatedly been found critical to nest success of ground-nesting birds, including sage-grouse. For example, recent studies such as those by Beck and Mitchell (2000), Watters, McLash, Aldridge, and Brigham (2002), Halloran et al. (2005), and Moynahan, Lindberg, Rotella, and Thomas (2007) support this point. Various components of herbaceous and woody vegetation shield a ground nest from the eyes of predators and thus contribute to nest success; grass height and density often dominate as factors in this relationship.

It is also highly significant from almost any perspective that nearly all the remaining sage-grouse habitat in western North America is located on public lands that are subjected to livestock grazing (Figure 10.2). Clearly, Figure 10.1 illustrates not all public lands that are grazed by livestock are sage-grouse habitat. However, it does illustrate that nearly all remaining sage-grouse habitat exists on public lands that are grazed by livestock. The impact of livestock grazing on shrub-steppe vegetation can be especially acute (Figure 10.3).

The recent publication of a major tome on sage-grouse (Knick and Connelly 2011) provides a reference that synthesizes and interprets the sage-grouse literature and introduces some new material into the mix. The fact that this 646-page behemoth contains twenty-four chapters, but not one dealing specifically with the effects of

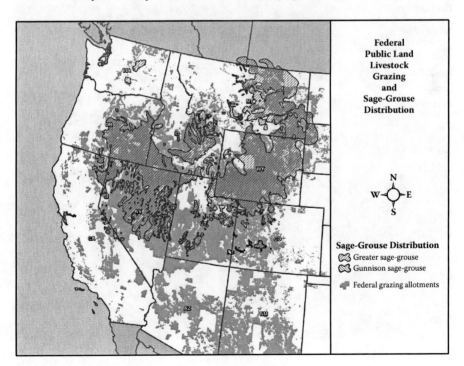

FIGURE 10.2 Geographic distribution of sage-grouse in relation to public land in the American West. Map courtesy of Mark Salvo, WildEarth Guardians.

(a)

(b)

FIGURE 10.3 (a and b) Livestock grazing can have acute and significant impacts on shrub-steppe vegetation that provides sage-grouse habitat. Photo courtesy of Katie Fite.

livestock grazing on sage-grouse numbers is astonishing. Such an omission has raised eyebrows in the conservation community and among some of the researchers themselves.

Grazing is—in my observation—grandly sidestepped as a central issue by Knick and Connelly (2011) and the other authors of this very useful and obviously hard-won reference. The sentence of greatest relevance to this sidestep comes in the last chapter, "Conservation of Greater Sage-grouse: a Synthesis of Current Trends and Future Management," which lists twenty co-authors from the sage-grouse community of managers and researchers. It reads, "Thus, the direct effect of livestock grazing expressed through habitat changes to population-level responses of sage-grouse cannot be addressed using existing information" (Connelly et al. 2011:553). This awkward but careful bit of English seeks to seal the coffin on our present state of understanding of the effects of livestock grazing, but it is largely undone in bits and pieces found elsewhere in Knick and Connelly (2011) and beyond. Based on my own observations in sage-grouse habitat and the observations and publications of others, I must argue that this is a classic example of disconnect between research and management, and it is probably the result of politics more than anything else.

There has been no shortage of livestock on public lands of the American West. In the earliest days for which there are data (1870), some 4,100,000 cattle and 4,800,000 sheep were grazed in the West; by 1900, that number had swelled to 19,600,000 cattle and 25,100,000 sheep. The Taylor Grazing Act of 1934 was meant to bring some control to grazing on public lands in the American West.

Some time in the 1930s, the magical but obfuscating term, "Animal Unit Month" or AUM was introduced and has become central to the lexicon of grazing on public lands. I have tried in vain for two decades to root out the origin of this clever term, but have failed. The most common definition is: "AUM: the amount of forage required to sustain a cow and her calf for one month" (USDA 1976). Knick et al. (2011:221) put it this way: "Numbers of livestock grazed on public lands are indexed by animal unit month (AUM), which is the amount of forage required to feed one 454-kg (1,000-lb) cow and her calf, one horse, five sheep, or five goats for one month." Livestock numbers are usually expressed in terms of AUMs today, obligating one interested in real quantification to make a calculation (even though that calf can be up to a year old). The pernicious quality of this term is manifest in its definition by Knick et al. (2011:221): "AUM = amount of forage to support one livestock unit per month." Now, that definition obfuscates the real *number* of livestock by up to 100%, which should be something of a problem for a quantitative scientist! Granted, the second individual is not likely the equivalent of the cow, but it can be (by definition of AUM) up to a year old, and that means it will be consuming a significant amount of forage.

It turns out that cows weigh a good deal more than 1,000 pounds these days, and it follows that their calves are heftier as well. Forage requirements could be expected to track this figure. Cattle have changed dramatically over the last two or more decades; they are probably 200 pounds bigger than two decades ago (Olson 2011). Until the definition of AUM is reset, the number of livestock currently grazing on sage-grouse habitat will continue to be understated, and the confusion inherent in the term "Animal unit" will continue to complicate any evaluation of the effects of grazing on sage grouse.

When it comes to the condition of public rangelands in the American West, it is very popular to imply (or simply state) that "it's not as bad as it once was," that there has been recovery, that current regulations mean a healthier "range." While it is true that numbers and even intensity of grazing has diminished from the "bad old days," there is one factor missing from that equation—one that has great relevance to range-wide sage-grouse populations: The area accessible to livestock today is far greater than what was available to graze 50 to 100 years ago. This is because water is a limiting factor to animals that require daily water and, over the decades, private owners and public agencies (primarily the U.S. Forest Service and the Bureau of Land Management) have gone beyond the original natural sources of water to drill wells and bring pumps, pipes, and water troughs to landscapes that earlier were without free surface water and hence livestock.

This is how the subject is treated in a new book by Knick et al. (2011:232): "We cannot conclude the effect of grazing across the sage-grouse range has been reduced because fewer numbers of livestock may still exert a larger influence over a broader region on lands that now have lower productivity." Other factors degrading sage-grouse habitat include various forms of fragmentation, roads, power lines, and—especially—fences. Beyond that, only 10% of current sage-grouse habitat is considered "in excellent condition." According to this estimate, 76% is in "good" condition, and 15% is "poor" (Knick et al. 2011).

Additionally, Knick et al. (2011:232) state, "The question of effects of livestock grazing at large spatial scales is difficult to assess because we lack control areas sufficiently large to include landscape processes." This is doubtful but difficult to test—certainly, there are large areas that are less damaged than others are, and these might lend themselves to comparison (Bock, Bock, and Smith 1993). In the mid-1990s, for example, livestock were removed from nearly a half million acres of prime sage-grouse habitat on Hart Mountain and Sheldon National Antelope Refuges, but it appears any opportunity to study the effects of that removal were missed by the agency managing the refuge—the U.S. Fish & Wildlife Service. In something of a final fallback position, Knick et al. (2011:232) stated, "Our inability to test for an effect does not demonstrate that livestock grazing has no effect or is a compensatory use of sagebrush habitats and therefore should be ignored."

FERAL HORSES AND BURROS

Feral horses and burros have occupied some 366,690 km² of semi-arid habitat in the West, and are "managed" on about a third of that area, which constitutes about 18% of the currently occupied sage-grouse range. So, despite the fact that feral horses and burros exert a potential impact that is only one-fifth of that posed by livestock (in terms of area—some of the effects of horses exceed the damage by cattle; horses have incisors above as well as below), this group warrants its own chapter (Beever and Aldridge 2011) in Knick and Connelly (2011). Cattle, for some strange reason, do not. Moreover, Knick and Connelly (2011) treat equids much more straightforward than they do cattle. Much of the literature reviewed and analyzed by Beever and Aldridge (2011) has to do with comparisons between areas occupied by feral horses and areas where horses have been removed; no such literature shows up in the various treatments of domestic livestock.

It is pointed out that domestic livestock consume an estimated 7,100,000 AUMs of forage annually within the current range of sage-grouse, the equivalent figure for free-roaming equids is 315,000 to 433,000 AUMs. The ratio of livestock to free-roaming equids over the public lands of western North America averaged 23:1 in 1982 (Wagner 1983). However, in Nevada, which holds over half of the equids and the highest state numbers of sage-grouse, that ratio was only 4.8:1.

POLITICS VERSUS SCIENCE

The disconnect between wildlife research and wildlife management is usually a manifestation of the rule, "politics trumps science," and that holds with special tenacity in the case of sage-grouse. Listing this species under the Endangered Species Act has broad conservation potential, and the public lands ranching community in particular fears such a move. Andy Kerr (personal communication), a prominent conservation activist, has put it this way: "Listing the sage-grouse has the potential of making it the Spotted Owl of the shrub-steppe."

Sometime late in the first decade of our new century, word circulated that the sage-grouse research community was working on a monograph that would provide managers and conservationists with a heap of "new research findings" about the bird. So momentous seemed the prospect of this new information, that the U.S. Fish & Wildlife Service used its pending publication to postpone their decision on listing.

On March 3, 2010, the U.S. Fish & Wildlife Service published a 103-page book of their findings in the Federal Register. Working from a draft or advance copy of the monograph by Knick and Connelly (2011), they found that the greater sage-grouse listing was "warranted but precluded." Listing in this case was determined to be a lower priority than listing other plant and animal species. Therefore, the decision cut both ways, but the impact was one of reinventing the status quo. In theory, at least, more protection would be encouraged, but mandates were not triggered.

In their findings, the Service treats grazing at some length, but in the end, they find (Federal Register, p. 32) that "currently there is little direct evidence linking grazing practices to population levels of Greater Sage-Grouse." However, "given the widespread nature of grazing, the potential for population level impacts cannot be ignored." Therefore, there seems to be a hazy acknowledgment that grazing has a negative impact here and there, but not enough for action.

Almost every recent sage-grouse research paper nods vaguely in the direction of recognizing the importance of grazing as a central question related to the sustainability of sage-grouse, but few come as close to confronting the issue head-on as Halloran et al. (2005:648): "Although little direct evidence associating livestock grazing practices with greater sage-grouse population levels exists, our results suggest annual grazing in nesting habitat, regardless of timing, could negatively impact the following year's nesting success."

My favorite definition of science comes from Robert MacArthur (1972:1) the father of modern ecology: "To do science is to search for patterns in nature …." All this sage-grouse research has clearly revealed a pattern, in my view. If, on the one hand, effects at the population level are necessary to trigger action on grazing, then action at the population level needs to be the mandate. Apparently, there is an effort currently underway to compare livestock stocking rates with degrees of decline in sage grouse, and I'm told that the outcome is liable to allow a judgment that codifies the thesis of this chapter.

CONCLUSIONS

Wildlife biology is largely a production-driven discipline, and that driving force is felt in management. In my opinion, it is not a primarily conservation or preservation driven field. It follows that managers begin with the premise that "wildlife needs help," with the assumption that there is some limiting factor in the environment that—if augmented or otherwise made more available—will lead to more birds, mammals, or fishes. The prescriptions that issue from these perceived needs often take the form of efforts to increase food. At the other end of the spectrum, they may result in campaigns to control or "manage" predators. Much less often managers might seek to increase cover for wildlife, or—theoretically—manage in such a way as to preserve habitat.

But all these planned strategies *assume* that the wildlife in question *need* help, and that is by no means true; the deer, ducks, or in this case, sage-grouse, may not be short anything in their environment, and the possibility of increasing their numbers through active management might be miniscule or even absent. In the case of sage-grouse, there have been efforts to show that grazing benefits the species by increasing forbs in meadows, and there was a trend in the late part of the last century to prescribe and create optimal habitat characteristics like percent shrub cover and grass density; these have largely faded away in light of recent research. Predator control has been rejected as a routine "management tool," but, ironically, increased nest predation (often an outcome of livestock grazing) and increases in numbers of their major nest predator, Common Ravens (*Corvus corax*), have led to some experimental control programs and may in the future foster control programs.

I predict that wildlife biology will inevitably move from a production-oriented discipline to one that emphasizes preservation and beauty (Herman 2002; Schmidly 2005) and that such a change will have a profound effect on managed landscapes, providing that the target landscapes have not been permanently and irreversibly damaged. With the ascendancy of nongame wildlife in the late twentieth century, this shift was initiated. However, today most wildlife managers still think of wildlife in terms of deer, ducks, elk, and pheasants, instead of frogs, warblers, hawks, and beetles. This orientation bears heavily on the disconnect between research findings and their application with respect to grazing and sage-grouse. The idea that a healthy landscape will support healthy wildlife populations is more often trumpeted than it is implemented; managers have to manage, and the idea that the best management is sometimes no management at all (Herman 2002) is still likely to remain foreign to most of the management community.

REFERENCES

Beck, J. L., and D. L. Mitchell. 2000. Influences of livestock grazing on sage grouse habitat. *Wildlife Society Bulletin* 28:993–1002.

Beever, E. A., and C. L. Aldridge. 2011. Influences of free-roaming equids on sagebrush ecosystems, with a focus on sage-grouse. *Studies in Avian Biology* 38:273–292.

Bent, A. C. 1932. Life histories of gallinaceous birds. Smithsonian Institution. *United States National Museum Bulletin*. 162.

Bock, C. E., J. H. Bock, and H. M. Smith. 1993. Proposal for a system of federal livestock exclosures on public rangelands in the Western United States. *Conservation Biology* 7: 731–733.

Connelly, J. W., S. T. Knick, C. E. Braun, W. L. Baker, et al. 2011. Conservation of greater sage-grouse: a synthesis of current trends and future management. *Studies in Avian Biology* 38:549–564.

Crawford, J. A., R. A. Olson, N. E. West, J. C. Mosley, M. A. Schroeder, et al. 2004. Ecology and management of sage-grouse and sage-grouse habitat. *Journal of Range Management* 57:2–19.

DeLong, A. K., J. A. Crawford, and D. C. DeLong, Jr. 1995. Relationships between vegetational structure and predation of artificial sage grouse nests. *Journal of Wildlife Management* 59:88–92.

Gregg, M. A., J. A. Crawford, M. S. Drut, and A. K. DeLong 1994. Vegetational cover and predation of sage grouse nests in Oregon. *Journal of Wildlife Management* 58:162–166.

Halloran, M. J. M., J. B. J. Heath, A. G. Lyon, S. J. Slater, J. L. Kuipers, and S. H. Anderson. 2005. Greater sage-grouse nesting habitat selection and success in Wyoming. *Journal of Wildlife Management* 69:638–649.

Herman, S. G. 2002. Wildlife biology and natural history: time for a reunion. *Journal of Wildlife Management* 66:933–946.

Johnsgard, P. A. 1973. *Grouse and Quails of North America*. Lincoln, NE: University of Nebraska Press.

Knick, S. T., and J. W. Connelly. Editors. 2011. Greater sage-grouse: ecology and conservation of a landscape species. *Studies in Avian Biology* No 38. Berkeley, CA: Cooper Ornithological Society and University of California Press.

Knick, S. T., S. E. Hanser, R. F. Miller, D. A. Pyke, M. J. Wisdom, S. P. Finn, E. T. Rinks, and C. J. Henny. 2011. Ecological influence and pathways of land use in sagebrush. *Studies in Avian Biology* 38:203–252.

MacArthur, R. H. 1972. *Geographical Ecology*. Princeton, NJ: Princeton University Press.

Moynahan, B. J., M. S. Lindberg, J. J. Rotella, and J. W. Thomas. 2007. Factors affecting nest survival of greater sage-grouse in northcentral Montana. *Journal of Wildlife Management* 71:1773–1783.

Olson, K. L. 2011. Relationship between cow size and nutrient requirements. *Tri-State Livestock News* 49: A–1, A–5.

Rasmussen, D. I., and L. A. Griner. 1938. Life histories and management studies of the sage grouse in Utah, with special reference to nesting and feeding habits. *Transactions of the North American Wildlife Conference* 3:852–864.

Schmidly, D. J. 2005. What it means to be a naturalist and the future of natural history studies at American universities. *Journal of Mammalogy* 86:449–456.

Schroeder, M. A., J. R. Young, and C. E. Braun. 1999. Sage grouse (*Centrocercus urophasianus*). In: *The Birds of North America*, No. 425, A. Poole and F. Gill, Eds. Philadelphia, PA: The Birds of North America, Inc.

USDA. United States Department of Agriculture. 1976. *National Range Handbook*. Washington, DC.

Wagner, F. H. 1983. Status of wild horse and burro management on public rangelands. *Transactions of the North American Wildlife and Natural Resources Conference* 48:116–133.

Watters, M. E., T. L. McLash, C. L. Aldridge, and R. M. Brigham. 2002. The effect of vegetation structure on predation of artificial greater sage-grouse nests. *Ecoscience* 9:314–319.

11 Ecological Goals, not Rigid Management Plans, are the Key to Effective Grassland Bird Conservation

Robert A. Askins
Department of Biology
Connecticut College
New London, Connecticut

CONTENTS

Yet there is still time to save many of the "living dead"—those [species] so close to the brink that they will disappear soon even if merely left alone. The rescue can be accomplished if natural habitats are not only preserved but enlarged, sliding the numbers of survivable species back up the logarithmic curve that connects quantity of biodiversity to amount of area. Here is the means to end the great extinction spasm. The next century will, I believe, be the era of restoration in ecology.[*]

—**Edward O. Wilson (1992)**

---

[*] Reprinted by permission of the publisher from *The Diversity of Life* by Edward O. Wilson, p. 340, Cambridge, MA: The Belknap Press of Harvard University Press, Copyright © 1992 by Edward O. Wilson.

ABSTRACT

Traditional approaches to wildlife management often focused on boosting populations of a few target species, and general ecosystem processes were not always taken
into account. The importance of understanding the role of ecosystem processes is
illustrated by examples from grassland bird conservation. An understanding of natural disturbances and ecological drivers is crucially important for understanding
how to manage habitats for particular species of grassland birds. This approach is
effective for both game and nongame species, and for preserving general biodiversity in a wide range of grassland ecosystems (shortgrass prairie, desert grassland,
coastal grassland, and wet meadows). To apply this approach successfully, wildlife
managers must use adaptive management to restore or simulate natural ecosystem
processes. This requires thinking like an ecologist, and not like a civil engineer who
is molding the environment according to a set of standard specifications or highly
prescribed plans.

INTRODUCTION

I was initially surprised by the theme of this volume because I don't normally perceive a major gulf between wildlife research and wildlife management. This may
reflect my interest in ecology and conservation of songbirds, a field in which collaboration between ecologists and land managers has been fostered at the international, national, and state levels by the Partners in Flight program (Hagan 1992).
During the past twenty years, I have met frequently with land managers from federal
and state agencies and nonprofit organizations, so I have had many opportunities
to explain the implications of my research and the research of my colleagues, and
to learn directly about the practical challenges and research needs of people who
manage public and private conservation lands. I developed close working relationships with biologists in the Connecticut Department of Energy and Environmental
Protection (DEEP), The Nature Conservancy, and Audubon Connecticut, and I also
frequently discuss conservation issues with wildlife biologists in federal agencies.
During the week I started writing this chapter, for example, I met with Shannon
Kearney-McGee, a state wildlife biologist who is working on conservation of birds
in early successional habitats. We discussed two studies my students and I completed
on nest success and distribution of birds in clearcuts and powerline corridors. Both
studies were partially funded by the Connecticut DEEP Wildlife Division, and they
had explicit conservation goals (determining what types of habitats support productive populations of declining species of early successional birds). We discussed the
results of the recently completed powerline study, focusing particularly on data indicating that sites in more heavily developed landscapes have a low abundance and
diversity of early successional birds even when the immediate habitat is favorable.
Not only was there no disconnect in this case, but the connection between research
and management was so immediate that it came with the proviso that the results are
tentative, pending peer review of a manuscript on the study. This is not an isolated
example; wildlife managers in Connecticut are in direct contact with scientists who
are doing research on a variety of conservation issues, such as ecology of threatened

saltmarsh sparrows (*Ammodramus caudacutus*), the spread of invasive common reed (*Phragmites australis*) in coastal marshes, and the ecological requirements of woodland amphibians. As time permits, they also participate in their own research and monitoring projects.

My experience may reflect the role of Partners in Flight Program in bringing people from diverse organizations together to discuss the mandate of saving migratory songbird populations (a goal that was later expanded to include terrestrial birds in general). The close collaboration among agencies, nonprofit organizations, and academic researchers was later reinforced and expanded during meetings to develop a comprehensive wildlife conservation strategy for the state. The main barriers to implementation of new approaches to conservation were usually financial limitations or a lack of understanding and support on the part of the public, not the lack of communication between wildlife researchers and wildlife managers.

LIMITATIONS OF STANDARD MANAGEMENT PROTOCOLS

Applying the results of ecological research is often a major challenge for land managers, however, who are forced to look at ecosystems in new, and sometimes counterintuitive, ways. This is particularly true when there is not a strong connection between people who manage land and people directly involved in ecological research. It is not surprising that some people resist the implications of research when the results point to the need for a major change in conservation priorities developed by experienced land managers. A classic example is the original resistance to the evidence that fire is not always an agent of destruction and environmental degradation, but is sometimes an agent of renewal that sustains biological diversity (reviewed in Askins 2002). Another challenge is that wildlife biologists and land managers need to develop a broader, more holistic understanding of ecosystems to successfully apply conclusions from research in landscape ecology and restoration ecology. Convenient handbooks with standard wildlife management protocols will no longer suffice. To manage for a target species such as grasshopper sparrow (*Ammodramus savannarum*), for example, there is no general protocol (such as "plant warm season grasses and mow annually during the late winter") that is likely to be effective at all sites. It is important to know that grasshopper sparrows require sparse bunch grass interspersed with open ground for foraging, and that they tend to be absent from sites that are not located within more than 100 ha of continuous grassland habitat (Askins 1993). The vegetation, soil fertility, and landscape context all must be considered in order to determine how (or whether) a site should be managed for this species. In addition, the effects of management on other species of plants and animals and on ecosystem dynamics must be considered. The result is that land managers are forced to think more like ecologists and less like civil engineers.

This transition in the field of wildlife biology not only facilitates new scientific perspectives, but also undermines widely accepted management practices for target species that have never been adequately tested and that may even have negative environmental effects. For example, recent analyses indicate that standard methods for installing in-stream structures (dams, deflectors, and cover structures) to enhance streams for trout not only damage stream ecosystems, but also probably do not increase trout populations (Thompson and Stull 2002; Thompson 2006). Designs for in-stream structures were developed more than seventy years ago with little systematic testing (Thompson and Stull 2002). These designs, which were replicated with minimal change in manuals and handbooks on fisheries management from the 1930s to the 1990s, guided the installation of tens of thousands of structures. In the meantime, the ultimate causes of low trout populations (overfishing and forest destruction) did not receive sufficient attention. In an analysis of data from seventy-nine publications on the effects of in-stream structures, Thompson (2006) found little conclusive evidence that these structures increase trout populations. In many cases, reported increases in trout catches following installation of structures may merely reflect increased fishing pressure by people attracted to the structures or even increases in trout stocking near the structures. Thompson recommends that fishery biologists address the ultimate reason for declines in trout populations (overfishing and changes in land use) rather than the symptom (lower catch rates). He suggested that this could best be accomplished by applying insights derived from research on natural stream channels to replicate natural processes in managed streams.

UNDERSTANDING ECOSYSTEM PROCESSES

An emphasis on understanding natural ecosystem processes is a key tenet of restoration ecology. For example, an understanding of the frequency and scale of natural disturbances such as fires and seasonal floods can guide land management activities in landscapes where these disturbances have been suppressed. Ecologists often use this approach when managing grassland ecosystems. In many grassland systems, natural disturbances are so frequent—and so critical for sustaining the original association of plant and animal species—that the term "disturbance" is considered a misnomer, and natural ecological processes such as fire, seasonal flooding, or grazing are called "drivers." An understanding of the role of disturbances or drivers in sustaining biological diversity in grasslands has led to surprising, often counterintuitive conclusions that do not support traditional approaches to either range management or natural area preservation.

ROLE OF GRAZING IN DRY PRAIRIES

The dry grasslands of the Great Plains are ecologically distinct from the wetter tallgrass prairies found east of the 100th meridian (Figure 11.1). The primary driver in tallgrass prairies is fire, and these prairies tend to be replaced with shrubs and trees in the absence of fire. In contrast, the drivers in shortgrass prairies are low rainfall and grazing (Knopf 1996; Askins et al. 2007). Most of the biomass of the dominant grasses is below the ground in the roots, so there is little fuel to sustain fires. Originally the shortgrass prairies were intensively grazed by a combination of bison (*Bison bison*) and black-tailed prairie dogs (*Cynomys ludovicianus*), and many shortgrass prairie species such as mountain plover (*Charadrius montanus*) and McCown's longspur (*Rhynchophanes mccownii*) are adapted to heavily grazed, close-cropped lawns found in prairie-dog colonies (Tipton, Dougherty, and Dreitz 2009).

FIGURE 11.1 Shortgrass prairie in Pawnee National Grassland, Weld County, Colorado. Photograph by Fritz L. Knopf.

Recent conservation efforts in the Great Plains have largely ignored the key ecological processes of the presettlement shortgrass prairie. The main regional grassland conservation effort is the Conservation Reserve Program (CRP) in which farmers are paid to take cropland out of production to reduce soil erosion and create habitat for wildlife. Most of the CRP land has been planted with tall grasses rather than the short, native grasses that originally covered most of this region (Askins et al. 2007). More importantly, grazing generally is not permitted on CRP lands. As a result, the specialized, shortgrass prairie-bird species that have declined in recent decades generally do not use CRP land.

Another challenge is that most environmental organizations oppose grazing on public land because of the well-documented negative impacts of heavy grazing in riparian woodlands, deserts, and other sensitive habitats. Completely removing grazers from shortgrass prairies will remove one of the key ecological drivers in this ecosystem; however, and the habitat will soon become unsuitable for most of the bird species that are endemic to the shortgrass prairies and are major targets for conservation (Knopf 1996; Askins et al. 2007). Cattle grazing can be carefully managed to sustain habitat for these species (Figure 11.2). If cattle are removed, then it is imperative to ensure that native grazers (bison and prairie-dogs) are reintroduced. Land managers must decide which of these approaches (or mix of approaches) is most appropriate for a particular expanse of prairie. Hence, there are no simple protocols for sustaining populations of shortgrass-prairie bird species. Management should be guided by the more general goal of creating the "grazing lawn" with occasional islands of tall grass that characterized the Great Plains when prairie-dog colonies stretched for hundreds of kilometers and massive bison herds moved across the landscape.

FIGURE 11.2   Mountain Plover, a shortgrass prairie specialist, in Pawnee National Grassland, Colorado. Photograph by Fritz L. Knopf.

The desert grasslands of the southwestern United States and northwestern Mexico present a somewhat different conservation challenge. Large expanses of desert grassland have been replaced by desert scrub dominated by mesquite (*Prosopis glandulosa* and *P. velutina*). Saab, Bock, Rich, and Dobkin (1995) hypothesized that fire was the primary driver in desert grassland and that suppression of fire led to encroachment by woody species. Migratory herds of bison occasionally reached the desert grassland, but this grazer was much less important than in the shortgrass prairies farther north. Despite this, grazing rather than fire may have been the primary driver in this system. Black-tailed prairie dogs were abundant in vast areas of grassland from West Texas to southeastern Arizona (Parmenter and Van Devender 1995; Askins et al. 2007). Prairie dogs constantly clip back woody plants in their colonies, preventing the spread of shrubs and trees (Koford 1958; Weltzin, Archer, and Heitshmidt 1997). This may be the reason that prairie-dog colonies in Mexico are still open grassland rather than mesquite scrub (Ceballos, Mellink, and Hanebury 1993; Manzano-Fischer, List, and Ceballos 1999). The almost complete eradication of prairie dogs in the southwestern United States by the 1960s may largely account for the extensive replacement of desert grassland by mesquite scrub. Because the spread of mesquite scrub reduces both the productivity of rangeland for cattle and the diversity of grassland (as opposed to shrubland) bird species, restoration of open grassland is favored by both ranchers and conservationists. A key element of restoration might involve reintroduction of prairie dogs to prevent the growth of mesquite seedlings. This would require major changes in understanding by both ranchers (who perceive prairie dogs as severe competitors of cattle rather than as part of a potentially integrated system of grazers) and conservationists (who often assume that any intense grazing is detrimental to the stability of arid grasslands). Thus, habitat restoration may depend on land managers who understand the ecological history and interspecific dynamics of grazing lawns.

Effective application of our understanding of ecosystem processes in dry grasslands also requires further research, however. We need large-scale experiments on how different types of grazers and other natural processes affect biodiversity in shortgrass prairies and desert grasslands (Askins et al. 2007). This would provide the sort of detailed ecological understanding that is already available for tallgrass prairies, the wetter "true prairies" that are found east of the shortgrass prairie, where intensive research in places like Konza Prairie has shown how prescribed burns, grazing, and mowing affect grassland birds and other organisms.

EASTERN GRASSLANDS AND NUTRIENT MINING

Much of northeastern North America was originally covered with deciduous forest or mixed hardwood/conifer forest, but we know from accounts of early explorers and settlers that there were open areas of grassland along the coast (Askins 1999, 2002). These "plains" or "barrens" were typically found on porous, sandy soil that tended to dry out, making the vegetation prone to fire. Most fires were probably set by indigenous people, but people had been a component of northeastern ecosystems for thousands of years, since the end of the last glacial period (Gill et al. 2009). In fact, the true "presettlement" landscape (the landscape that prevailed before human influence) is most accurately described as a 2-km thick continental glacier bordered by arctic steppe. The most extensive coastal grasslands in the Northeast at the time of European settlement were the Hempstead Plains (about 24,000 ha of virtually treeless, little bluestem [*Schizachyrium scoparium*] prairie on Long Island) and the blueberry barrens of eastern Maine. These coastal grasslands supported a variety of prairie plant species and specialized grassland birds, including the heath hen (*Tympanuchus cupido cupido*, an extinct coastal subspecies of the greater prairie-chicken), upland sandpiper (*Bartramia longicauda*), and several species of grassland sparrows (Askins 2002). Where the grassland grew on well-drained, sandy soil, it was dominated by little bluestem and other grass species that grow in clumps with interspersed spaces. This is the preferred habitat of a number of grassland bird species, such as the upland sandpiper and grasshopper sparrow (Figure 11.3).

Another important requirement of these and many other specialized grassland bird species is large, continuous areas of appropriate habitat (Renfrew and Ribic 2008; Ribic et al. 2009). Regional bird surveys in both the blueberry barrens of Maine and grasslands in Illinois showed that both upland sandpiper and grasshopper sparrow tend to be infrequent in patches of grassland smaller than 100 ha (250 acres) (Herkert 1994; Vickery, Hunter, and Melvin 1994). Hence, these species usually are not found in small pastures and meadows surrounded by forest.

Both the upland sandpiper and the grasshopper sparrow are listed as endangered or threatened species in northeastern states because nearly all of the natural coastal grassland has been developed, and the extensive areas of fallow field and pasture that supported these species after European settlement have steadily disappeared (Askins 1993, 2002). These fields and pastures either reverted to forest or were converted to intensively managed farmland that does not provide appropriate habitat. Consequently, land managers in northeastern states have attempted to restore or even create habitat for these species. This involves planting warm-season grasses (often

FIGURE 11.3 Sandplain grasslands on Kennebunk Plains, York County, Maine. Grasshopper sparrow, upland sandpiper, and other grassland birds breed in this habitat. Photograph by Peter D. Vickery.

after removing introduced cool-season grasses that were planted for livestock forage) and maintaining the site with mowing or prescribed burning (Jones and Vickery 1997). Frequently, however, the outcome is disappointing because the result is a dense growth of tall grass, not the sparse, low, patchy grass needed by the target species. Even if managers have taken into account the requirements of these species for large areas of grassland, native species of grass, and periodic disturbance, the restored habitat may look nothing like the coastal grasslands or mixed grass prairies that originally supported these species. Although more frequent disturbance of the site or application of a particular type of disturbance (such as burning) may help address this problem, it may also be important to consider soil fertility. Part of the problem may be that the soil is too rich to support bunchgrass vegetation, which typically grows on relatively dry, nutrient-poor soil. This is especially problematic if the site does not have sandy soil, but it can even be a problem on sandy soil if the site was enriched by application of manure and artificial fertilizers during centuries of farming and by deposition of nitrogen due to atmospheric pollution (Walker et al. 2004; Wamelink, van Dobben, and Berendse 2009). This is a problem in England, where restoration of dry grasslands is important for preserving a large number of threatened species (Walker et al. 2004). English land managers prepare an area for grassland restoration with several years of "nutrient stripping," which involves intensive haying (including removal of the hay from the site) without fertilization. The soil is then sufficiently impoverished to support dry grassland. Another, more intensive approach that was used in restoration of plant diversity in fen meadows in The Netherlands is turf removal (Van der Hoek and Heijmans 2007). Several other methods are used to specifically reduce nitrogen levels in soil for ecological resto-ration (Perry et al. 2010). Nutrient mining apparently has not been used to restore grasslands in northeastern North America so it is not clear that the European meth-ods will work in superficially similar North American ecosystems (Ernie Steinauer, Massachusetts Audubon Society, personal communication). Although eventually it may be possible to provide a standard recipe for accomplishing grassland restoration using methods of this sort, I suspect that sites vary enough so that successful man-agers will use adaptive management based on a goal of creating and maintaining a particular type of vegetation (such as sparse bunchgrass dominated by native species of grasses and forbs) rather than depending on a set of highly standardized manage-ment procedures.

WET MEADOWS AND ECOSYSTEM ENGINEERS

Before the extensive use of maize agriculture began about 1,200 years ago, the most abundant type of grassland in eastern North America was probably wet meadow. The two main sources of wet meadow were beaver (*Castor canadensis*) activity and seasonal floods (Askins 2002). Beavers dam streams to create ponds, but even-tually move on after they exhaust the local food supply. Without maintenance, a beaver dam leaks and breaks down, draining the pond. The drained pond typically becomes a marsh and then either a shrub swamp or a "beaver meadow," which is a wet meadow (Figure 11.4). Similar habitat was originally created without help from

FIGURE 11.4   Beaver meadow near Harwinton, Connecticut. Photograph by Paul J. Fusco, CT-DEP-Wildlife.

beavers along large rivers and creeks because of severe seasonal flooding of the floodplain and meandering of the river channel. In the most frequently and intensely flooded sections of the floodplain, riparian forest was replaced with wet meadow. Early European visitors to New England describe large areas of wet meadow along many of the major rivers. Even in the 1850s, Henry Thoreau observed extensive wet meadows along the Sudbury and Concord rivers in Massachusetts (Foster 1999).

By the end of the nineteenth century, the two major sources of wet meadows had been removed from most of the Northeast, however. Beavers had been extirpated due to the fur trade, and most rivers were dammed and subject to flood control. Bobolinks (*Dolichonyx oryzivorus*) and other grassland species that depend on wet meadows survived in artificial grasslands (hay meadows and fallow fields). Later the intensification of farming (particularly more frequent mowing of hay meadows) made these fields unsuitable or, worse, turned them into "ecological traps" that attracted nesting birds and then destroyed their eggs and nestlings (Bollinger and Gavin 1992). To sustain populations of these species, we need to think in terms of their original habitats. In larger, wilder natural areas, dams could be removed and beavers could be permitted to resume their role of "ecosystem engineers" (Jones, Lawton, and Shachak 1994). Unregulated beavers initiate a successional sequence of habitats—open pond to marsh to wet meadow to shrub swamp to forested wetland—that support a large array of aquatic and early successional species of plants and animals. Because it is not usually practical to reintroduce seasonal flooding or unrestricted beaver activity in settled areas, however, wet meadows will need to be managed by simulating these natural disturbances with the goal of maintaining wet meadow vegetation. Fortunately, the bird species found in wet meadows do not require extensive areas of unbroken grassland, probably because they evolved in beaver meadows and patches of floodplain terrace that were relatively small openings in the forest (Askins et al. 2007). Consequently, relatively small meadows are often valuable for these species.

MANAGING GAME AND NONGAME SPECIES

Much of the funding for wildlife conservation is derived from fees and taxes on people who hunt or fish, so it is understandable that wildlife agencies emphasize management of game species. Consequently, changing the funding structure for wildlife conservation may be one of the most important ways to improve management of natural ecosystems to sustain biodiversity. Nongame species would have a higher priority if wildlife programs received more funding from people who enjoy wildlife but do not pay for hunting and fishing licenses. Eventually this would help improve communication between wildlife researchers (especially those who do not focus specifically on game species) and wildlife managers.

Because of the emphasis on propagating game species, highly artificial habitat modifications have frequently been used to boost populations of target species. Hence, we have in-stream structures and hatcheries to increase the number of trout; impoundments in salt marshes to create freshwater pools for dabbling ducks; and cornfields planted on prairie grassland to provide feeding areas for geese. These approaches to wildlife management frequently conflict with more general goals of sustaining biological diversity and natural ecosystem processes. However, the goals for managing nongame and game species are not always in conflict and are often congruent. Many of the ecological insights that have been derived from studies of landscape ecology can be applied to game species. Why, for example, do American woodcocks (*Scolopax minor*), which depend on dense, young forest for their food supply and nesting sites, need forest openings for their aerial courtship flights and summer roosts (Keppie and Whiting 1994)? Where were these forest openings when most of the eastern North America was blanketed with forest? The answer is obvious in places like Algonquin Provincial Park in Ontario and Voyageurs National Park in Minnesota, where beaver populations are not suppressed or highly regulated (Askins 2002). The forest canopy in these parks is repeatedly interrupted by beaver ponds, beaver meadows, and patches of young forest growing on former beaver meadows. Before beavers were eradicated in much of eastern North America, there was a close association between young riparian woodland where woodcocks forage and wet meadows where males court females with flight displays. Management of wet meadows may benefit woodcocks as well as bobolinks, common yellowthroats (*Geothlypis trichas*), alder flycatchers (*Empidonax alnorum*), and other meadow species. Similarly, the ecology of greater prairie-chickens and northern bobwhites (*Colinus virginianus*) are best understood in terms of the ecological disturbances and drivers that originally molded their habitats. As long as the goal is not to maximize production of a game species at the cost of degrading biological diversity in general, the management of game and nongame populations can be compatible. This is particularly true if the overriding goal is to manage natural ecosystems that are similar to the habitats in which these species evolved.

CONCLUSIONS

Ecological research has shown that it is difficult to manage wildlife populations effectively without a deep understanding of the natural ecosystem processes that support them. Traditional approaches to game management that focus on nesting habitat, food supply, and predators of target species don't always take these processes into account. It is particularly important to understand the various ecological processes such as persistent ecological drivers or occasional natural disturbances that sustain biological diversity, as well as the history of changing land use within particular ecosystems. It is also important to take into account important components of natural ecosystems that have been eliminated by people. Populations of many species were once highly dependent on natural processes (floods, fires, beaver activity, or grazing) that are now absent or infrequent. Compensating for these missing ecosystem components requires application of well-supported scientific theories combined with careful testing of the effectiveness of the resulting management plans. It also requires flexibility as new research refines or even displaces earlier theories. Manuals of wildlife management methods can be useful as starting points, but they are not sufficient for achieving an effective management goal that depends on a clear vision of the structure and functioning of the restored landscape (Samson and Knopf 2001). This vision should be based on the most conclusive ecological research available, and on a clear understanding of the history and ecology of a particular site. It should be the ultimate guide to conservation. Approved conservation plans for particular sites and standardized habitat management and restoration protocols are only provisional methods for achieving this ecological vision. They should be modified whenever they are not achieving the ultimate goal of restoring or simulating the natural ecological processes that sustain biological diversity.

ACKNOWLEDGMENTS

Funding for research related to this chapter was provided by the State Wildlife Grants Program administered by the Connecticut Department of Energy and Environmental Protection Wildlife Division, and by the Connecticut Light and Power Company, Northeast Utilities Transmission. My understanding of issues addressed in this chapter benefited greatly from discussion with Fritz Knopf (ecology of dry grasslands), Shannon Kearney-McGee (wildlife management and research), Ernie Steinauer (coastal grassland ecosystems), and Douglas Thompson (trout management). I appreciate editorial suggestions for improving the chapter from Fritz Knopf and the editors of this volume.

REFERENCES

Askins, R. A. 1993. Population trends in grassland, shrubland, and forest birds in eastern North America. In: *Current Ornithology,* Volume 11, D. Power, Ed. New York: Plenum Press, pp. 1–34.

Askins, R. A. 1999. History of grassland birds in eastern North America. In: *Ecology and Conservation of Grassland Birds in the Western Hemisphere,* P. D. Vickery and J. R. Herkert, Eds. Studies in Avian Biology No 19. Washington, DC: American Ornithologists' Union, pp. 60–71.

Askins, R. A. 2002. *Restoring North America's Birds: Lessons from Landscape Ecology,* 2nd ed. New Haven, CT: Yale University Press.

Askins, R. A., F. Chávez-Ramírez, B. C. Dale, et al. 2007. Conservation of grassland birds in North America: understanding ecological processes in different regions. *Ornithological Monographs* 64:1–46.

Bollinger, E. K., and T. A. Gavin. 1992. Eastern bobolink populations: ecology and conservation in an agricultural landscape. In: *Ecology and Conservation of Neotropical Migrant Landbirds,* J. M. Hagan, III and D. W. Johnson, Eds. Washington, DC: Smithsonian Institution Press, pp. 497–506.

Ceballos, G., E. Mellink, and L. R. Hanebury. 1993. Distribution and conservation status of prairie dogs *Cynomys mexicanus* and *Cynomys ludovicianus* in Mexico. *Biological Conservation* 63:105–112.

Foster, D. R. 1999. *Thoreau's Country. Journey through a Transformed Landscape.* Cambridge, MA: Harvard University Press.

Gill J. L., J. W. Williams, S. T. Jackson, K. B. Lininger, and G. S. Robinson. 2009. Pleistocene megafaunal collapse, novel plant communities, and enhanced fire regimes in North America. *Science* 326:1100–1103.

Hagan, J. M. 1992. Conservation biology when there is no crisis—yet. *Conservation Biology* 6:475–476.

Herkert, J. R. 1994. The effects of habitat fragmentation on Midwestern grassland bird communities. *Ecological Applications* 4:461–471.

Jones, A., and P. Vickery. 1997. *Managing Large Grasslands for Grassland Birds.* Lincoln, MA: Massachusetts Audubon Society, Center for Biology Conservation.

Jones, C. G., J. H. Lawton, and M. Shachak. 1994. Organisms as ecosystem engineers. *Oikos* 69:373–386.

Keppie, D. M., and R. M. Whiting, Jr. 1994. American Woodcock (*Scolopax minor*). In: *The Birds of North America Online*, A. Poole, Ed. Ithaca, NY: Cornell Lab of Ornithology; Birds of North America Online. http://bna.birds.cornell.edu/bna/species/100 (accessed August 10, 2010).

Knopf, F. L. 1996. Perspectives on grazing nongame bird habitats. In: *Rangeland Wildlife*, P. R. Krausman, Ed. Denver, CO: The Society of Range Management, pp. 51–58.

Koford, C. B. 1958. Prairie dogs, whitefaces, and blue grama. *Wildlife Monographs* 3:1–78.

Manzano-Fischer, P., R. List, and G. Ceballos. 1999. Grassland birds in prairie-dog towns in northwestern Chihuahua, Mexico. In: *Ecology and Conservation of Grassland Birds in the Western Hemisphere*, P. D. Vickery and J. R. Herkert, Eds. Studies in Avian Biology No 19. Washington, DC: American Ornithologists' Union, pp. 263–271.

Parmenter, R. R., and T. R. Van Devender. 1995. Diversity, spatial variability, and functional roles of vertebrates in the desert grassland. In: *The Desert Grassland*, M. P. McClaran and T. R. Van Devender, Eds. Tucson, AZ: University of Arizona Press, pp. 196–229.

Perry, L. G., D. M. Blumenthal, T. A. Monaco, M. W. Paschke, and E. F. Redente. 2010. Immobilizing nitrogen to control plant invasion. *Oecologia* 163:13–24.

Renfrew, R., and C. Ribic. 2008. Multi-scale models of grassland passerine abundance in a fragmented system in Wisconsin. *Landscape Ecology* 23:181–193.

Ribic, C. A., R. R. Koford, J. R. Herkert, et al. 2009. Area sensitivity in North American grassland birds: patterns and processes. *Auk* 126:233–244.

Saab, V. A., C. E. Bock, T. D. Rich, and D. S. Dobkin. 1995. Livestock grazing effects in western North America. In: *Ecology and Management of the Neotropical Migratory Birds*. Martin, T. E. and D. M. Finch, Eds. New York: Oxford University Press, pp. 311–353.

Samson, F. B. and F. L. Knopf. 2001. Archaic agencies, muddled missions and conservation in the 21st century. *BioScience* 51:869–873.

Thompson, D. M. 2006. Did the pre-1980 use of in-stream structures improve streams? A reanalysis of historical data. *Ecological Applications* 16:784–796.

Thompson, D. M., and G. N. Stull. 2002. The development and historic us of habitat structures in channel restoration in the United States: the grand experiment in fisheries management. *Géographie physique et Quaternaire* 56:45–60.

Tipton, H. C., P. F. Dougherty, Jr., and V. J. Dreitz. 2009. Abundance and density of mountain plover (*Charadrius montanus*) and burrowing owl (*Athene cunicularia*) in eastern Colorado. *Auk* 126:493–499.

Van der Hoek, D., and M. M. P. D. Heijmans. 2007. Effectiveness of turf stripping as a measure for restoring species-rich fen meadows in suboptimal hydrological conditions. *Restoration Ecology* 15:627–637.

Vickery, P. D., M. L. Hunter, Jr., and S. M. Melvin. 1994. Effects of habitat area on the distribution of grassland birds in Maine. *Conservation Biology* 8:1087–1097.

Walker, K. J., P. A. Stevens, D. P. Stevens, J. O. Mountford, S. J. Manchester, and R. F. Pywell. 2004. The restoration and re-creation of species-rich lowland grassland on land formerly managed for intensive agriculture in the UK. *Biological Conservation* 119:1–18.

Wamelink G. W. W., H. F. van Dobben, and F. Berendse. 2009. Vegetation succession as affected by decreasing nitrogen deposition, soil characteristics and site management: a modelling approach. *Forest Ecology and Management* 258:1762–1773.

Weltzin, J. F., S. Archer, and R. K. Heitshmidt. 1997. Small-mammal regulation of vegetation structure in a temperate savanna. *Ecology* 78:751–763.

Wilson, E. O. 1992. *The Diversity of Life*. New York: W.W. Norton & Co.

12 A Historical Perspective of the Connectivity between Waterfowl Research and Management

Christopher K. Williams
Department of Entomology and Wildlife Ecology
University of Delaware
Newark, Delaware

*Paul M. Castelli**
New Jersey Division of Fish and Wildlife
Nacote Creek Research Station
Port Republic, New Jersey

CONTENTS

* Current address: U.S. Fish and Wildlife Service, Forsythe National Wildlife Refuge, Oceanville, New Jersey.

Research must keep ahead of, not lag behind the need for facts. Game yields can be greatly increased and the costs and risks of management decreased, by more research.

—**Aldo Leopold**
Chair of the American Wildlife Policy Committee
reporting to the 1929 American Game Conference

Sound waterfowl management practices should be based on and modified by the results of careful scientific research.

—**Harrison F. Lewis**
Superintendent of Wildlife Protection for Canada
(now the Canadian Wildlife Service) at the 1946 meeting of the
International Association of Game, Fish and Conservation Commissioners

ABSTRACT

Over the last century of managing for wildlife resources, a disconnect has emerged between research (both theoretical and empirical) and on-the-ground practical management. However, we introduce the idea that waterfowl biologists have historically taken an inclusive and landscape level approach to management, and therefore sought lofty goals for application of the information base that was available through research. In this chapter, we first discuss the historical background for macroscale thinking within the waterfowl community. Second, we introduce the foundations of North American Waterfowl Management Plan, its joint ventures, and their recent efforts to better integrate research and management. Lastly, we discuss three case studies that illustrate the current integrated and adaptive approaches taken by wildlife managers and researchers to address waterfowl information needs.

INTRODUCTION

Over the last century of managing for wildlife resources, a disconnect has emerged between research (both theoretical and empirical) and on-the-ground practical management. Although these two disciplines were commonly considered to be closely linked to each other, the reality did not always meet the expectation. However, some areas of wildlife research and management have been more closely linked than others. In this chapter, we introduce the idea that early waterfowl biologists took an inclusive and landscape-level approach to management, and therefore sought lofty goals for application of the information base that was available through research. Perhaps the reason for pursuing such an extensive information base was the need to include partners from across North America in management decisions due to the migratory nature of the waterfowl resource, the small number of federal personnel, and the interest of numerous partners. A common theme was the desire to get an equitable share of the waterfowl harvest. Data and the results of scientific research was often the avenue to reaching consensus among partners as to how to allocate harvest opportunity and management effort while ensuring the perpetuation of the resource. A high demand for research on waterfowl provided many opportunities for graduate student training, and these graduate students in turn provided a pool of highly trained employees, who understood and valued research, for the agencies charged with managing waterfowl.

Consequently, there exist today multiple examples in waterfowl conservation that represent an effective integration of research and management. Early efforts by sportsmen to establish duck hatcheries and protect habitat have matured into important research programs. State and federal entities both developed research and monitoring efforts. The development of the Waterfowl Flyway Council System during the late 1940s and early 1950s led to a long research and management collaboration between federal, state, and provincial governments as well as universities and non-governmental organizations. These collaborations had a synergistic effect, allowing researchers to address questions whose answers might require work during multiple seasons of the year in a variety of locations across the continent. The North American Waterfowl Management Plan (NAWMP) in the mid-1980s and subsequent evolution of an associated system of joint venture initiatives (hereafter termed joint ventures), provided a unique and new starting point for productive dialogue between waterfowl researchers and managers that continues today. Accordingly, in this chapter we first discuss the historical background for macroscale thinking within the waterfowl community. Second, we introduce the foundations of NAWMP, its joint ventures, and their recent efforts to better integrate research and management. Lastly, we briefly discuss three case studies that illustrate the current integrated and adaptive approaches taken by wildlife managers and researchers to address waterfowl information needs, including: (1) the adaptive harvest management of mid-continent mallards (*Anas platyrhynchos*), (2) the need for winter carrying capacity estimates of American black ducks (*Anas rubripes*), and (3) the need for metapopulation structural information for Pacific Flyway lesser snow geese (*Chen caerulescens*).

THE ORIGINS OF THE WATERFOWL
RESEARCH-MANAGEMENT CONNECTION

North America's massive growth during the nineteenth century came from many fronts. To support a growing population seeking economic opportunity, the North American landscape supplied ample resources in the form of minerals, forests, fertile lands for agriculture, and wildlife. Although wildlife in North America provided food and fur for families building a new future, the lure of economic expansion and exploitation was to change many people's attitude toward the use of wildlife. The stories are familiar: the extinction of animals such as passenger pigeons (*Ectopistes migratorius*), heath hens (*Tympanuchus cupido*), Carolina parakeets (*Conuropsis carolinensis*), and near extinction of many others such as the wood duck (*Aix sponsa*) and American bison (*Bison bison*).

The great market-hunting era of the late nineteenth century eventually came to an end. This occurred for three main reasons. First and foremost, the depletion of the wildlife resources themselves effectively disrupted the sustainable supply in relation to demand. Second, amateur ornithologists, who were appalled at the millinery trade's use of bird feathers as well as the obvious impact on bird populations, spoke out vigorously on the behalf of conservation. Third, a growing call from sportsmen and sporting organizations questioned the reckless destruction of wildlife and other natural resources. In 1848, Henry William Herbert promoted a new ethic of sportsmanship that encouraged a growing sporting press, such as *American Sportsman* (1871) and *Field and Stream* (1874) as well as rod and gun clubs, such as the Boone and Crockett Club (1887) founded by George Bird Grinnell and Teddy Roosevelt (Trefethen 1975; Reiger 2000).

The result of these growing concerns caused a myriad of states to enact legislation to limit the impacts of market hunting on their wildlife resources (Matthews 1986; Anderson 2002). However, it was becoming clear that federal legislation was needed to support state regulation efforts for management of all species, with special emphasis on migratory birds, which could not be regulated by any one state (Matthews 1986; Baldassarre and Bolen 2006). In 1900, the Lacey Act was passed to prevent interstate transport of illegally harvested wildlife, but the Act also delegated power to the Secretary of Agriculture to include "the preservation, distribution, introduction, and restoration of game birds and other wild birds." The Migratory Bird Treaty Act of 1918, which ratified the Migratory Bird Convention of 1916, provided the federal government oversight of migratory bird conservation, including harvest. Although a few states fought such centralized management all the way to the U.S. Supreme Court, the legislation's footing was solid, as affirmed by Chief Justice Oliver Wendell Holmes's statement,

> The state of Missouri claims to have exclusive authority over birds that yesterday had not arrived; tomorrow may be in another state; and in a week a thousand miles away. Here a national interest of great importance is at stake. But for the treaty and the statute there soon might be no birds for any powers to deal with. We see nothing in the constitution that compels the government to sit by .... It is not sufficient to rely upon the states. The reliance is vain. (State of Missouri v. Holland 252 U.S. 416, 1920)

With these laws, a foundation was laid to ultimately promote a unique system where research and management would be integrally embedded and necessary. By the 1920s, there was increasing recognition that better scientific information was needed to improve management of game animals and especially waterfowl populations (Meine 1988:259). Aldo Leopold, as Chair of the American Wildlife Policy committee, reported at the 1929 American Game Conference that for migratory birds "the management measures most needed are the public acquisition of (wetland) habitats threatened with drainage, a continental system of ... refuges and a more adequate program of fact finding." The report concluded that the knowledge base of waterfowl was inadequate and that "research must keep ahead of, not lag behind the need for facts. Game yields can be greatly increased and the costs and risks of management decreased, by more research" (Hawkins et al. 1984). Leopold later joked, "if only [waterfowl] could feed on the broad acres of paper we have dedicated to guesses about their welfare" (Meine 1988:315). These concerns by the foremost wildlife ecologist of the day led the way for increased applied waterfowl research on waterfowl in subsequent years.

When waterfowl populations crashed during the drought of the 1930s, sportsmen responded with several initiatives of their own that eventually led to significant waterfowl research efforts. A Minneapolis sportsman, James Ford Bell, acquired approximately 5,000 acres of Delta marsh, located at the foot of Lake Manitoba, for waterfowl hunting (Delta Waterfowl 2011). An ethical hunter, he established his own limits when none existed and followed them when they were established. In an effort to give back to the resource, he created a duck hatchery during the 1930s and released almost 10,000 hand-reared ducks of seven species (Delta Waterfowl 2011). However, Bell and his hatchery manager, Edward Ward, came to realize there was much to learn about duck biology and a better understanding of their life cycle was needed. Bell believed the best hope for sustaining waterfowl populations was through a science-based understanding of the birds' behavior. Bell's efforts to establish a waterfowl research facility almost failed because when he tried to interest Aldo Leopold in organizing a waterfowl research program at the Delta Marsh, Leopold initially turned him down (Meine 1988). Leopold was under the impression that Bell was interested only in the artificial production of canvasbacks. Not easily deterred, Bell convinced Dr. Miles D. Pirnie from Michigan State University to visit the Marsh (Delta Marsh). Pirnie was so impressed by the potential for waterfowl research he convinced Leopold to give Bell another chance. After a second meeting in 1938, Leopold threw his support to the idea, helped garner support and partners, and provided the first graduate student at the new Delta Research Station, H. Albert Hochbaum (Meine 1988). Since 1938, the Delta Research Station has supported over 340 graduate students representing seventy-five colleges. These students produced over 750 research papers on waterfowl ecology, habitat, and management (Delta Waterfowl 2011). Among the most notable work at Delta was the Marsh Ecology Research project, a detailed long-term study of wetland ecology and management (Murkin, van der Valk, and Clark 2000).

In 1930, a group of concerned sportsmen formed the More Gamebirds in America Foundation (Meine 1988). The organization was primarily interested in restoring duck populations via incubators and releasing them to the flight lanes; something with which Aldo Leopold strongly disagreed (Meine 1988). However, the group was important because it raised awareness of the plight of waterfowl during the Dustbowl Era and then ultimately explored the possibility of using aircraft to survey and monitor waterfowl and the success of their efforts. During August 1935, the group coordinated the first international duck census. Coordinated ground counts were conducted in the more settled areas of Canada and the United States, while aerial counts were made in the northernmost areas. The resultant publication, *The 1935 International Wild Duck Census,* documented almost 43 million ducks and estimated the continental duck population at 65 million. This work was the precursor to modern waterfowl inventory work and is remarkable in that their estimated population is deemed "in the ball park" based on subsequent modern surveys (Leitch 1984).

The More Gamebirds in America Foundation went on to form Ducks Unlimited (DU) in 1937 and Ducks Unlimited Canada (DUC) in 1938 and shifted its focus more to habitat acquisition and preservation than its precursor (Gavin 1964; Meine 1988). While best known for decades of raising funds and delivering habitat conservation, both DU and DUC eventually established waterfowl research programs. DU initiated its research group to guide the evolution of conservation practices with changing climate, land uses, and needs of waterfowl. DU has partnered with various research agencies to address these issues while training graduate students and producing scientific reports. In 1991, DUC established the Institute for Wetland and Waterfowl Research (IWWR), as the scientific research arm of its organization to conduct research on the problems facing North America's waterfowl (Baldassarre and Bolen 2006). During the past twenty years, IWWR has led or participated in nearly 300 studies throughout North America, while training graduate students from over forty universities and collaborating with more than forty-five external research agencies.

Another early waterfowl research center was the Illinois Natural History Survey. The river bottoms in Illinois were famous for the concentrations of mallards and other ducks and when the number of ducks declined precipitously during the severe drought of the 1930s, duck research became a priority. While most famous for the experimentation by Frank Bellrose that led to the development of wood duck boxes, many other subjects were addressed (Baldessarre and Bolen 2006). Wetland ecology studies documented the decline of aquatic habitats caused by sedimentation. Waterfowl food habits, migration, lead poisoning, and the effects of waterfowl regulations were studied. Intensive studies revealed the population structure of Canada geese in the Mississippi Flyway. Long-term survey and banding programs were initiated and many techniques for studying waterfowl were developed.

Federal biologists responsible for waterfowl harvest regulations also initiated research to inform their regulatory decisions. One of their earliest research questions about waterfowl was their spatial and temporal movements and leg banding was the primary technique developed to address this question. Building on the early work of Wells W. Cooke, who recognized the overwhelming trend of most waterfowl to migrate in a north-south pattern (Cooke 1906), Frederick C. Lincoln reviewed years of accumulated banding data and introduced the concept of "flyways" to the waterfowl management community. Lincoln defined flyways as "those broad areas into which certain migration routes ... (of ducks, geese, and swans) ... blend or come together in a definite geographic region" (Lincoln 1935). Although the biological data did show some overlap, Lincoln was able to define four biological flyways: Atlantic, Mississippi, Central, and Pacific.

A milestone in federal research occurred in1935, when Jay "Ding" Darling, then Chief of the U.S. Bureau of Biological Survey, was able to establish nine wildlife research cooperative units at universities in various parts of the United States whose purpose was to conduct fundamental investigations, apply the results to local wildlife management conditions, and carry them to both public and private landowners (Baldassarre and Bolen 2006). This action was perfectly timed as the new research results and graduate students trained were needed in the rapidly growing field of waterfowl biology. Over the decades, it is likely no other entity has produced more research, trained more graduate students, and been the source of more wildlife professionals than the Cooperative Wildlife Research Program.

The federal governments of the United States and Canada further provided leadership in waterfowl research by establishing wildlife research centers. Early waterfowl research was conducted at the Denver Wildlife Research Center in Colorado and the Patuxent Wildlife Research Center located in Maryland. Each of these had several substations to accommodate work in various parts of the country (Patuxent National Research Center 1989). In 1965, the United States Fish and Wildlife Service established the Northern Prairie Wildlife Research Center in Jamestown, North Dakota, and field stations were soon established in Minnesota, South Dakota, and North Dakota. Northern Prairie gained an international reputation for excellence in waterfowl research and may have been the premier waterfowl research facility in the world. During the 1960s, the Canadian Wildlife Service established the Prairie Migratory Bird Research Center, in Saskatoon, Saskatchewan where it conducted similar work, often in collaboration with the Northern Prairie Center. Overall, the centers worked with a variety of partners on many research projects and produced many seminal works on all aspects of waterfowl ecology and management. The permanent staff of researchers was often able to conduct long-term and geographically extensive projects that addressed large problems through a series of conceptually linked studies (Patuxent National Research Center 1989).

THE STATES' CONTRIBUTION AND THE WATERFOWL FLYWAY COUNCIL SYSTEM

Not many states had a waterfowl research program during the 1930s. The ability of states to participate in waterfowl management was bolstered by the Pittman-Robertson Federal Aid in Wildlife Restoration Act of 1938, which provided reliable funding for state waterfowl biologists and state-based research programs (Kallman 1987). With a steady supply of funding, and highly trained students from the newly formed Cooperative Wildlife Research Units, state programs in waterfowl research and management developed quickly. Armed with their own data and analyses, states began seeking a greater voice in waterfowl regulatory decisions and began to informally organize regionally to discuss waterfowl management. Soon it became clear a framework to formalize the growing demand for state participation was needed. In 1951, the International Association of Game, Fish and Conservation Commissioners passed Resolution #10, which proposed that Frederick Lincoln's biological flyways serve as a model for four administrative flyways (Figure 12.1). The recommendation was formalized and four Flyway Councils were established in 1952, with a National Flyway Council (NFC) established in 1953 (Hawkins et al. 1984) consisting of each state's wildlife administrator. Importantly, Technical Sections (consisting of official membership by state biologists and associate membership by private conservation organizations, universities, and federal agencies) existed below the NFC to advise it and discuss data and analysis related to waterfowl populations and their habitats.

Even as Flyway Councils and Technical Sections were being established and state participation in waterfowl management was increasing, there was the recognition that harvest decisions were being made without specific data related to the annual status and number of waterfowl, waterfowl hunters, or their harvests. Although much progress had been made since Bell founded the Delta Research Station, there was still a continuing need to expand the existing knowledge about the basic biology of waterfowl species and their relationship to various habitats throughout the annual cycle.

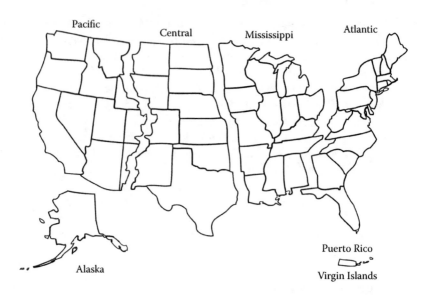

FIGURE 12.1 Administrative waterfowl flyways in the United States.

Experimental aerial monitoring of breeding and wintering waterfowl across the North American continent had begun during the 1930s and 1940s. By 1955, after assessment and refinement of their design, the breeding waterfowl survey of the mid-continent region and mid-winter survey became operational (Blohm 1989). Both these surveys continue to this day, with improvements, making them among the largest and longest-running wildlife surveys in the world. Similarly, leg-banding programs established by Cooke and Lincoln, as well as nationwide harvest surveys instituted during the 1950s to enumerate the species, number, and location of the harvest of migratory waterfowl also continue (Trost and Carney 1989). These long-term, large-scale data collection efforts require more money and personnel than any one agency can muster. However, the Waterfowl Flyway System provided a forum to find the many state and federal partners necessary to carry out the ambitious continent-wide monitoring and research necessary to manage waterfowl populations.

Breeding, wintering, and harvest surveys, combined with banding data, provided the basis for Flyway Councils and the United States Fish and Wildlife Service (USFWS) to begin to manage waterfowl species based on data. Initially, annual population-based harvest regulations were developed based on managers' beliefs as to how environmental conditions, waterfowl populations, and harvest were related. Some of these beliefs were widely held, yet largely untested (Romesburg 1981). Over the years, the accumulation of data and improvement in analytical techniques permitted testing and refinement of these relationships. This allowed waterfowl management to develop *a priori* predictions about the effect of various annual harvest regulations on a population's size given current environmental conditions and the management goal for that species. In some cases, where adequate data existed, it has been possible to set up species management in an adaptive framework (see Case Study #1) and conduct management and research concurrently as suggested by MacNab (1983). In recent years, Flyway Councils have also begun to research the human dimension issues of waterfowl management (e.g., hunter recruitment and retention) in a more formal way (Witter et al. 2006).

THE NORTH AMERICAN WATERFOWL MANAGEMENT PLAN

The decline of waterfowl populations and the loss of wetlands were of increasing concern in the 1970 and 1980s. Because of these growing concerns, as well as the infrastructure of federal oversight of waterfowl issues, an ambitious wetland and waterfowl recovery plan was designed in the form of the North American Waterfowl Management Plan (NAWMP). Signed jointly by the United States and Canada in 1986 and by Mexico in 1994, the plan's framework and philosophy was to design and implement regional strategies to reverse waterfowl decline by improving habitat quantity and quality. The goal was to restore waterfowl populations to a level representative of the 1970s by improving and securing long-term protection of 2.4 million ha of habitat in thirty-four geographic areas of major concern (U.S. Department of the Interior and Environment Canada 1986).

Despite this broad habitat management goal, several important constructs were developed that paved the way for improved connection between waterfowl management and research. First, NAWMP promoted regional partnerships among all federal, state, provincial, territorial, tribal, and private organizations to facilitate the understanding of population and habitat declines. Second, NAWMP proposed several specific research initiatives to help clarify population and habitat relationships, including investigation of: (1) the effects of land-use practices on the breeding success of waterfowl, (2) factors affecting the carrying capacity of migration and wintering areas, (3) evaluation of subpopulation relationships between mallards and black duck populations, and (4) the effects of hunting mortality on waterfowl populations. Third, NAWMP recommended a series of formal joint ventures to parse out and oversee specific efforts to connect research and management within specific regions deemed important for waterfowl. Additionally, NAWMP identified priority species lacking basic information or data sufficient for management.

DEVELOPMENT OF THE JOINT VENTURE PARTNERSHIPS

Under the organization of the 1986 NAWMP, multiple habitat joint ventures were created to meld resources for maximizing financial, organizational, and other in-kind support toward common waterfowl research and management objectives in a geographic region. Today, twenty-one habitat joint ventures exist, including eighteen in the United States: Appalachian Mountains, Atlantic Coast, Central Hardwoods, Central Valley, Eastern Gulf Coastal Plain, Gulf Coast, Intermountain West, Lower Mississippi Valley, Northern Great Plains, Oaks and Prairies, Pacific Coast, Playa Lakes, Prairie Pothole, Rainwater Basin, Rio Grande, San Francisco Bay, Sonoran, and Upper Mississippi River/Great Lakes (Figure 12.2). There are three additional habitat joint ventures in Canada: Eastern Habitat, Prairie Habitat, and Inter-Mountain West. Within each, an organizational construct evolved to include a Management

1) Appalachian Mountains   7) Intermountain West   13) Prairie Pothole
2) Atlantic Coast   8) Lower Mississippi Valley   14) Rainwater Basin
3) Central Hardwoods   9) Northern Great Plains   15) Rio Grande
4) Central Valley   10) Oaks and Prairies   16) San Francisco Bay
5) Eastern Gulf Coastal Plain   11) Pacific Coast   17) Sonoran
6) Gulf Coast   12) Playa Lakes   18) Upper Mississippi River/
                                                      Great Lakes

FIGURE 12.2   Eighteen habitat Joint Ventures in the United States.

Board and a Technical Committee. The Management Board establishes priorities and direction for each joint venture, while the Technical Committee, consisting of federal, provincial, state, private partners, and universities, works to carry out projects at the local level. Although each joint venture has different strategies for accomplishing its stated objectives, all depend on multiple partnerships to protect, restore, and enhance targeted habitat for migratory birds.

In addition to the tewnty-one habitat joint ventures, the NAWMP has sanctioned three species joint ventures. In contrast to habitat joint ventures, which direct efforts to projects on the ground, species joint ventures were established to address critical information gaps for individual species or species groups through the development of monitoring programs and directed research. The species joint ventures work by identifying necessary research and monitoring activities, assigning priorities from a continental perspective, promoting and encouraging funding and participation in priority research, and facilitating timely dissemination of information. Currently, three species joint ventures exist: the Black Duck Joint Venture (1986), the Arctic Goose Joint Venture (1989), and the Sea Duck Joint Venture (1999).

The American black duck (*Anas rubripes*), once the most abundant freshwater duck in eastern North America (Longcore et al. 2000a), reached a population low in the 1980s after a thirty-year decline. Habitat loss and degradation, competition with mallards, and hunting mortality are believed to have contributed to this decline. Consequently, the mission of the Black Duck Joint Venture (BDJV) is to coordinate and promote data gathering (through monitoring and research) among Flyway Councils, universities, and federal, provincial, state, and tribal conservation agencies to improve population and habitat management (Black Duck Joint Venture Management Board 2008).

The Arctic Goose Joint Venture (AGJV) facilitates research and monitoring of arctic goose populations. Prior to the formation of the AGJV, goose management was based heavily upon research and monitoring of goose populations in migrating and wintering areas. However, because (1) wintering ground mixing of northern-nesting geese was increasing and greatly complicated population assessment and management of all goose populations, and (2) knowledge regarding the breeding ground population distribution, status, and demographics continued to be limiting, it became apparent that traditional means of coordinating field studies were no longer suitable to resolve these challenges. Beginning in 1991, the AGJV and its partners set in motion a coordinated approach for meeting information needs for the management of northern-nesting geese in North America. Today the AGJV now encompasses twenty-eight northern-nesting goose populations, ranging from the Aleutian Islands to Labrador. The AGJV works cooperatively to provide a coordinated and cost-effective approach to meeting high priority information needs for the management of northern-nesting geese in North America. This partnership approach is especially valuable for conducting arctic research where logistics are more costly and where maximum return from available funds is highly desirable. Attention is focused on subspecies of the brant (*Branta bernicla*), the greater white-fronted goose (*Anser albifrons*), the Canada goose (*Branta Canadensis*), the cackling goose (*Branta hutchinsii*), and greater and lesser snow geese (Arctic Goose Joint Venture Technical Committee 2008).

Lastly, the Sea Duck Joint Venture (SDJV) promotes the conservation of fifteen species of North American sea ducks through partnerships by seeking greater knowledge and understanding for more effective management. The SDJV is focused on four key pillars of emphasis, including science, communications, funding, and conservation actions. In an effort to link management and research, the science focus of SDJV initially concentrated much of its effort on population delineation in an effort to provide a firm foundation for subsequent monitoring programs. Efforts that are more recent have begun to focus on developing these effective monitoring programs for sea duck populations as trends for most species are based on limited information, and abundance estimates are not feasible. Additionally, SDJV is working to improve harvest surveys, identify important habitats, and document biological impacts of contaminants, parasites, and disease (Sea Duck Joint Venture Management Board 2008).

2004 NAWMP REVISIONS STRENGTHEN RESEARCH AND MANAGEMENT CONNECTIVITY

In 2004 a NAWMP revision subtitled *Strengthening the Biological Foundation* was written to reflect the Plan Committee's continuing belief that a strong scientific base is critical to the Plan's success in conservation (North American Waterfowl Management Plan, Plan Committee 2004a). It was recognized that significant gaps remained in scientific information on the ecology, abundance, and trends of many waterfowl populations, especially for sea ducks and resident ducks in Mexico. However, because there had been a strong scientific history and extensive practical management experience concerning waterfowl, NAWMP partners believed conservation could be strengthened by expressly stating it was a priority for researchers and managers to work together in an adaptive management framework.

Therefore, NAWMP partners were challenged to improve the scientific knowledge on which conservation decisions depend and to continuously improve their work through adaptive management. Specifically, a series of recommendations were promoted. First, because programs to track population trends had been lacking or inadequate for several species, population monitoring capabilities needed to be enhanced to detect meaningful changes in waterfowl abundance and gauge those changes against objectives. Second, to better understand how specific habitat changes affect waterfowl populations, habitat joint ventures were encouraged to develop and maintain monitoring and assessment systems capable of discerning habitat changes over time and space. Third, because the Plan was designed to encompass continental, regional, and local issues, adaptive management and strategic planning were encouraged to occur at multiple spatial scales. Fourth, the Plan Committee encouraged continued commitment to expanding scientific information where it was lacking and integrating the best possible science into the Plan's decision-support systems from continental to local project scales.

Perhaps more important than simply promoting an increase in the connectivity of research and management within the 2004 NAWMP revisions, there was also a general call for embracing adaptive management (North American Waterfowl Management Plan, Plan Committee 2004b). Although adaptive management is a broad concept, and one that had been successfully implemented a decade earlier for managing waterfowl harvests (see Case Study #1), NAWMP considered that it should be used for cyclic planning, implementation, and evaluation to improve management performance. Implicit in this charge was the necessary and continued connection between research and management. Therefore, within future joint venture and NAWMP conservation programs, there must be consideration of: (1) clear, quantifiable objectives, (2) specific predicted biological outcomes of alternative management actions, (3) monitoring procedures to measure the outcome variables defined in the objectives, (4) an evaluation process to compare outcomes with original objectives, and (5) a commitment to use the lessons learned to adjust future decisions. For example, if waterfowl population goals are set and models are designed to predict how they will respond to habitat changes and management actions, ultimately the implementation and monitoring of such actions will improve future decisions. Importantly, this strategic planning incorporates a biological and research foundation. Although adaptive management requires discipline by all partners to work within an information structure, NAWMP recognized that the ultimate payout for biological response and cost-effectiveness was strong.

CASE STUDY #1: ADAPTIVE HARVEST MANAGEMENT IN MID-CONTINENT MALLARDS

The legal authority of the federal government to regulate the hunting of migratory birds was rooted in the 1918 Migratory Bird Treaty Act. These regulations serve as a framework each year for state governments, which may adopt federal guidelines without change or introduce changes that are more restrictive. Such frameworks establish beginning and ending dates for the hunting season, the season lengths within those dates, and daily allowable bag limits. An essential component of this regulatory process is to demonstrate that the harvest allowed is sustainable. Thus, each year biologists estimate breeding populations, harvest levels, productivity, migration, and other population characteristics (Martin, Pospahala, and Nichols 1979; Smith et al. 1989) with the ultimate goal of informing managers about any meaningful changes in population size and growth (Williams and Nichols 1990).

Unfortunately, because all such variables have a great deal of uncertainty over time and space, there is reasonable limitation to our capacity to predict outcomes using models and other assessment tools (Conroy 1993; Nichols, Johnson, and Williams 1995). Four such sources of uncertainty are associated with the regulatory process for hunting waterfowl and other migratory game birds each year. First, "environmental variation" occurs and alters the yearly availability of breeding habitat and reproductive success. Second, "partial controllability" is the uncertainty that results from the inability of managers to precisely control harvest rates. Third, "partial observability" causes uncertainty from the limited precision in estimating population size, reproduction, and harvest rates. Fourth, "structural uncertainty" is uncertainty resulting from the incomplete understanding about the ways biological systems actually work, such as the degree of additive versus compensatory mortality at work during the annual cycle (Williams 1997).

In 1995, following a number of years of discussion and development, the USFWS, with the support of the four Flyway Councils, applied the concept of adaptive resource management to the regulation of mid-continent mallard harvests, using a process called Adaptive Harvest Management (AHM) that was designed to account for the uncertainties in the response of waterfowl populations to various harvest regulations. The key components of AHM included four considerations (Johnson et al. 1993; Williams and Johnson 1995). First, a limited number of regulatory alternatives were considered that described Flyway-specific season lengths, bag limits, and framework dates. Second, several competing population models, describing various hypotheses about the effects of harvest and environmental factors on waterfowl abundance, were formulated. Four models were developed for mid-continent mallards, including consideration of additive versus compensatory mortality and weak versus strong effects of density dependence on recruitment. Third, measures of reliability (or "weight") for each competing population model were assigned. Fourth, a mathematical description was assigned to the objective of harvest management by which alternative regulatory strategies could be compared. Once these components were assigned, a stochastic optimization procedure was conducted to derive an optimal regulatory strategy with respect to the stated management objective, for each possible combination of breeding population size, environmental conditions, and model weights (Johnson et al. 1997).

Once the adaptive framework was established, the setting of annual hunting regulations then involved an iterative process to include monitoring. Each year, an optimal regulatory choice was identified based on population estimates, environmental conditions (number of ponds as a surrogate), and current model weights. For mid-continent mallards, the wetland condition in the Prairie Pothole Region defined the environmental condition. After the regulatory selection was made, model-specific predictions for subsequent breeding population size were determined. When monitoring data for the estimated breeding population became available, model weights could be increased or decreased to the extent that observations of population size agreed with predictions. The new model weights would then be used to start another iteration of the process in the following year. By iteratively updating model weights and optimizing regulatory choices, the process should eventually identify which model was the best overall predictor of changes in population abundance (Figure 12.3). In the case of mid-continent mallards, by 2009, updated model weights suggested preference for the weakly density-dependent reproductive hypothesis (88%) and evidence for additive mortality (62%) (U.S. Fish and Wildlife Service 2009). Since 1995, consideration of yearly population sizes and numbers of available breeding ponds has allowed for liberal harvest regulations for all of the years the AHM structure has been in place.

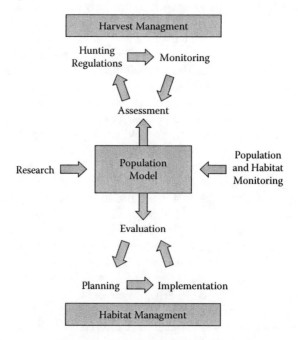

FIGURE 12.3   Analysis, assessment, and research are linked iteratively to management, leading to improved understanding of waterfowl populations over time as conservation goals are achieved.

AHM is adaptive in the sense that the harvest strategy "evolves" to account for new knowledge generated by a comparison of predicted and observed breeding population sizes. Importantly, this process establishes a quantitative connection between population monitoring and regulation framework and allows for continued research into the system to better understand the biological processes that drive the system. Ultimately, AHM has encouraged the increased connection between research and management to iteratively reduce uncertainty and improve predictions (Lancia et al. 1996) while better informing management decisions.

CASE STUDY #2: BLACK DUCK JOINT VENTURE AND A PRIORITY TO ESTIMATE HABITAT CARRYING CAPACITIES

As the first species joint venture, established in 1986, the Black Duck Joint Venture's (BDJV) mission has been to implement and coordinate cooperative population monitoring, research, and communication programs to provide information required to manage black ducks and restore numbers to the NAWMP goal (Black Duck Joint Venture Management Board 2008). Specifically, the joint venture has three goals. First, there is the mission to develop, support, and guide monitoring programs to provide information on the distribution, population growth trends, and other demographic parameters of black ducks throughout their annual range. Second, the joint venture supports and guides research on black duck population dynamics, habitat ecology, monitoring programs, and management techniques to identify and mitigate limiting factors. Third, results of monitoring and research programs are communicated in a timely and effective manner to associated habitat joint ventures, natural resource agencies, policymakers, and stakeholders to support black duck conservation and management efforts.

To achieve these three main mission components, the BDJV consists of two bodies: a Management Board and a Technical Committee. The Technical Committee works with the BDJV Science Coordinator to advise the Management Board about black duck research and management needs and is responsible for development of annual work plans, project implementation, and progress evaluation. To maximize the Technical Committee's efficiency and productivity, members are organized into working groups, representing each BDJV program area (i.e., population monitoring, research, communications, and evaluation). The goals of the research program are to identify and prioritize research and information needs, facilitate the exchange of ideas and data among cooperators and stakeholders, coordinate and support research, and serve as a one-stop clearinghouse of current scientific knowledge of black duck ecology and management.

Black duck populations have declined by >50% between the late 1950s and 1980s, potentially due to over-harvest, competition and hybridization with mallards, decrease in quality and quantity of wintering and breeding habitat, and environmental contaminants (Conroy, Costanzo, and Stotts 1989; Rusch et al. 1989; Longcore et al. 2000a, b; Merendino, Ankney, and Darrell 1993; Nudds, Miller, and Ankney 1996; Conroy, Miller, and Hines 2002; McAuley, Clugston, and Longcore 2004; Zimpfer and Conroy 2006). To address these declines, the BDJV research program identifies priority research needs, and updates them annually. In the 2008–2013 strategic plan, four broad categories exist, including population monitoring, population ecology, habitat ecology, and integration of population dynamics and habitat (Table 12.1). Across all nineteen priorities, the top priority identified was to quantify regional nutritional carrying capacity on seasonal ranges.

As an example of such work, Cramer (2009) set out to estimate the current carrying capacity of winter black duck habitats in 3 million ha of wetlands in southern New Jersey, which traditionally has held the largest proportion of the wintering population. Cramer used bioenergetics modeling to calculate energetic carrying capacity in terms of duck-use days (DUD) and found estimates of carrying capacity were below the five-year mean (75,190 black ducks) counted during the midwinter waterfowl survey, indicating a possible shortage of energy during the core of the winter. He further hypothesized that reductions to estimates of DUD may be attributed to interspecific competition for resources, avoidance of developed areas resulting in an overestimate of available habitat, or the temporary suspension of food availability during winter freeze events.

The ultimate value of Cramer's (2009) research, in combination with the other funded studies (e.g., Plattner, Eichholz, and Yerkes 2010) under the auspices of the BDJV, was twofold. First, such studies can identify priority habitats where future management efforts could be taken to increase the carrying capacity. For instance, in the case of Cramer (2009) and Plattner et al. (2010), mudflat habitat had the largest energy value across all habitat types. This information allows for local management actions, such as (1) the acquisition and enhancement of ecologically functioning salt marsh habitat with an emphasis on mudflat habitat, and (2) restoration of full tidal exchange to formerly tidally restricted areas, such as impoundments, salt hay farms, and dense stands of common reed (*Phragmites australis*). Second, these studies produced estimates of energy availability for all coastal habitat types across a wide spatial area. These combined values can be applied to landscape level land-use maps (such as the National Wetland Inventory) to make predictions of black duck carrying capacity across their wintering range. Such results reflect the interface of research and management that is promoted within the BDJV and helps to ensure the development of science-based population goals and realistic habitat management recommendations.

CASE STUDY #3: ARCTIC GOOSE JOINT VENTURE, PACIFIC FLYWAY, AND UNDERSTANDING METAPOPULATION STRUCTURE OF WESTERN ARCTIC AND WRANGEL ISLAND LESSER SNOW GEESE

The primary goal of the Arctic Goose Joint Venture (AGJV) is to "foster greater research and monitoring of northern-nesting geese for the purpose of improving and refining population management from a breeding ground perspective" (Arctic Goose Joint Venture Technical Committee 2008). The strategy of the AGJV is to achieve this general goal by planning, facilitating, communicating, and coordinating activities directed at improving the information base for northern-nesting populations of

TABLE 12.1

Description and Categorization of Priority Research Needs for the American Black Duck as Identified by the Black Duck Joint Venture During the 5-Year Planning Horizon, 2008–2013

Category	Issue	Priority Within Category	Overall Priority
Population monitoring	Identify sources of and quantify heterogeneity in band reporting rates.	1	3
	Evaluate genetic methods to monitor changes in black duck population size and structure.	2	9
	Quantify and identify sources of variation and bias in estimates of fall age ratios.	3	15
	Quantify effect of "availability" bias in aerial surveys	4	16
	Evaluate and quantify influence of behavioral responses of black ducks to survey related disturbances.	5	17
	Integration of ground breeding population surveys with aerial surveys.	6	18
	Development of post-season banding and aging techniques.	7	19
Population ecology	Development of Adaptive Harvest Management models.	1	2
	Quantify the rate and identify causes of apparent decline in black duck productivity.	2	4
	Quantify regional differences in and factors influencing black duck productivity.	3	5
	Evaluate and quantify population structure and sub-units.	4	10
	Test hypotheses of range contraction versus population decline.	5	13
Habitat ecology	Quantify regional nutritional carrying capacity on seasonal ranges.	1	1
	Evaluate and quantify influence of habitat patch size and human disturbance (e.g., road density, perforation, fragmentation, and recreation) on black duck habitat selection, use, and quality.	2	6
	Identify and quantify habitat factors influencing productivity.	3	8
	Evaluate the quality and availability of healthy coastal marshes and other large (>10 ha) marsh/open water complexes during winter and spring.	4	14
Integration of population dynamics and habitat	Develop synthetic models integrating black duck habitat and population dynamics as recommended by the Joint Task Group.	1	7
	Develop synthetic models predicting the effects of climate change of black duck population dynamics and habitat use.	2	11
	Identify migration routes and factors influencing time budgets and distribution during migration.	3	12

geese. The activities of the AGJV include both short-term and long-term information-gathering programs directed at measuring basic population parameters, such as population growth, production, harvest, and survival rates. To identify priority information needed to facilitate effective population management, the AGJV has subsequently categorized seven categories of "Information Needs" into high, medium, and low designation for the twenty-eight goose populations (Table 12.2 illustrates snow geese). A high priority indicates an immediate need for information; a medium priority recognizes a demonstrated need for the information, but other information is required first; and a low priority suggests that the information is relevant but other information is presently deemed more important.

The Beringia region of the Arctic contains two colonies of lesser snow geese breeding on Wrangel Island, Russia, and Banks Island, Canada, that winter in North America. The Wrangel Island population is composed of two subpopulations from a sympatric breeding colony but with separate wintering areas on the Fraser/Skagit River Deltas and the Central Valley of California. The Banks Island population shares a sympatric wintering area in California with one of the Wrangel Island subpopulations. Although intensive cooperative marking efforts were undertaken on several major breeding colonies in the 1980s, including Wrangel Island, to delineate population units and document interchange rates, multiple management concerns still relate to the metapopulation dynamics of these populations. First, the Wrangel Island colony represents the last major snow goose population in Russia and has fluctuated considerably since 1970, whereas the Banks Island population has more than doubled. The reasons for these changes are unclear, but hypotheses include independent population demographics (survival and recruitment) and immigration and emigration among breeding or wintering populations. These demographic and movement patterns have important ecological and management implications for understanding goose population structure, harvest of admixed populations, and gene flow among populations with separate breeding or wintering areas. Second, the northern and southern wintering segments have experienced important shifts in recent decades: the proportion of the Wrangel population wintering on the Fraser and Skagit River Deltas (the northern component) has increased steadily from approximately 20% to 30% in the 1950s to 1960s to >50% in recent years. The Fraser-Skagit winter population has grown to the point where the geese are now causing problems with crop depredation on local farms, air traffic safety issues at the Vancouver International Airport, nuisance concerns in urban areas, and overconsumption of bulrush rhizomes on the foreshore marshes. Third, the harvest of lesser snow geese cannot be currently apportioned by population units. Mixing of western arctic geese with other populations with different harvest goals further complicates harvest management. Population growth in some breeding areas may necessitate the development of strategies to increase population-specific harvest.

To address these concerns, the Pacific Flyway Council developed a Management Plan in 1992 (and further updates in 2006) for Wrangel Island breeding lesser snow geese (Pacific Flyway Council 2006). The Council recommended the following efforts: (1) analyze existing banding data to evaluate occurrence and harvest of Wrangel Island snow geese in migration and wintering areas where they mix with other populations, and (2) use existing demographic data collected on Wrangel

TABLE 12.2
Short-Term Information Needs for Snow Goose Populations Included in the Arctic Goose Joint Venture

Genus, Species, Population	Population Definition or Delineation	Population Status or Assessment	Population Dynamics	Population Biology and/or Ecology	Harvest Assessment	Habitat Concerns	Parasites, Disease, and/or Contaminants
				Information Need			
Snow Goose (*Chen caerulescens*)							
Greater	Low	High	High	Low	Med	Med	Low
Midcontinent Lesser	Low	Med	Med	Low	High	High	Low
Western Central	Low	Med	High	Low	Med	High	Low
Flyway							
Western Arctic Lesser	High	Low	High	Low	Low	Med	Med
Wrangel Island Lesser	Med	High	High	Low	Med	Low	Low

Note: A high priority indicates an immediate need for information; a medium priority indicates a demonstrated need for the information, but other information is required first; and a low priority suggests that the information is relevant but other information is presently deemed more important.

Island and on the Fraser-Skagit deltas to develop a reliable population model to forecast trends and identify significant factors influencing population dynamics of that wintering segment. Recognizing these management needs, the AGJV put forth multiple research charges described generally in the 1991 prospectus (Arctic Goose Joint Venture Technical Committee 1991) and more specifically in the 2008–2012 strategic plan (Arctic Goose Joint Venture Technical Committee 2008) to initiate, expand, or update research on factors influencing metapopulation mixing (such as is occurring on Banks and Wrangel Islands), including population size, distribution, migrations corridors, and production.

As part of a larger study to estimate the demography and management of the Wrangel Island-Banks Island metapopulation of lesser snow geese, funded in part by the Arctic Goose Joint Venture and the Pacific Flyway, Williams et al. (2008) neck-banded molting birds at their breeding colonies and resighted birds on the wintering grounds between 1993 and 1996. Using multistate mark–recapture models to evaluate apparent survival rates, resighting rates, winter fidelity, and potential exchange among these populations, they found similar apparent survival rates between subpopulations of Wrangel Island snow geese and lower apparent survival, but higher emigration, for the Banks Island birds. Importantly, transition between wintering areas was low (<3%), with equal movement between northern and southern wintering areas for Wrangel Island birds and little evidence of exchange between the Banks and northern Wrangel Island populations. Their findings suggested that northern and southern Wrangel Island subpopulations should be considered collectively as a metapopulation to assist in understanding and managing Pacific Flyway lesser snow geese. Furthermore, the absence of a strong population connection between Banks Island and Wrangel Island geese suggests that these breeding colonies can be managed as separate but overlapping populations. Finally, the authors concluded that winter population fidelity may be more important in lesser snow geese than in other species, and both breeding and wintering areas are important components of population management for sympatric wintering populations.

An understanding of arctic goose population structures (including breeding, wintering, and migration areas) and admixing among populations is critical for their conservation and management. The characteristics of population structuring help to accurately define appropriate population units, identify constraints on population dynamics, establish appropriate harvest regulations, and determine interactions among goose metapopulations (Esler 2000). Through the Pacific Flyway identifying management needs and the AGJV making those needs a high priority research need, they both facilitated the initiation of necessary research and provided management recommendations for managers.

SUMMARY

During the twentieth century, the management of waterfowl made great strides forward due to a close link with research. Together, the powerful combination of federal legislation of migratory birds along with the formulation of a new discipline of wildlife science ultimately allowed an informed, adaptive, and careful management of waterfowl. The wide and varied distribution of waterfowl and the need to convince the public and partners across the length and breadth of North America that management was productive, led to an early understanding that management of waterfowl populations would only be successful if actively connected to research. Waterfowl management provided an example for many other species and areas of applied ecology. To this day, personnel with federal, state, private, and university organizations continue to dedicate their time and efforts to implementation of the ideals of connectivity of research and management. NAWMP and the development of the joint ventures and their adaptive approach to improving management have asked difficult questions of researchers. Data gathering and analytical techniques for harvest management are generally well developed; however, the fundamental question of setting goals is a socio-political one without a simple answer. Researchers are now being asked to turn their sights on the human dimension aspects of waterfowl management. In addition, researchers will also continue to be challenged to inform the integration of harvest and habitat management into a coherent waterfowl management strategy (Runge et al. 2006).

ACKNOWLEDGMENTS

We thank Robert Blohm, Lenny Brennan, Steve DeMaso, Patrick Devers, James Ringleman, Joseph Sands, and Robert Trost for providing valuable review and comments that improved this chapter.

REFERENCES

Anderson, S. H. 2002. *Managing our Wildlife Resources*. Upper Saddle River, NJ: Prentice Hall.

Arctic Goose Joint Venture Technical Committee. 1991. Arctic Goose Joint Venture: a prospectus. Unpublished report (c/o AGJV Coordination Office, CWS, Edmonton, Alberta, Canada).

Arctic Goose Joint Venture Technical Committee. 2008. Arctic Goose Joint Venture Strategic Plan: 2008–2012. Unpublished report (c/o AGJV Coordination Office, CWS, Edmonton, Alberta, Canada).

Baldassarre, G. A., and E. G. Bolen. 2006. *Waterfowl Ecology and Management*. Malabar, FL: Krieger Publishing Company.

Black Duck Joint Venture Management Board. 2008. Black Duck Joint Venture Strategic Plan 2008–2013. U.S. Fish and Wildlife Service, Laurel, MD; Canadian Wildlife Service, Ottawa, Ontario, Canada.

Blohm, R. J. 1989. Introduction to harvest: understanding surveys and season setting. *International Waterfowl Symposium Proceedings* 6:118–133.

Conroy, M. J. 1993. The use of models in natural resource management: prediction, not prescription. *Transactions of the North American Wildlife and Natural Resources Conference* 58:509–519.

Conroy, M. J., G. R. Costanzo, and D. B. Stotts. 1989. Winter survival of female American black ducks on the Atlantic coast. *Journal of Wildlife Management* 53:99–109.

Conroy, M. J., M. W. Miller, and J. E. Hines. 2002. Identification and synthetic modeling of factors affecting American black duck populations. *Wildlife Monographs* No. 150.

Cooke, W. W. 1906. Distribution and migration of North American ducks, geese and swans. U.S. Department of Agriculture, Biological Survey, Bulletin 26, Washington, DC.

Cramer, C. 2009. Estimating habitat carrying capacity for American black ducks wintering in southern New Jersey. M.S. Thesis, University of Delaware, Newark.

Delta Waterfowl. 2011. Delta's research program. http://www.deltawaterfowl.org/research/index.php Accessed August 15, 2011.

Esler, D. 2000. Applying metapopulation theory to conservation of migratory birds. *Conservation Biology* 14:366–372.

Gavin, A. 1964. Ducks unlimited. In: *Waterfowl Tomorrow*, J. P. Linduska and A. L. Nelson, Eds. Washington, DC: United States Department of the Interior, pp. 545–553.

Hawkins, A. S., R. C. Hanson, H. K. Nelson, and H. M. Reeves, Eds. 1984. *Flyways: Pioneering Waterfowl Management in North America*. Washington, DC: U.S. Government Printing Office.

Johnson, F. A., C. T. Moore, W. L. Kendall, J. A. Dubovsky, D. F. Caithamer, J. R. Kelley, Jr., and B. K. Williams. 1997. Uncertainty and the management of mallard harvests. *Journal of Wildlife Management* 61:202–216.

Johnson, F. A., B. K. Williams, J. D. Nichols, J. E. Hines, W. L. Kendall, G. W. Smith, and D. F. Caithamer. 1993. Developing an adaptive management strategy for harvesting waterfowl in North America. *Transactions of the North American Wildlife and Natural Resources Conference* 58:565–583.

Kallman, H., Ed. 1987. *Restoring America's Wildlife, 1937–1987: The First 50 Years of the Federal Aid in Wildlife Restoration (Pittman-Robertson) Act*. Washington, DC: United States Department of Interior, Fish and Wildlife Service.

Lancia, R. A., C. E. Braun, M. W. Collopy, R. D. Dueser, J. G. Kie, C. J. Martinka, J. D. Nichols, T. D. Nudds, W. R. Porath, and N. G. Tilghman. 1996. ARM! For the future: adaptive resource management in the wildlife profession. *Wildlife Society Bulletin* 24:436–442.

Leitch, W. G. 1984. Response of the private sector. In: *Flyways: Pioneering Waterfowl Management in North America*, A. S. Hawkins, R. C. Hanson, H. K. Nelson, and H. M. Reeves, Eds. Washington, DC: U.S. Government Printing Office.

Lincoln, F. C. 1935. The Waterfowl Flyways of North America. U.S. Department of Agriculture, Circular 342. Washington, DC.

Longcore, J. R., D. G. McAuley, G. R. Hepp, and J. M. Rhymer. 2000a. American Black Duck (*Anas rubripes*). In: *The Birds of North America*, No. 481, A. Poole and F. Gill, Eds. Philadelphia, PA: The Birds of North America, Inc..

Longcore, J. R., D. G. McAuley, D. A. Clugston, C. M. Bunck, J. Giroux, C. Ouellet, G. R. Parker, P. Dupuis, D. B. Stotts, and J. R. Goldsberry. 2000b. Survival of American black ducks radiomarked in Quebec, Nova Scotia, and Vermont. *Journal of Wildlife Management* 64:238–252.

Martin, F. W., R. S. Pospahala, and J. D. Nichols. 1979. Assessment and population management of North American migratory birds. In: *Environmental Biomonitoring, Assessment, Prediction, and Management—Certain Case Studies and Related Quantitative Issues*, J. Cairns, Jr., G. P. Patil, and W. E. Walters, Eds. Fairland, MD: International Cooperative Publishing House, pp. 187–239.

Matthews, O. P. 1986. Who owns wildlife? *Wildlife Society Bulletin* 14:459–465.

MacNab, J. 1983. Wildlife management as scientific experimentation. *Wildlife Society Bulletin* 11:397–401.

McAuley, D. G., D. A. Clugston, and J. R. Longcore. 2004. Dynamic use of wetlands by black ducks and mallards: evidence against competitive exclusion. *Wildlife Society Bulletin* 32:465–473.

Meine, C. 1988. *Aldo Leopold: His Life and Work*. Madison: University of Wisconsin Press.

Merendino, T. D., C. D. Ankney, and D. G. Darrell. 1993. Increasing mallard, decreasing American black ducks: more evidence for cause and effect. *Journal of Wildlife Management* 57:199–208.

Murkin, H. R., A. G. van der Valk, and W. R. Clark, Eds. 2000. *Prairie Wetland Ecology: The Contribution of the Marsh Ecology Research Program.* Ames: Iowa State University Press.

Nichols, J. D., F. A. Johnson, and B. K. Williams. 1995. Managing North American waterfowl in the face of uncertainty. *Annual Review of Ecology and Systematics* 26:177–199.

North American Waterfowl Management Plan, Plan Committee. 2004a. North American Waterfowl Management Plan 2004. Strategic Guidance: Strengthening the Biological Foundation. Canadian Wildlife Service, U.S. Fish and Wildlife Service, Secretaria de Medio Ambiente y Recursos Naturales, Washington, DC.

North American Waterfowl Management Plan, Plan Committee. 2004b. North American Waterfowl Management Plan 2004. Implementation Framework: Strengthening the Biological Foundation. Canadian Wildlife Service, U.S. Fish and Wildlife Service, Secretaria de Medio Ambiente y Recursos Naturales, Washington, DC.

Nudds, T. D., M. W. Miller, and C. D. Ankney. 1996. Black ducks: harvest, mallards, or habitat? *Proceedings of the International Waterfowl Symposium* 7:50–60.

Pacific Flyway Council. 2006. Pacific Flyway management plan for the Wrangel Island population of lesser snow geese. White Goose Subcommittee, Pacific Flyway Study Committee. United States Fish and Wildlife Service, Portland, OR.

Patuxent National Research Center. 1989. Patuxent Wildlife Research Center 50th anniversary: wildlife conservation through scientific research 1939–1989. Laurel, MD.

Plattner, D., W. Eichholz, and T. Yerkes. 2010. Food resources for wintering spring staging black ducks. *Journal of Wildlife Management* 74:1554–1558.

Reiger, J. F. 2000. *American Sportsmen and the Origins of Conservation.* Corvalis: Oregon State University Press.

Romesburg, H. C. 1981. Wildlife science: gaining reliable knowledge. *Journal of Wildlife Management* 45:293–313.

Runge, M. C., F. A. Johnson, M. G. Anderson, M. D. Koneff, E. T. Reed, and S. E. Mott. 2006. The need for coherence between waterfowl harvest and habitat management. *Wildlife Society Bulletin* 24:1231–1237.

Rusch, D. H., C. D. Ankney, H. Boyd, J. R. Longcore, F. Montalbano, III, J. K. Ringleman, and V. D. Stotts. 1989. Population ecology and harvest of the American black duck: a review. *Wildlife Society Bulletin* 17:379–406.

Sea Duck Joint Venture Management Board. 2008. Sea Duck Joint Venture Strategic Plan 2008–2012. USFWS, Anchorage, Alaska; CWS, Sackville, New Brunswick, Canada.

Smith, R. I., R. J. Blohm, S. T. Kelly, and R. E. Reynolds. 1989. Review of databases for managing duck harvests. *Transactions of the North American Wildlife and Natural Resources Conference* 54:537–544.

Trefethen, J. B. 1975. *An American Crusade for Wildlife.* New York: Winchester Press.

Trost, R. E., and S. M. Carney. 1989. Measuring the waterfowl harvest. *International Waterfowl Symposium Proceedings* 6:134–147.

United States Department of the Interior and Environment Canada. 1986. North American Waterfowl Management Plan, Washington, DC.

United States Fish & Wildlife Service. 2009. Adaptive Harvest Management: 2009 Hunting Season. U.S. Department of Interior, Washington, DC. http://www.fws.gov/migratorybirds/CurrentBirdIssues/Management/AHM/AHM-intro.htm Accessed October 5, 2011.

Williams, B. K. 1997. Approaches to the management of waterfowl under uncertainty. *Wildlife Society Bulletin* 25:714–720.

Williams, B. K., and F. A. Johnson. 1995. Adaptive management and the regulation of waterfowl harvests. *Wildlife Society Bulletin* 23:430–436.

Williams, B. K., and J. D. Nichols. 1990. Modeling and the management of migratory birds. *Natural Resource Modeling* 4:273–311.

Williams, C. K., M. D. Samuel, V. Baranyuk, E. Cooch, and D. Kraege. 2008. Winter fidelity and apparent survival of lesser snow goose populations in the Pacific Flyway. *Journal of Wildlife Management* 72:159–167.

Witter, D., D. J. Case, J. H. Gammonley, and D. Childress. 2006. Social factors in waterfowl management: Conservation goals, public perceptions, and hunter satisfaction. *Transactions of the 71st North American Wildlife and Natural Resources Conference.*

Zimpfer, N. L., and M. J. Conroy. 2006. Models of production rates in American black duck populations. *Journal of Wildlife Management* 70:947–954.

13 Deer in the Western United States

James R. Heffelfinger
Regional Game Specialist
Arizona Game and Fish Department
and
Adjunct Professor
University of Arizona
Tucson, Arizona

CONTENTS

Everybody is ignorant, only on different subjects.

—**Will Rogers**

ABSTRACT

The wildlife profession began with the arms of research and management in full embrace. Through time, the connection between the two has weakened and in some cases started to diverge. Managers may not have the technical training or convenient access to the scientific literature. They often work in remote locations without large university libraries or even access to online journals. Managers who are familiar with pertinent, recent research with direct management applicability must still consider logistical, political, fiscal, and sociological issues that constrain their ability to implement management changes. Researchers may not conduct studies that answer the most pressing needs of managers because funding sources or personal interests play a larger role in developing studies. When management-applicable research is conducted, there still may be barriers to its use and implementation. These barriers fall into four main categories: temporal, technical, spatial, and logistic. Although the disconnect between research and management specialists is common, means exist to enhance the relationships and improve the connection. An example of this is taken from the recent work of the Mule Deer Working Group (MDWG) assembled and sponsored by the Western Association of Fish and Wildlife Agencies. The MDWG has been very successful in its efforts to compile past and present research and decades of agency management experience to inform decision-makers and thus improve management of mule deer. Disconnects between the research and management disciplines can be repaired, but a conscious effort, starting with an increase in communication, is required. Large-scale efforts to increase collaboration between research and management disciplines can be successful for many conservation challenges, just as they were for mule deer in the West. Improved recognition and understanding of the different worlds in which researchers and managers live is an initial step.

INTRODUCTION

Since its early beginnings, the wildlife management profession has experienced incredible advances in the ability to identify research hypotheses, analyze data, improve technology, and answer questions. Management activities are increasingly implemented in effective manners that incorporate evaluations that improve future actions. Unfortunately, the evolution of both research and management has occurred along increasingly dichotomous paths. Research has headed off in an increasingly esoteric direction in the minds of many managers, and researchers express frustration when managers do not immediately adopt their findings. This growing disconnect is most obvious in species with great economic, aesthetic, recreational, and cultural importance.

Western deer, especially the iconic black-tailed and mule deer (*Odocoileus hemionus*), are an integral part of western culture (Figure 13.1). In a 2006 survey of outdoor activities, the U.S. Fish and Wildlife Service (USFWS) reported that nearly 3 million people hunted in the nineteen western states (USFWS 2007). Although this included hunters that pursued other species, deer have traditionally been the most important big game animals in the West. In 2006 alone, hunters were afield for almost 50 million days and spent more than $7 billion in local communities across the West on lodging, food, fuel, and hunting-related equipment.

Hunters have contributed billions of dollars through license fees and excise taxes that finance wildlife management and benefit countless wildlife species (Heffelfinger, Geist, and Wishart 2012). These funds support wildlife management agencies, which manage all wildlife species, not just those that are hunted. Deer are the most avidly sought game animals, and thus are responsible for supporting a wide variety of conservation activities valued by the public, including law enforcement, habitat management and acquisition, and wildlife population management.

FIGURE 13.1 Research information has always been important for proper management of mule deer winter range such as this in Middle Park, Colorado. Photo courtesy of David J. Freddy, Colorado Parks and Wildlife.

The social and economic interest goes far beyond hunters and wildlife management agencies. Deer are valued as an integral part of the western landscape by hunters and non-hunters alike. According to the 2006 USFWS survey, 25.6 million residents in nineteen western states spent more than $15.5 billion that year "watching wildlife" (USFWS 2007). Thus, the proper management of deer in the West is critical to western culture in general and wildlife management agencies specifically (Heffelfinger and Messmer 2003).

To manage such a critically important group of animals, both management experience and relevant research are needed to inform decision-makers. Effective wildlife research can be applied to management under the current fiscal, logistic, and social constraints that are the modern manager's reality. Western deer serve as a good example of the growing disconnect between research and management in wildlife science today. Fortunately, they also serve as an example of how this gap can be reduced to the mutual benefit of researchers, managers, and most importantly, the wildlife resources we all care so much about.

FACTORS CONTRIBUTING TO A DICHOTOMOUS RELATIONSHIP

RESEARCH PERSPECTIVE

Many researchers believe their role is solely to provide facts and publish those facts in peer-reviewed literature. Some wildlife research being conducted is esoteric with little or no management applicability (see, for example, Heffelfinger 2000). Research topics might stem from the researcher's personal interest or be directed at topics that are popular with funding sources. A decade ago, a grant request for deer research merely had to intimate a connection (even a weak one) to Chronic Wasting Disease in deer to be nearly guaranteed funding. Today, "Global Climate Change" is the nexus that is getting a lot of research funded. Federal funding for climate change increased from $2.4 billion in 1993 to $5.1 billion in 2004 (116%; GAO 2005).

Even when research is done with management in mind, there are still barriers to implementation. When management-applicable research is conducted, researchers may not have funding or feel an obligation to provide any follow-up to managers who would benefit from (or may have even commissioned) the research. There is sometimes an assumption that everyone will immediately read the research. Further, there is an expectation that managers will immediately redesign management programs to implement all the suggestions published in the latest manuscript. Researchers may have a firm understanding of the biological implications of the work, but may not fully appreciate the social or economic implications that may make some recommendations inappropriate for managers.

MANAGEMENT PERSPECTIVE

Many managers do not have advanced graduate degrees because they were interested in pursuits other than research. Without the scientific training that comes with graduate-level education or convenient access to the scientific literature, the manager is unlikely to be familiar with the latest research. Managers often work in remote locations without large university libraries or even access to online journals. The modern wildlife manager is inundated with daily management tasks and a flood of information reaching his or her "inbox" on a daily basis. They may have insufficient time to read journals compared with academics or researchers who are required to stay current with the latest research.

Managers who are familiar with pertinent, recent research with direct management applicability must still consider logistical, political, fiscal, and sociological issues that constrain their ability to implement management changes. Results in terms of treatments or population monitoring that are accomplished with a full-time student and their technicians on a discrete study area may be difficult for a manager to implement on a larger scale, such as statewide.

Another important cause of the disconnect between research and management is simply the loss of institutional memory as long-term employees retire or move to different jobs. Young wildlife managers often start positions with enthusiasm and expertise in general science, but familiarity with the last five decades of research is difficult to develop. Thus, the chasm of current management actions and past research widens. Conversely, long-term employees can be neophobic and resist management change. It is certainly easier to keep doing something "the way we've always done it."

DIVISION OF DISCONNECT

Disconnect between research and management specialists is largely because researchers think managers lack adequate scientific rigor, and managers perceive that researchers are not producing information that can be applied to the real world. The issues elucidated previously can be used to categorize this disconnect into four divisions: (1) temporal, (2) technical, (3) spatial, and (4) logistic.

TEMPORAL

As years pass, management and research personnel retire or move out of positions that are then filled by younger biologists. Thus, research published decades previously may fall from the institutional memory of an agency. In addition, management personnel may not write down many things they have learned from years in the field or, if they do, it may be published in gray literature such as proceedings and transactions.

TECHNICAL

Research is not always published in a format that will be seen and then used by managers. Research results published in obscure journals, which are not widely available, can create barriers to managers wishing to access this information. Even with full access, managers may not have the technical training to understand some research. In addition, technical jargon of some fields may introduce additional barriers to acceptance.

SPATIAL

Research conducted on a small study area with several full-time workers is difficult or impossible to replicate statewide by managers who have other simultaneous responsibilities. For example, research on the effects of predation in a small (~1 square mile) enclosure may be impossible to apply at the same intensity to larger areas (Aikens 2004).

LOGISTIC

Managers may not have the financial or logistical ability to implement improved management activities highlighted by research (Keegan et al. 2011). In addition, funding sources must be developed for management-applicable research. Major sources of funding (e.g., National Science Foundation) frequently request proposals for more esoteric research, making it hard to find funds for projects that will answer managers' questions and contribute to improving on-the-ground wildlife conservation.

RECONNECTING: A WESTERN DEER EXAMPLE OF COOPERATIVE CONNECTIONS

Although the disconnect between research and management specialists is common, means exist to enhance the relationships and improve the connection. Here I describe one such example focusing on mule deer population management.

FIGURE 13.2   The Mule Deer Working Group. Photo courtesy of Jim Heffelfinger Collection.

The size of black-tailed and mule deer populations throughout North America fluctuates in response to various factors in different ecoregions (Heffelfinger et al. 2003). Distinct populations sometimes increase and decline in near synchrony (Workman and Low 1976). During the 1990s, the size of most mule deer populations was declining and a general concern existed that the nearly ubiquitous nature of the declines might be due to some new and irreversible factor.

To address this range-wide deer decline, the Western Association of Fish and Wildlife Agencies (WAFWA) chartered the MDWG in 1997 (Figure 13.2). The MDWG, composed of representatives of all twenty-three WAFWA member agencies, was established to accomplish three specific tasks:

1. Develop solutions to common mule deer management challenges.
2. Identify and prioritize cooperative research and management activities in the western states and provinces.
3. Increase communication between agencies and the public who are interested in mule deer, and among those in agencies, universities, and non-governmental organizations who are interested in mule deer management.

At the first MDWG meeting, state wildlife agency representatives and researchers with relevant expertise (e.g., predator/prey relationships, nutrition, and habitat management) identified issues considered important to mule deer management. These topics included short- and long-term changes to habitat, differences in mule deer ecology among ecoregions, changes to nutritional resources, effects of different hunting strategies, competition with elk (*Cervus elaphus*), inconsistent collection and analyses of data, deer-predator relationships, weather patterns, disease impacts, and interactions among other factors.

Of topics identified, predation was not deemed the most important for influencing deer abundance, but it was the topic upon which state wildlife agencies were spending a disproportionate amount of time. A vocal public minority perceived predation to be an important contributor to low mule deer abundance. Because of this sensitivity and pressure, the MDWG believed it was most important to summarize the state of our knowledge about effects of predation on black-tailed and mule deer populations. This summary, titled "Deer-Predator Relationships: A Review of Recent North American Studies with Emphasis on Mule and Black-tailed Deer" was published as a technical paper and has been cited frequently since publication (Ballard et al. 2001). This publication was timely and appeared when agencies throughout the West were struggling with public pressure on both sides of this issue. This example demonstrated how research and management needs could be incorporated to affect a positive result to field managers struggling with the sometimes volatile mix of sociobiology.

With the topic of predation addressed, the MDWG turned to the complete list of issues originally identified. Each of these issues formed the basis of a series of chapters that culminated in a book (*Mule Deer Conservation: Issues and Management Strategies*) addressing the major issues affecting mule deer (deVos, Conover, and Headrick 2003). This book summarized the status of research and management experience into a single volume to inform discussions and management decisions regarding all the species. This book referenced the original literature, providing a source for the most current available science, and offered a technical foundation upon which to build future MDWG products.

Because of public passion regarding mule deer, the readability of this information for non-biologists was important. Therefore, the MDWG commissioned a writer who worked in the natural resources field, but was not a biologist, to render the content of the book into an easily read brief synopsis of the main points without technical jargon. This was accomplished with great success and more than 35,000 copies of that magazine-format publication were distributed to the public at every agency office and related venue across North America (WAFWA 2003).

In order to develop an over-arching strategic document to guide conservation of black-tailed and mule deer, the MDWG wrote and published the North American Mule Deer Conservation Plan (NAMDCP). The NAMDCP provided goals, objectives, and strategies for implementing coordinated activities to benefit mule deer throughout their range (WAFWA 2004). The overall goal of the NAMDCP was to provide "ecologically sustainable levels of black-tailed and mule deer throughout their range through habitat protection and management, improved communication, increased knowledge, and ecoregional-based decision making." To garner support from all state, provincial, and federal agencies with jurisdiction over mule deer populations or their habitat, the MDWG obtained an accompanying Memorandum of Understanding (MOU) signed by the Directors of the Bureau of Land Management, U.S. Forest Service, U.S. Fish and Wildlife Service, U.S. Geological Survey, Natural Resources Conservation Service, and the president of WAFWA representing all twenty-three member agencies. The MOU assured commitment by all agencies that they would work to accomplish the goals and objectives outlined in the NAMDCP.

This conservation plan provided the strategic umbrella document, below which specific operational guidance could be nested. The most important documents tiering off of this conservation plan were a complete set of habitat management guidelines for each of the seven North American ecoregions identified by the MDWG: Southwest Deserts, California Woodland and Chaparral, Colorado Plateau, Coastal Rainforest, Northern Forest, Intermountain West, and the Great Plains. These guidelines assembled research on all aspects of habitat management and provided comprehensive and specific recommendations to private, tribal, state, provincial, and federal land managers for maintaining and improving mule deer habitat. These strategies are already being used to improve black-tailed and mule deer habitat on a landscape scale by allowing federal, state, local, private, and tribal land managers to incorporate mule deer habitat requirements into land management plans.

Habitat guidelines and other products of the MDWG are already providing the information for improving mule deer habitat on a landscape scale throughout many areas of the West. Benefits to current populations include:

- Western Landscape Conservation Initiative in Southwest Wyoming is using information the MDWG produced.
- Statewide Mule Deer Initiatives in Idaho, Wyoming, and New Mexico are all based heavily on documents and information the MDWG has produced.
- Murphy complex sagebrush restoration in Nevada/Idaho is using the combination of research and management information provided in the habitat guidelines and the strategies in the North American Mule Deer Conservation Plan.
- All states are working when possible to interject the mule deer habitat guidelines and NAMDCP strategies into revisions of upcoming resource, forest, and allotment management plans of federal land management agencies.
- Agencies involved in a cooperative effort in Idaho, Utah, and Nevada to manage the wildfire/cheat grass cycle are also using the information assembled and disseminated by the MDWG.

In accordance with the original desire for the MDWG to increase communication among member agencies, the group took part in a workshop in Reno, Nevada in 2005 to discuss differences in deer population monitoring techniques. Agencies across the West often collect mule deer population data differently. In some jurisdictions, different monitoring approaches are needed, but in others, the lack of consistency may be due to an unfamiliarity with information wildlife research has provided. The result of the workshop was a published paper that summarized how agencies were collecting and storing these data with recommendations for improvement (Mason et al. 2006).

Following up on one of the recommendations from that workshop, the MDWG developed a document summarizing research available on mule deer population monitoring techniques (Keegan et al. 2011). Clearly, standardization is difficult across all jurisdictions because of the biological, logistical, and fiscal constraints each state/province faces with regard to selecting a specific monitoring approach. Nevertheless, the MDWG has assembled a document that lists and discusses what should be monitored and the best methods to collect those data in the least biased manner. This is yet another example of the melding of available research and management experience into a cooperative document to help wildlife managers implement the best available science.

To prioritize research and management efforts and to aid in managing mule deer and their habitat on a landscape scale, the MDWG completed an interactive Geographic Information Systems (GIS) map of North America to identify mule deer distribution and important mule deer habitat features. For ease of use, GIS files were converted to files used by Google Earth™ software so those in remote duty stations with Internet access could use this information as a reference or in habitat project planning.

The MDWG leads coordination of the biennial Western States Deer/Elk Workshop and standardization of long-term deer population trend data to further cooperation and communication between researchers and managers. This workshop is an opportunity for those working with black-tailed and mule deer to share information and data. Full workshop proceedings have been published since the second meeting in 1972 and represent a valuable collection of mule deer information. Papers presented at these workshops often provide information unavailable outside these proceedings. The MDWG collected and scanned all past Deer/Elk Workshop Proceedings (1972–present) into a searchable PDF format. This complete series, as well as all publications mentioned previously, are available to anyone on the MDWG website (www.muledeerworkinggroup.com). This free exchange of information serves to bring research and management entities closer together for their mutual benefit.

The MDWG developed strategies to improve mule deer management throughout western North America, and has effectively increased communication among mule deer managers, researchers, administrators, and the public. Increased communication among agency biologists allows managers to face new resource challenges with the best available science and techniques. This ecoregional and range-wide approach to mule deer conservation allows natural resource administrators to make science-based decisions and provide up-to-date and accurate information to their stakeholders.

CONCLUSIONS

The MDWG has been very successful in their efforts to compile past and present research and decades of agency management experience to inform decision-makers and thus improve management of mule deer. This effort, and the MDWG in general, has been touted by western directors as the most productive working group ever sponsored by WAFWA.

For its efforts and accomplishments, the MDWG was awarded the Wildlife Management Institute's "Touchstone Award" in 2006. The Touchstone Award recognizes the achievement of a natural resource management program, professional, or group of professionals in the public or private sector. The future of deer and our western wildlife populations depends on a cohesive network of wildlife professionals working together with all conservation agencies and organizations toward a common goal. The MDWG would serve as a solid model for enhancing collaborations among researchers and managers.

Disconnects between the research and management disciplines can be repaired, but a conscious effort, starting with an increase in communication, is required.

Before a study is designed, researchers will optimize the relevance by asking, "How can I make this useful to those conserving wildlife on the ground?" Researchers who do not think of how their research can be used until they begin to write the "Management Implications" section of their paper will be less useful to the manager. Graduate programs producing our future researchers will have to do a better job of instilling an understanding of the logistical, social, and fiscal realities managers face.

Managers who understand how to incorporate research results into their management programs will be more successful. Undergraduate programs can help future wildlife managers incorporate research into their programs by emphasizing: (1) the scientific method, (2) collecting unbiased data, (3) interpreting research results, and (4) filtering important and applicable information out of the overwhelming overload of research results produced in hundreds of journals. Successful managers will ask, "How can I add scientific rigor to what I do?" Opportunities abound to conduct cooperative long-term "management experiments" using data collected as part of normal management activities and then applying adaptive management principles to improve effectiveness.

Large-scale efforts to increase collaboration between research and management disciplines can be successful for many conservation challenges, just as they were for mule deer in the West. Improved recognition and understanding of the different worlds in which researchers and managers live is an initial step. For this to happen all wildlife professionals, regardless of vocation, must commit to actively investigate opportunities to pool their expertise to achieve common goals.

REFERENCES

Aikens, R. K. 2004. Deer, predators, and drought. *Arizona Wildlife Views*, July/August issue.

Ballard, W. B., D. Lutz, T. W. Keegan, L. H. Carpenter, and J. C. deVos, Jr. 2001. Deer-predator relationships: a review of recent North American studies with emphasis on mule and black-tailed deer. *Wildlife Society Bulletin* 29:99–115.

deVos, J. C., Jr., M. R. Conover, and N. E. Headrick, Eds. 2003. *Mule Deer Conservation: Issues and Management Strategies*. Logan, UT: Jack H. Berryman Institute Press, Utah State University.

GAO. 2005. United States Government Accountability Office, Report to Congressional Requesters, Climate Change: Federal reports on climate change funding should be clearer and more complete, GAO-05-461. http://www.gao.gov/cgi-bin/getrpt?GAO-05-461 Accessed on 1 September 2011.

Heffelfinger, J. R. 2000. Status of the name Odocoileus hemionus crooki (Mammalia: Cervidae). *Proceedings of the Biological Society of Washington* 113:319–333.

Heffelfinger, J. R., L. H. Carpenter, L. C. Bender, G. L. Erickson, M. D. Kirchhoff, E. R. Loft, and W. M. Glasgow. 2003. Ecoregional differences in population dynamics. In: *Mule Deer Conservation: Issues and Management Strategies*, J. C. deVos, Jr., M. R. Conover, and N. E. Hedrick, Eds. Logan, UT: Berryman Institute Press, Utah State University, chapt. 2.

Heffelfinger, J. R., V. Geist, and W. Wishart. 2012. The role of hunting in North American wildlife conservation. *International Journal of Environmental Science* in press.

Heffelfinger, J. R., and T. A. Messmer. 2003. Introduction. In: *Mule Deer Conservation: Issues and Management Strategies*, J. C. deVos, Jr., M. R. Conover, and N. E. Hedrick, Eds. Logan, UT: Berryman Institute Press, Utah State University, pp. 1–12.

Keegan T. W., B. B. Ackerman, A. N. Aoude, L. C. Bender, T. Boudreau, L. H. Carpenter, B. B. Compton, M. Elmer, J. R. Heffelfinger, D. W. Lutz, B. D. Trindle, B. F. Wakeling, and B. E. Watkins. 2011. Methods for monitoring mule deer populations. Mule Deer Working Group, Western Association of Fish and Wildlife Agencies.

Mason, J. R., L. H. Carpenter, M. Cox, J. C. deVos, J. Fairchild, D. J. Freddy, J. R. Heffelfinger, R. H. Kahn, S. M. McCorquodale, D. F. Pac, D. Summers, G. C. White, and B. K. Williams. 2006. The case for standardized ungulate surveys and data management in the western United States. *Wildlife Society Bulletin* 34:1238–1242.

U.S. Fish and Wildlife Service (USFWS). 2007. 2006 National Survey of Fishing, Hunting, and Wildlife-Associated Recreation. http://library.fws.gov/Pubs/nat_survey2006_final.pdf (Accessed April 8, 2011).

WAFWA. 2003. Mule Deer: Changing landscapes, changing perspectives. Mule Deer Working Group, Western Association of Fish and Wildlife Agencies.

WAFWA. 2004. North American Mule Deer Conservation Plan. Mule Deer Working Group, Western Association of Fish and Wildlife Agencies.

Workman, G. W., and J. B. Low, Eds. 1976. *Mule Deer Decline in the West: A Symposium.* Logan: Utah State University.

14 Collaboration among Scientists, Managers, Landowners, and Hunters
The Kinzua Quality Deer Cooperative

David deCalesta
Wildlife Analysis Consulting
Hammondsport, New York

CONTENTS

> In the field of game, however, it seems doubtful whether theories and plans alone, no matter how well supported by evidence, are nearly so useful as samples or demonstrations of how those theories and plans work out in practice.
>
> —**Aldo Leopold (1933:386)**

ABSTRACT

After being nearly extirpated in Pennsylvania by the late 1800s, white-tailed deer (*Odocoileus virginianus*) populations expanded dramatically in the early 1900s as a result of elimination of natural predators, severely restricted hunting opportunities, deer reintroductions from other states, and vast increases in forage created by state-wide clearcutting. Hunters became accustomed to seeing deer densities in excess of twenty deer per square kilometer and came to expect that as the norm. By the late 1930s, it was apparent that high deer densities were devastating understory vegetation and habitat and preventing forest regeneration. Decades of attempts by the Pennsylvania Game Commission (PGC) to reduce deer densities to levels associated with ecosystem sustainability (~eight deer per square kilometer) achieved only limited success. Hunters successfully lobbied the Pennsylvania Game Commissioners and legislators to water down science-based recommendations of deer managers for lowering deer density via increasing harvest of antlerless deer. The straight-line relationship between scientists and managers was circumvented by hunter-influenced decisions made by administrators, commissioners, and legislators. This situation changed dramatically in the early 2000s when the PGC revitalized its deer management program. A dynamic and effective deer program leader, Dr. Gary Alt, developed and delivered an effective, science-based education and research program to hunters, commissioners, legislators, and other groups that emphasized restoration of ecosystem and deer health by reducing deer to densities established as sustainable by science. Alt's 3-point deer management program, endorsed and enacted by the PGC, featured restricting harvest of young antlered deer, providing additional antlerless licenses to hunters through a Deer Management Assistance Program (DMAP), and allowing hunters to harvest antlerless as well as antlered deer during the regular twelve-day rifle season. Concurrently, a consortium of public and private landowners, hunters, biologists, foresters, and natural resource funding groups initiated a ten-year deer demonstration program on a 30,000-ha northern hardwood forest in northwestern Pennsylvania. Guided by the Sand County Foundation, a not-for-profit institution dedicated to working with landholders to improve natural habitats, the Kinzua Quality Deer Cooperative (KQDC) employed the new PGC deer management regulations in an adaptive management mode to effect a halving of deer density, significant reductions in deer impact, and significant increases in deer weights and antler characteristics within three years. Reduced deer density and impact were maintained for the following four years; improvements in deer weights and antler characteristics continued throughout. The pairing of a renewed scientist-manager relationship with a facilitated partnership of hunters, landowners, biologists, and natural resource entities provides a remarkable example of how science-based wildlife management can be successful.

INTRODUCTION

This book explores the relationship between wildlife scientists and wildlife managers as it affects wildlife management. This chapter provides an example of how politics can interfere with the relationship between science and management with negative repercussions for management of white-tailed deer. In counterpoint, the chapter also describes a demonstration project, the Kinzua Quality Deer Cooperative (KQDC), wherein scientists, managers, landowners, and hunters teamed up to produce a model for adaptive management of white-tailed deer (*Odocoileus virginianus*) in Pennsylvania. The KQDC provides a case history where wildlife science and management came together in partnership with hunters and landowners to reduce the environmental impact of deer overabundance in Pennsylvania. However, despite the successes of the KQDC, political pressure continues to exert influence that is counterproductive to science-based management of white-tailed deer in Pennsylvania.

THE SCIENCE-MANAGEMENT INTERACTION MODEL FOR WILDLIFE MANAGEMENT

Ever since Aldo Leopold introduced the practice of wildlife management to America in 1933, the management model has focused on the interactions among hunters (or other stakeholders such as livestock growers with predator problems), wildlife managers, wildland managers (landowners who own/manage habitat with direct effects on wildlife), and wildlife scientists. The basic adaptive management model (Figure 14.1) was simple. Wildlife managers, wildland managers, stakeholders (including hunters), and scientists developed plans to meet desired goals for management and conservation of species. They then developed and implemented management activities to achieve their goals.

Results were assessed by monitoring populations and habitats of the species being managed. Such monitoring efforts were often developed jointly by scientists and managers, and during the course of implementing the plan, corrective adjustments in the plan or goals were often made. If goals were met, or progress toward goals was acceptable, all was considered well and good. If not, goals and actions were usually modified to meet the needs of stakeholders, including landowners. When stakeholder

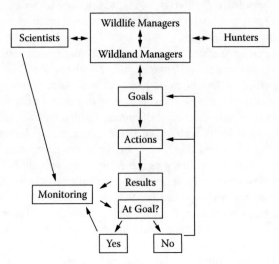

FIGURE 14.1   Typical adaptive management flow chart for managing wildlife species.

goals were unrealistic, managers depended on scientists to provide the research and technology to justify the goals that were at odds with stakeholders' expectations.

Wildlife management entails manipulating wildlife populations and their habitats and occurs on two or more levels: administrative (management by administrators of the agency responsible for manipulating wildlife populations), biological (management by biologists and landowners of species and their habitats), and, in states with commissions, political (management of people by politicians). Administrators implement and enforce regulations and manage budgets, biologists manage species and their habitats (in cooperation with landowners), and, where they exist, commissioners hire and fire administrators and biologists and vote on proposed policies and regulations. Biologists can recommend regulatory actions, such as shortening or lengthening hunting seasons to affect species' numbers, or recommend size and placement of administrative management units, but the administrators create the exact dimensions of such regulations and units, and the commissioners vote to enact these actions. Wildland managers manipulate habitats in ways that affect wildlife populations, but their activities generally are not affected by politicians or commissioners, except in those instances where regulations are enacted to protect or enhance habitats.

Unfortunately, in states like Pennsylvania, recommendations for manipulating wildlife abundance by biologists and administrators from wildlife agencies are sometimes ignored by commissions that have the power to accept, change, or ignore such recommendations. When commissions instead press for management of wildlife abundance that hunters want, such management may not be realistic, sustainable, or scientifically sound. In some cases, as in the example provided by this chapter, the disconnect between science and management lies in the bureaucratic layers (administrators and commissions) that come between scientists and managers, and in the influence stakeholder groups may exert over administrators and commissioners. However, it should be noted that the conflicts relate to regulation of wildlife abundance. It is rare for politicians or stakeholders to interfere with the communication between scientists and wildland managers related to implementation of advances in the science of habitat management.

For wildlife species that affect one or more stakeholder groups [e.g., ruffed grouse (*Bonasa umbellus*) and hunters; endangered species like spotted owls (*Strix occidentalis*) and conservation biologists; pests like pocket gophers (*Thomomys talpoides*) and alfalfa farmers], the simple model worked, in part because there were few differing human groups involved—scientists, the managers, and the affected stakeholders. Hunters generally went along with advances in habitat manipulation, seeing them as ways to improve quality of the habitat and their intended quarry. Regulations for manipulating population abundance generally were straightforward and non-controversial, in most cases, with a notable exception being the ongoing acrimony among various stakeholder groups and spotted owl management.

The model is vastly more complex with white-tailed deer (Figure 14.2) because this species affects multiple classes of stakeholders, who in turn interact with additional, non-stakeholder groups (e.g., politicians, government agencies, etc.) that can influence decisions of commissions about actions for regulating wildlife abundance, which are taken by managers. The once straight-line relationships between managers

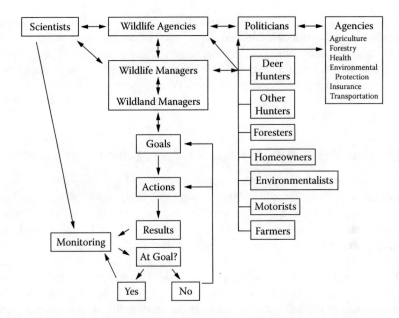

FIGURE 14.2 Adaptive management flow chart for managing deer.

and scientists and between managers and single stakeholder groups become entangled with additional intertwined and interfering groups, including commissions, politicians, and agencies representing other stakeholders. However, even with deer management, there is virtually no controversy or interference with application of science for manipulating wildlife habitat.

In large part, the complexity of deer management relates to the impact deer have on many different stakeholders. Deer hunters view deer positively, and having grown up in Pennsylvania in a culture that experienced high deer densities for generations, they view high deer densities as the expected norm. Other stakeholders want fewer deer—the other hunters whose preferred species is negatively impacted by habitat degradation by deer; the forest landowners who are unable to harvest trees because replacement seedlings are greatly reduced or eliminated by deer; the farmers whose crops are destroyed by deer; the homeowners whose landscaping is damaged by deer and whose health is endangered by Lyme disease spread by deer; the motorists whose cars are damaged by collisions with deer; and the conservation biologists concerned about deer reducing vegetation diversity and endangering rare plants. Stakeholders have learned that lobbying wildlife managers tasked with managing deer is not nearly as effective as lobbying wildlife agencies, other agencies representing their interests, or commissioners and politicians who influence wildlife management by their control over policies, regulations, funding, and budgets.

When it comes to managing deer, all these conflicting and interacting layers of stakeholders, agencies, commissioners, and politicians create barriers between managers and scientists. Administrators of wildlife agencies charged with supervising the work of managers are often beset by political pressures that force them to compromise management options formulated by the managers. Pennsylvania provides an example of how the complex management of deer in the context of multiple, highly engaged stakeholder groups influenced effective scientist-manager interaction to produce sustainable, responsible deer management. The result was unsustainable management of deer that resulted in decades of habitat degradation. Re-establishment of the tie between scientists and managers in Pennsylvania by Gary Alt and by the KQDC serve as a case study of how scientists and managers can work together and produce sustainable management of white-tailed deer.

DEER MANAGEMENT IN PENNSYLVANIA PRE-2000 AND PRE-KQDC

Management of white-tailed deer has a storied and controversial history in Pennsylvania (Latham 1950; Frye 2006). By the end of the nineteenth century, deer had been nearly eliminated by overharvest in Pennsylvania. The PGC was formed in 1895 with restoration of the deer herd as a primary goal (Kosack 1995). Initial steps included restriction of deer harvest to antlered deer and reintroduction of approximately 1,200 deer transplanted from a number of states including Kentucky, Maine, Michigan, New Hampshire, North Carolina, and Ohio—all designed to increase the deer herd. As these management actions were being implemented, concurrent forest clearcutting of the entire state resulted in production of vast quantities of prime deer forage (Redding 1995). In response to these factors—which were essentially protection of populations from overharvest and a huge increase in food supply—the deer population skyrocketed during the early 1900s. Aldo Leopold, among others, reported deer overpopulation and habitat degradation as an emerging threat to wildlife habitat and the deer themselves by the 1920s and 1930s (Leopold, Sowls, and Spencer 1947).

In response to overpopulation by the late 1930s, and at the urging of its biologists and administrators, the PGC held an antlerless only deer season in 1938, which coincided with a severe winter with deep snow persisting into late spring. The much-reduced forage from deer overbrowsing and maturation of early succession forest created at the turn of the century resulted in a massive die-off of deer due to starvation. However, hunters blamed the resulting low deer herd exclusively on antlerless hunting, which as a consequence has been a hard sell in Pennsylvania ever since.

The deer herd gradually increased during the 1940s and 1950s because of increased timber harvest, which caused an increase in deer forage production. In 1950 Roger Latham, Chief of the Division of Wildlife Research for the PGC, released his report (Latham 1950) on the deer situation. He stated that there were too many deer; he called for a large reduction in herd density and he chastised hunters for wanting more deer than the habitat could sustain. The report was met with outrage by hunters, who steadily increased the political pressure on the commission, which resulted in Latham's dismissal in 1957. As the deer herd continued to increase from the late 1950s through the 1970s, foresters and farmers began to complain about deer damage to forest regeneration and crops. Because of these complaints and those from other stakeholders, and at the urging of biologists, the PGC administrators and commissioners allowed hunting of antlerless deer. However, the size of the herd continued to creep upwards until the late 1970s when two harsh winters resulted once again in massive winter die-offs from starvation. From the 1950s to the 1970s, many of the predictions and warnings by Aldo Leopold and Roger Latham had come true. The forests of Pennsylvania had changed in many areas into maturing forests with an understory of ferns (Horsley 1993) that provided little to no food for white-tail deer populations, resulting in greatly reduced deer abundance, corroborating the predictions of those pioneers that carrying capacity would be cut in half (Leopold et al. 1947).

An emerging body of science concerned with the effect of deer overabundance on ecosystem resources began in the 1980s and culminated in the 1990s with a set of important environmental consequences that occurred when deer density in northern hardwood forests in the eastern United States exceeded thirty-eight to fifty-two deer per square kilometer (fifteen to twenty deer per square mile):

- Tree regeneration was stifled or altered (Tilghman 1989; Healy 1997; Horsley, Stout, and deCalesta 2003).
- Diversity and abundance of native bird species and sensitive herbaceous species were reduced (Whitney 1984; deCalesta 1994; deCalesta and Stout 1997; Augustine and Frelich 1998; Rooney and Dress 1997; McShea and Rappole 2000).
- Deer health suffered (Jones, deCalesta, and Chunko 1993; deCalesta 1997; Frye 2006).

Pennsylvania's deer managers embraced these results as evidence that supported their continuing attempts to reduce deer density to levels that did not have wide-spread negative impacts on the habitat as well as on other vertebrates. They urged administrators to adjust hunting regulations to permit more aggressive reduction of deer density and, again, commissioners complied by setting goal densities of deer in the 7.7 deer per square kilometer range (twenty-one deer per square mile), and initiating a "bonus" antlerless deer program that allowed hunters to harvest additional antlerless deer by obtaining special licenses.

However, little improvement occurred in the Pennsylvania deer situation during the 1990s. The antlerless programs did not result in reduction of deer density sufficient to reduce impact on forests and agricultural crops. Hunters resisted harvesting antlerless deer, and commissioners began cutting back on the number of antlerless licenses issued. Deer management became more controversial as hunters who wanted more deer were pitted against an increasing number of stakeholders who were negatively impacted by the large deer herd. Deer managers and PGC administrators knew they had to reduce deer density but they were hindered by the power and growing influence of that sector of the deer-hunting public that wanted higher deer density and lobbied commissioners to scale back efforts that would reduce deer density. The situation was exacerbated by the nature of the commission. In Pennsylvania, PGC commissioners are nominated by hunting clubs, selected by the governor, and confirmed by the state senate. However, few PGC commissioners had training in the biological or wildlife sciences. Some were deer hunters with a sympathetic ear for higher deer densities as requested by many deer hunters.

The situation became untenable during the 1990s with the rising tensions and conflicts among the various stakeholder groups, state and federal agencies, the administration, the commission, legislators, and lobbyists. Pressure by deer hunt-ers on commissioners resulted in suppression of levels of antlerless permits needed to reduce deer density to achieve population goals. In 1999, PGC administrators restructured deer management in Pennsylvania. They began by hiring Dr. Gary Alt as director of the deer management program. Alt, who had an international reputa-tion as a bear researcher and manager with the PGC, assembled scientific knowl-edge about deer and their impact on the environment. He perceived that the key to successful deer management was educating hunters, commissioners, legislators, and stakeholders about the scientifically proven environmental impacts of overabundant deer and of the need for control of deer density for the sake of the environment and the deer herd itself. He developed a comprehensive educational program, which he delivered to tens of thousands of hunters and other stakeholders, including his own administrators, commissioners, and state legislators. He oversaw a research program developed to fill in gaps (e.g., impact of predators on deer recruitment, distance hunt-ers travel from roads to hunt).

Key messages in Alt's program were that habitat health influenced deer health and antler size, and that changes in hunting regulations combined with reductions in the deer herd would improve habitat health and deer health, including improved antler characteristics. Concurrent with Alt's education campaign, the PGC convened a group of stakeholders to establish goals for the PGC deer management plan. Three goals were established for the 2003–2007 PGC Deer Management Plan: manage deer for a healthy and sustainable herd; manage deer-human conflicts at levels con-sidered safe and acceptable to Pennsylvanians; and manage deer impacts for healthy and sustainable forest habitat.

Because of initial positive reactions to the program by hunters and Alt's impec-cable reputation with the commissioners, regulations recommended by deer manag-ers, scientists, and administrators to achieve the goals set forth by the PCG Deer Management Plan were passed and implemented. These regulations are as follows:

- In 2002, antler restrictions for harvesting bucks were instituted to reduce harvest of yearling bucks and increase the proportion of mature bucks in the population.
- In 2003, the Deer Management Assistance Program (DMAP) was passed into law. The program was designed to increase harvest of antlerless deer by allowing landowners experiencing high deer impact to obtain permits to harvest antlerless deer during the regular hunting season and distribute them to hunters.
- In 2003, regulations were changed to permit hunters to harvest antlered and antlerless deer (concurrent harvest) during the general rifle season for the entire length of the twelve-day season.

These three initiatives temporarily reduced the influence of the commissioners, hunters, and legislators who wanted fewer antlerless deer killed and allowed manag-ers to harvest what research indicated was required to balance deer populations with habitat. For a short time, Alt revitalized the process by which wildlife scientists and wildlife managers collaborate to produce responsible, science-based, and sustainable management of deer.

Into this window of opportunity stepped the KQDC program. The KQDC was developed in 2001 by a consortium of foresters, public and private forest landowners, hunters, biologists, scientists, local business interests, and wildlife managers. Initiation and coordination of the KQDC were orchestrated by the Sand County Foundation (SCF), a private nonprofit organization dedicated to working with private landholders to improve natural habitats on their land by advancing the use of ethical and scientifically sound land management practices and partnerships for the benefit of people and the ecological landscape. SCF served as a conduit and facilitator for sharing ecological and management information between and among private individuals, scientists, wildlife and wildland managers and landowners, hunters, and local communities.

Primary KQDC goals were to reduce deer density and improve deer and habitat health. The success of the KQDC program demonstrates how collaboration among hunters, forest landowners, wildlife managers, and volunteers, as fostered by the SCF under the aegis of a functioning wildlife scientist/wildlife manager relationship, can develop and deliver a management strategy that improves a managed species and habitat while addressing stakeholder needs.

THE KQDC PROGRAM

The KQDC program began with identification of goals, developed jointly by a team comprised of scientists, forest and wildlife managers, SCF administrators, local hunters, foresters, PGC wildlife conservation officers, and the local recreational bureau (Allegheny National Forest Vacation Bureau). Support for the program was provided by public (Allegheny National Forest) and private (Bradford Water Authority, Collins Pine, Forest Investment Associates, RAM Forest Products) landowners, volunteers, and foundations, notably the SCF, Bradley Fund for the Environment, the Heinz Foundation, and the National Fish and Wildlife Foundation. Initial goals were:

- Reduce deer density to 5.8 to 7.0 deer per square kilometer (15 to 18 deer per square mile and later amended to 4.0 to 5.8 deer per square kilometer [10.4 to 15.0 deer per square mile]).
- Reduce deer impact on regeneration to the point where fencing no longer is required to protect tree regeneration from deer.
- Improve deer health (body weight, antler characteristics).
- Monitor and improve hunter satisfaction.
- Improve habitat.

The KQDC leadership team realized that engaging hunters and soliciting their participation in the program, including deer harvest, were essential for accomplishing the goals. A kick-off information event was held at a local community college in 2001 with Dr. Alt as the main speaker to provide background for the KQDC program and attract hunters. Hundreds of local hunters attended.

The KQDC program was designed around a monitoring program that measured changes in deer density and impact, deer health, habitat condition, and hunter participation. Monitoring data were collected by volunteers (approximately fifty hunters or members of stakeholder groups annually) to keep down costs and to enhance stakeholder buy-in. The leadership team was aware of key requests hunters routinely asked of the PGC for deer management:

- Provide estimates of deer density.
- Provide evaluation of deer health through deer check stations.
- Reduce the size of deer management units to represent local landscapes and local hunting conditions for localized deer management.

Reports were issued to provide stakeholders, including hunters, with annual estimates of deer density and impact and deer and habitat health. The size of the KQDC management unit (30,000 ha [74,131 acres]) represented the smaller, localized management unit size hunters favored.

Involvement of hunters was recognized as a key factor in the KQDC program, and hunters participated in conducting roadside counts, working check stations, bringing deer to check stations, and attending deer density and impact workshops where they helped gather, analyze, and interpret deer density data. Hunters were rewarded for bringing deer to check stations by receiving invitations to hunter appreciation banquets and tickets for banquet raffles. A local outdoor writer played a key role in providing hunters (and other stakeholders) with informative and supportive newspaper stories about the project. The local recreational bureau and the information office of the surrounding Allegheny National Forest provided timely news releases and other sources of information about the project. A website (www.kqdc.com) was built to communicate with hunters and to provide monitoring updates and practical information (e.g., maps of hunting areas, instructions for obtaining DMAP licenses, etc.).

During the first year of the KQDC program (2002), the PGC enacted the 3-point antler restriction in the management unit encompassing the KQDC Project Area. This regulation restricted buck harvest to those with three or more antler points on at least one side. This spared many yearling bucks from harvest and resulted in an immediate and continuous improvement in antler characteristics (spread, beam diameter, total points), improvement in buck-to-doe ratio, and incentive for meat hunters to harvest antlerless deer. The second year of the project the DMAP program and concurrent buck-doe seasons were initiated, resulting in an immediate increase in harvest of antlerless deer and reduction in deer density. Because KQDC monitoring began prior to enactment of these two management changes, baseline data were available to gauge the success of both initiatives.

KQDC: DEER DENSITY AND IMPACT

Overwinter deer density, estimated by the pellet group technique (adapted from Eberhardt and Van Etten 1956), was initially in the ten to eleven deer per square kilometer range (twenty-seven to twenty-eight deer per square mile), then declined rapidly as the DMAP program was implemented (Figure 14.3) (deCalesta 2001–2010).

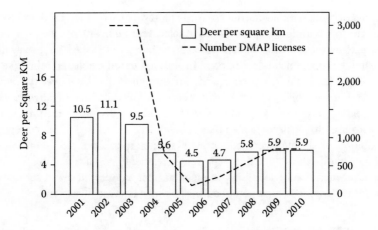

FIGURE 14.3 Changes in deer density and DMAP license availability on the KQDC demonstration area, 2001–2010.

Once density began to decline, the number of DMAP licenses available for the KQDC Project Area was reduced by consensus of the leadership team (2005–2007), increased as deer density increased (2008–2009), and then stabilized (2010) to maintain deer density within the target interval. These changes in DMAP license availability stabilized deer density and maintained it at target density. Involvement of hunters in annual workshops, where data on one of twenty-six data grids were collected, and contracting with a few other hunters to collect density data contributed to credibility of these results.

Impact of deer on understory vegetation similarly was surveyed prior to implementation of the management steps of 2002–2003 and documented the heavy impact of deer (Kirschbaum and Anacker 2005; deCalesta 2001–2003). As deer density declined, the impact of deer on vegetation declined (Figure 14.4). For example, the

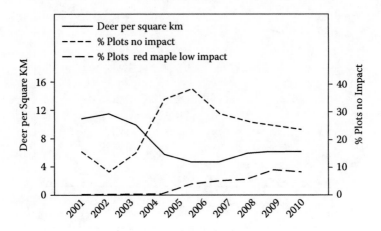

FIGURE 14.4 Relationships among deer density, percentage of plots with no impact and percentage of plots with low impact on red maple, an indicator species, 2002–2010.

percentage of plots with no impact on seedling regeneration increased and the percentage of plots with low impact on red maple (*Acer rubrum*), one of six indicator species, also increased. This demonstrated how changes in DMAP license availability resulted in reductions of deer impact. In addition to red maple, all other five indicators [American beech (*Fagus grandifolia*), striped maple (*Acer pensylvanicum*), birches (*Betula* sp.), black cherry (*Prunus serotina*), and hemlock (*Tsuga canadensis*)] responded similarly to reductions in deer density. A study on KQDC project plots by Royo et al. (2010) indicated that as deer density declined, size and reproductive condition of selected herbaceous plant species increased. Additionally, forest landowners within the KQDC Project Area were able to begin removing fences that protected regeneration of seedlings on recent timber harvest sites, and subsequently erected far fewer fences to protect seedlings on new harvest sites.

KQDC: DEER ROADSIDE COUNTS

Every summer, fifteen to twenty volunteer hunters counted deer by sex and age class (fawn, adult doe, spike buck, branch antlered buck, and unknown) along six roadside routes, traveling over 1,126 km (700 miles) of roadside. Roadside counts (deer counted per 166 km [100 miles]) roadside driven) were highly correlated with estimates derived from pellet group counts, corroborating the trend in deer density derived from pellet group counts (Figure 14.5).

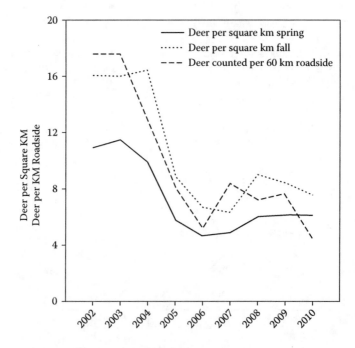

FIGURE 14.5 Relationships among spring and fall estimates of deer density from pellet group counts and fall roadside count density, 2002–2010.

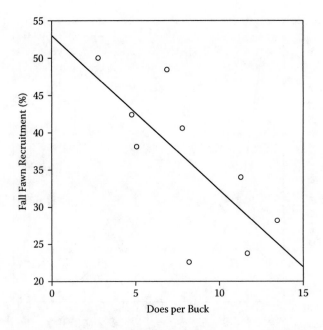

FIGURE 14.6   Plot of recruitment versus doe:buck ratio from previous fall.

Roadside counts, which produced estimates of buck-to-doe ratios and fall fawn recruitment rates, revealed a surprising relationship. As buck-to-doe ratio decreased (fewer bucks to more does), fall fawn recruitment rate the following fall decreased. In other words, recruitment was inversely related to the number of adult does per antlered buck in the population (Figure 14.6). This relationship indicated that as hunters increasingly harvested antlered deer in preference to antlerless deer, the resulting low ratio of bucks-to-does depressed recruitment and depressed deer density. This is the exact opposite of what hunters thought they were doing by avoiding harvesting does. It remains to be seen how hunters will interpret this relationship, but this finding, and the reduced and stabilized number of DMAP licenses available, may help explain why deer density stabilized after 2005 on the KQDC Project Area when hunters increasingly avoided harvesting antlerless deer.

KQDC: CHECK STATION OPERATION

As an immediate consequence of the 3-point antler restriction, age distribution of antlered deer switched primarily from 1½-year-old bucks in 2001 to primarily 2½-year-old bucks (with three points or more on at least one antler; Figure 14.7) from 2003 to subsequent years.

Subsequent increase in relative numbers of older bucks (>3½-year-old) indicates that the 3-point antler restriction regulation resulted in long-term improvement in the age structure of harvested antlered deer. Because deer density was cut in half, and remained at approximately half that prior to inception of the KQDC program, relatively more forage was available per deer.

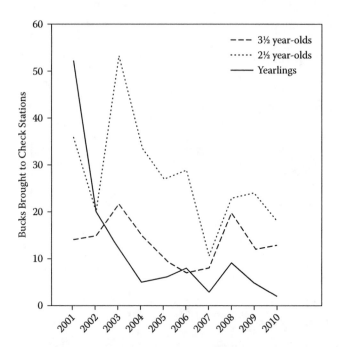

FIGURE 14.7 Relative numbers of antlered bucks by age class brought to check stations, 2001–2010.

Increase in relative forage availability was also reflected in a significant increase in body weights of all sex and age classes of deer [e.g., buck weights (Figure 14.8) and antler characteristics (Figure 14.9)].

KQDC: HUNTER PARTICIPATION

A protracted, annual effort to involve hunter participation and communication was conducted via the website noted previously, timely news releases, an annual hunter appreciation banquet, and facilitated participation of hunters in all phases of collecting monitoring data. Annual reports on the success of the program were provided to stakeholder groups including hunters, PGC biologists, managers, administrators, and commissioners in an effort to retain their long-term support for the KQDC project.

REASONS FOR CONCERN

Because of the regulations implemented by the PGC after 2000, deer density declined significantly across much of northern Pennsylvania. Hunter enthusiasm for the new regulations began to wane as they continued to notice the reduction in deer abundance during hunting season. In addition, an initial selling point in the KQDC program had been that as deer density declined and regeneration improved, resulting increased timber harvest would produce more forage for deer and pave the way for

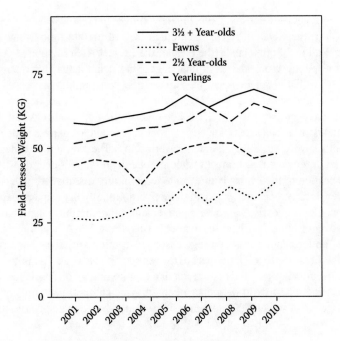

FIGURE 14.8   Field-dressed weights of bucks in four age classes, 2001–2010.

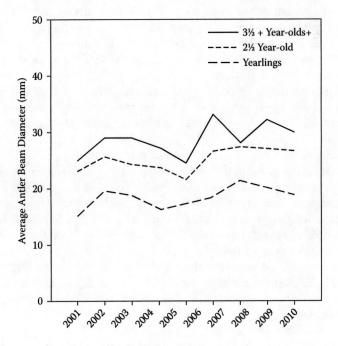

FIGURE 14.9   Average antler beam diameters of bucks in three age classes, 2001–2010.

increases in target deer density. Unfortunately, timber harvest was largely conducted on the 34% of the KQDC Project Area comprised of private timber land—timber harvest on National Forest lands (the 66% of the Project Area in public land) undergoes intense public input and legal challenges that delay actual harvest. Thus, on much of the KQDC Project Area, there was little concomitant increase in forage production as deer density declined. In addition, except where aggressive, sustainable forest management was practiced, tree regeneration by seedlings and increases in shrub and herbaceous vegetation were slower to respond to changes in deer density than hoped (Royo et al. 2010). Hunters reverted to harvesting antlered deer, which has resulted in reduced recruitment and continued depression of deer density.

As the portion of the Pennsylvania hunter community dissatisfied with the reduction in deer density became increasingly active politically, its impact was seen on regulations. Consequently, for some management units commissioners have scaled back numbers of antlerless licenses available (including DMAP) and rescinded the concurrent antlered/antlerless hunting season, restricting antlerless harvest (except with DMAP licenses) to the latter part of the regular rifle season. The initial triumph of science over perception, and restoration of a functioning wildlife science and management interaction for wildlife management, may be compromised in Pennsylvania if the resurgence of stakeholder influence over science-based management continues.

REASONS FOR SUCCESS

Success of the KQDC project was the result of a fortuitous combination of wildlife agency revitalization, cooperation and contributions of private and public timber-managing organizations comprising the KQDC Project Area, support by the SCF, and collaboration in development and delivery of a coordinated management program by a consortium of KQDC Project Area stakeholders that featured:

- Groundwork resulting from revitalized collaboration of wildlife scientists and managers, and education of administrators, politicians, and other stakeholders by the PGC deer management program under the guidance of Dr. Alt.
- Development of goals mutually acceptable to stakeholders.
- A monitoring program that provided quantitative justification for adjustments of management actions (i.e., DMAP licenses), and provided annual descriptions of deer herd density, health, impact, and habitat quality.
- A sustained, comprehensive effort to educate and engage hunters and other stakeholders through timely news releases, a website, workshops, and promotion by outdoor writers.
- A determined and comprehensive effort to involve hunters in all phases of the KQDC program, including membership on the KQDC leadership team, direct participation in all phases of monitoring (roadside counts, check stations, harvesting deer, estimation of deer density and impact) and recognition at an annual banquet for hunters and volunteers.
- Direct and ongoing two-way communication among field-experienced foresters, hunters, biologists, and managers at the scale of the management unit.

- Development and maintenance of a system for encouraging stakeholder feedback on management actions and responding to feedback.
- Reliance on volunteer hunters and other stakeholders for collection of annual monitoring data to minimize costs and increase hunter buy-in by involvement.
- Sustained efforts by cooperating KQDC Project Area landowners to improve and maintain condition of access roads into the KQDC Project Area for hunters, and to provide signage and other information to help hunters find and negotiate access roads.
- Conduct of the KQDC project within a management unit of a size and configuration providing consistent deer habitat, deer density, and access conditions that agreed with hunter perception of what a functioning management unit size should be.

In summary, the success of the KQDC project was a direct result of deer management being based on a science-driven model with emphasis on collaboration among primary stakeholders—hunters, foresters, biologists, and landowners—as facilitated by financial and philosophical support from a consortium of organizations (e.g., SCF) fostering enhanced, collaborative management of natural resources on public and private wildlands. It developed stakeholder-driven goals that integrated stakeholder desires with science and management; it minimized operational costs by relying on volunteers to collect monitoring data; it monitored results; it made justifiable and transparent management adjustments to make progress toward and maintain goal conditions; it engaged, informed, rewarded, and listened to stakeholders; it confined itself to an operational unit of a size that favored relatively consistent operational parameters; it was conducted over a period of time sufficiently long to develop believable and predictable trend information and management responses; and it successfully negotiated the complex political environment of deer management in Pennsylvania.

ACKNOWLEDGMENTS

Financial support and philosophical guidance for the Kinzua Quality Deer Cooperative were provided by the Sand County Foundation, Pennsylvania State University Cooperative Extension Service, Warren Forestry Sciences Laboratory, Northeast Research Station, USDA Forest Service, National Fish and Wildlife Foundation, Bradley Fund for the Environment, Heinz Endowment, the Wild Resources Conservation Program of the Pennsylvania Department of Conservation and Natural Resources, and the Allegheny National Forest Vacation Bureau. Financial support, consultation, volunteer help for data collection, and maintenance of access roads were provided by cooperating landowners: Allegheny National Forest, Bradford Watershed Authority, Collins Pine, Forest Investment Associates, and RAM Forest Products. Technical support was provided by Wildlife Conservation Officers, Pennsylvania Game Commission. A dedicated core of repeat deer hunters was the linchpin of success for the KQDC project—they reduced deer density to levels associated with healthy deer and habitat and their continued hunting support maintained deer density and impact at these levels. John Dzemyan (PGC), Kevin McAleese (SCF), and Susan Stout (U.S.D.A. Forest Service, Northeast Research Station) provided useful reviews of manuscript drafts.

REFERENCES

Augustine, D. J., and L. E. Frelich. 1998. Effects of white-tailed deer on populations of an understory forb in fragmented deciduous forests. *Conservation Biology* 12:995–1004.

deCalesta, D. S. 1994. Impact of white-tailed deer on songbirds within managed forests in Pennsylvania. *Journal of Wildlife Management* 58:711–718.

deCalesta, D. S. 1997. Deer and ecosystem management. In: *The Science of Over-Abundance: Deer Ecology and Population Management*, W. J. McShea, H. B. Underwood, and J. H. Rappole, Eds. Washington, DC: Smithsonian Institution Press, pp. 267–279.

deCalesta, D. S. 2001–2010. Annual report: Kinzua Quality Deer Cooperative. Unpublished annual reports.

deCalesta, D. S., and S. L. Stout. 1997. Relative deer density and sustainability: a conceptual framework for integrating deer management with ecosystem management. *Wildlife Society Bulletin* 25:252–258

Eberhardt, L., and R. C. Van Etten. 1956. Evaluation of the pellet group count as a deer census method. *Journal of Wildlife Management* 20:70–74.

Frye, R. 2006. *Deer Wars; Science, Tradition, and the Battle Over Managing Whitetails in Pennsylvania*. State College, PA: Pennsylvania State University Press.

Healy, W. M. 1997. Influence of deer on the structure and composition of oak forests in central Massachusetts. In: *The Science of Over-Abundance: Deer Ecology and Population Management*, W. J. McShea, H. B. Underwood, and J. H. Rappole, Eds. Washington, DC: Smithsonian Institution Press, pp. 249–266.

Horsley, S. B. 1993. Mechanisms of interference between hay-scented fern and black cherry. *Canadian Journal of Forest Research* 23:2059–2069.

Horsley, S. B., S. L. Stout, and D. S. deCalesta. 2003. White-tailed deer impact on the vegetation dynamics of a northern hardwood forest. *Ecological Applications* 13:98–118.

Jones, S. B., D. S. deCalesta, and S. E. Chunko. 1993. Whitetails are changing our woodlands. *American Forests*. Nov/Dec:20–25.

Kirschbaum, C. D., and B. L. Anacker. 2005. The utility of trillium and maianthemum as phytoindicators of deer impact in northwestern Pennsylvania, U.S.A. *Journal of Forest Ecology and Management* 217:54–66.

Kosack, J. 1995. *The Pennsylvania Game Commission, 1895–1995: 100 Years of Wildlife Conservation*. Harrisburg, PA: Pennsylvania Game Commission.

Latham, R. 1950. The Pennsylvania deer problem. *PA Game News*, Special Issue #1. PA Game Commission, Harrisburg, PA.

Leopold, A. 1933. *Game Management*. New York: Charles Scribner's Sons.

Leopold, A., L. K. Sowls, and D. L. Spencer. 1947. A survey of over-populated deer ranges in the United States. *Journal of Wildlife Management* 11:162–177.

McShea, W. J., and J. H. Rappole. 2000. Managing the abundance and diversity of breeding bird populations through manipulation of deer populations. *Conservation Biology* 14(4):1161–1170.

Redding, J. R. 1995. History of deer population trends and forest cutting on the Allegheny National Forest. In: *Proceedings of the Tenth Annual Hardwood Forest Conference*, K. W. Gottschalk and J. C. L. Fosbroke, Eds. U.S.D.A. Forest Service, Gen. Tech. Rep. NE-17, Northeastern Forest Experiment Station, Radnor, PA, pp. 214–224.

Rooney, T. P., and W. J. Dress. 1997. Species loss over sixty-six years in the ground-layer vegetation of Heart's Content, an old-growth forest in Pennsylvania, USA. *Natural Areas Journal* 17:297–305.

Royo, A. A., S. L. Stout, D. S. deCalesta, and T. G. Pierson. 2010. Restoring forest herb communities through landscape-level deer herd reductions—is recovery limited by legacy effects? *Biological Conservation* 143:2425–2434.

Tilghman, N. G. 1989. Impacts of white-tailed deer on forest regeneration in northwestern Pennsylvania. *Journal of Wildlife Management* 53:524–532.

Whitney, G. C. 1984. Fifty years of change in the arboreal vegetation of Heart's Content, an old-growth hemlock-white pine-northern hardwood stand. *Ecology* 65:403–408.

Section IV

Management and Policy Case Studies

15 Impacts of Wind Energy Development on Wildlife
Challenges and Opportunities for Integrating Science, Management, and Policy

Edward B. Arnett
Bat Conservation International
Austin, Texas

CONTENTS

Wind energy is one of the fastest growing forms of renewable energy worldwide. As demand for wind energy grows, environmental impacts must be considered because wind energy development is not environmentally neutral.

—**The Wildlife Society (2007)**

Now, clean energy breakthroughs will only translate into clean energy jobs if businesses know there will be a market for what they're selling. So tonight, I challenge you to join me in setting a new goal: by 2035, 80% of America's electricity will come from clean energy sources. Some folks want wind and solar. Others want nuclear, clean coal, and natural gas. To meet this goal, we will need them all—and I urge Democrats and Republicans to work together to make it happen.

—**President Barack Obama**
2011 State of the Union Address

ABSTRACT

The aforementioned State of the Union address by President Obama highlights the growing and politically charged demand for development of renewable energy sources, including wind. Wind energy is one of the fastest growing renewable energy sources under development and while representing a clean energy source, it is not environmentally neutral. The impacts of wind energy development on wildlife and their habitats have become an increasing concern and topic of research and political discussion. However, a lack of credible scientific information, poor or even nonexistent frameworks for decision making, lack of transparency, and weak policies or lack of policy and regulation altogether continue to fuel controversy over wind energy development and wildlife impacts. The issues and challenges presented in this chapter are not all-inclusive, but represent some key factors that must be overcome if wildlife and wind energy development conflicts are to be resolved such that responsible development that minimizes wildlife impacts can be achieved.

INTRODUCTION

Fossil fuels provide more than two-thirds of all energy consumed worldwide, including two-thirds of electricity and virtually all transportation fuels used in the United States (U.S. Department of Energy 2011). Conventional power generation from fossil fuels has a host of well-documented environmental impacts, most notable being emissions of carbon dioxide (CO_2) and particulate pollution such as nitrogen and sulfur oxides. Many climate change models predict that increased atmospheric CO_2 concentrations could contribute to additional pressures on flora and fauna to adapt to changing environmental conditions (Inkley et al. 2004). With rising costs and long-term environmental impacts from use of fossil fuels, the world community of scientists and engineers is increasingly exploring alternatives to supply emission-free electrical energy and fuel for transportation (McLeish 2002; Bernstein, Griffin, and Lempert 2006; Kunz et al. 2007a).

Wind power is one of the fastest growing renewable energy sources under development, in part due to recent technological advances, cost-competitiveness with conventional energy sources, and significant tax incentives in the United States (Bernstein et al. 2006). The U.S. Energy Information Administration (EIA 2011) estimated that world net electricity generation will increase from 18.8 billion megawatt hours (MWh) in 2007 to 25.0 billion MWh in 2020 and 35.2 billion MWh in

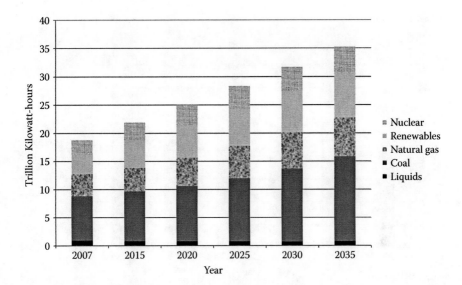

FIGURE 15.1   Projected world net electricity generation by fuel source (in trillion kilowatt-hours) from 2007–2035. (Modified from EIA 2011.)

2035 (Figure 15.1). In the United States alone, electricity demand is expected to grow by 39% from 2005 to 2030, reaching 5.8 billion MWh by 2030 (U.S. Department of Energy 2008). To meet 20% of that demand, U.S. wind power capacity would have to reach more than 300,000 megawatts (MW) (U.S. Department of Energy 2008), which would represent an increase of more than 260,000 MW (e.g., 130,000 2.0 MW turbines) over current installed capacity (U.S. Department of Energy 2011) within twenty years. Some energy analysts suggest, however, that while wind energy development is growing exponentially in the United States, fossil-fuel-burning power plants may continue to increase, which raises questions about the effectiveness of renewable energy sources to reduce greenhouse gas emissions over time. Indeed, the proportion of fossil fuels in the world's energy mix, currently at ~68%, is not projected to change significantly by 2035 (EIA 2011; Figure 15.2). However, some older coal-fired plants are being taken offline and replaced with natural gas or renewable sources, which theoretically will result in net reductions (J. Anderson, American Wind Energy Association, personal communication).

At regional to global scales, it is often postulated that the effects of wind energy on the environment is positive owing to the displacement of mining activities, air pollution, and greenhouse gas emissions associated with non-renewable energy sources (NRC 2007). Wind energy development, however, is not environmentally neutral. The impacts of wind energy development on wildlife and their habitats have become an increasing concern, owing to rapid development, new evidence indicating significant bat and bird fatalities and potential for habitat impacts, and a lack of scientific evidence upon which to base decisions regarding siting, permitting, and mitigating impacts (Arnett at al. 2007). There are two primary ways that wind-energy development may influence wildlife—through direct mortality impacts on

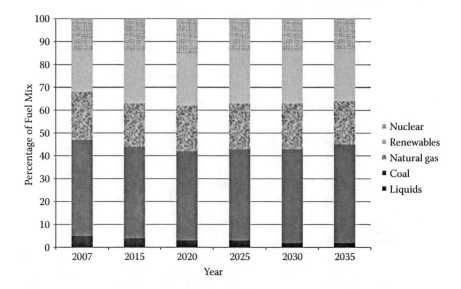

FIGURE 15.2 Percentage of projected world net electricity generation by fuel source from 2007–2035. (Modified from EIA 2011.)

individual organisms and through impacts on habitat structure and function (Arnett et al. 2007; Kunz et al. 2007a; NRC 2007). Fatalities of birds (e.g., Erickson et al. 2001; Drewitt and Langston 2008; Smallwood and Karas 2009) and bats (e.g., Dürr and Bach 2004; Johnson 2005; Arnett et al. 2008; Rydell et al. 2010) have been documented at wind facilities worldwide. Often overlooked are habitat impacts, both direct (e.g., resulting from turbine construction and increased human access) and indirect (e.g., habitat fragmentation and avoidance of habitats in proximity to turbines; Arnett et al. 2007; Kuvlesky et al. 2007; NRC 2007).

This chapter focuses on issues surrounding wildlife and wind energy development, and while I briefly discuss what we know and do not know from available studies, my primary focus centers on some key challenges that I believe impede integrating science, management, and policy to effectively site, permit, and operate wind energy projects consistently that minimize wildlife impacts. The data and discussions I present here are most specific to the United States and Canada, but certainly may apply to other regions of the world.

UNFORESEEN CONSEQUENCES

Given that many species of birds strike stationary objects, especially under inclement environmental conditions (e.g., Gehring et al. 2009), avian scientists may have predicted potential problems with wind turbines and birds long before they occurred. However, well before bat fatalities at wind facilities became a concern, few if any chiropteran scientists, myself included, might have predicted these fatality events and particularly at the level currently documented (Arnett et al. 2008; Kunz et al. 2007a). Habitat-related impacts, both direct and indirect, may have been more readily predicted by scientists if asked, but only based on surrogate disturbances (e.g., Robel et al. 2004; Pitman et al. 2005; Doherty et al. 2008; Naugle 2011) and none of which truly emulate potential impacts of wind turbines.

FATALITIES

Avian fatalities have been summarized by several authors (e.g., Erickson et al. 2001; Drewitt and Langston 2008; Arnett et al. 2007; Kuvlesky et al. 2007). Although fatalities of many bird species have been documented at land-based wind facilities, passerines by far dominate avian fatalities, but raptors have received the most attention (Arnett et al. 2007). Initial observations of dead raptors at the Altamont Pass Wind Resource Areas (APWRA) (Orloff and Flannery 1992) triggered concern from regulatory agencies, environmental groups, wildlife resource agencies, and wind and electric utility industries about possible impacts to birds from wind energy development. Raptor fatalities have been relatively low at most facilities studied, with the exception of APWRA (e.g., Smallwood et al. 2007; Smallwood and Thelander 2008). This site was developed nearly forty years ago when little was understood about wind energy and wildlife interactions, and to date appears to represent an anomaly that has not been replicated since. By 1998, APWRA consisted of about 5,400 turbines ranging from 40 kilowatt (kW) to 400 kW turbines, with most being 100 to 150 kW, with smaller rotor diameters and faster rotor speeds that likely attributed to such high raptor kills at this site (Orloff and Flannery 1992; Howell 1997; Smallwood and Karas 2009). Recent repowering indicates that raptor fatalities appear to be lower at new generation wind turbines (Smallwood and Karas 2009). Turbine characteristics, turbine siting, and bird behavior and abundance appear to be important factors determining raptor fatalities at wind power facilities. Nevertheless, the number of studies at utility-scale wind energy facilities is relatively small and most have occurred in areas with low raptor density.

In comparison with other sources, wind turbines appear to be a relatively minor source of passerine fatalities, particularly for migrants, at current levels of wind energy facility development. Although collision mortality may currently be relatively insignificant on passerine populations, it may have serious impacts on some raptor populations because raptors have longer life spans than passerines and consequently lower reproductive potential (Kuvlesky et al. 2007). In addition, other species of conservation concern may be impacted by wind facilities more so than most species of passerines. Thorough evaluation during the site selection process and site development plans that consider bird use and bird habitats at the site should allow development that reduces risk to raptors and other birds (e.g., Kunz et al. 2007b; Strickland et al. 2011). As turbine size increases and development expands into new areas with higher densities of passerines, the risk to passerines could increase and should continue to be evaluated, particularly concerning migration during inclement weather.

In recent years, large numbers of bat fatalities have been reported at some wind energy facilities in North America (Johnson 2005; Kunz et al. 2007a; Arnett et al. 2008). Several plausible hypotheses relating to possible sources of attraction, density and distribution of prey, and sensory failure (i.e., echolocation), for example, have

FIGURE 15.3 A hoary bat (*Lasiurus cinereus*) found having been killed by a wind turbine at a wind facility in West Virginia. Migratory tree roosting bats like the hoary bat are
the most frequently killed species by wind turbines in North America. (© Ed Arnett, Bat
Conservation International, www.batcon.org)

been proposed to explain why bats are killed by wind turbines (Kunz et al. 2007a;
Cryan and Barclay 2009). Estimates of bat fatalities from wind facilities in North
America range from 0.9 to 53.3 bats/MW/year (Arnett et al. 2008). Bat fatalities in
North America appear heavily skewed to migratory tree roosting species that include
the hoary bat (*Lasiurus cinereus,* Figure 15.3), eastern red bat (*Lasiurus borealis*),
and silver-haired bat (*Lasionycteris noctivagans*; Johnson 2005; Arnett et al. 2008).
In Europe, migratory and open-foraging species also dominate fatalities (Dürr and
Bach 2004; Rydell et al. 2010).

Bat fatalities are consistently higher during late summer and early fall (Arnett et al.
2008; Rydell et al. 2010). This period is when bats typically begin autumn migration
(Cryan 2003; Fleming and Eby 2003); although, to a lesser extent, fatalities during
spring have been reported (Arnett et al. 2008). One theory is that migratory tree bats
may follow different migration routes in the spring and fall (Cryan 2003), and behavioral differences between migrating bats during these two periods also may be related
to the different fatality patterns that have been reported (Johnson 2005). Bats have
been observed with thermal imaging cameras attempting to and actually landing on
stationary turbine blades and investigating turbine masts (Horn et al. 2008), and thus
may be attracted to turbines. A substantial portion of bat fatalities occurs during relatively low-wind conditions over a relatively short period during the summer-fall bat
migration period (Arnett et al. 2008). Operational adjustments under these conditions
and during this period of time have been proposed as a possible means of reducing
impacts to bats (Kunz et al. 2007a; Arnett et al. 2008). Indeed, recent results from
studies in Canada (Baerwald et al. 2009) and the United States (Arnett et al. 2011)
indicate that raising turbine "cut-in speed" (i.e., wind speed at which wind-generated
electricity enters the power grid) from the manufactured speed (usually 3.5 to 4.0 m/s

for modern turbines) to 5.0 m/s or higher results in significant reduction in bat fatalities (44% to 93%) compared to normally operating turbines.

Although population impacts are unknown, given the level of bat fatalities at some wind-energy facilities, biologically significant additive mortality is likely and must be considered for some species as wind power development expands and fatalities accumulate (Kunz et al. 2007a). At the end of 2010, there were ~4,062 MW of wind energy installed in Canada and 40,267 in the United States (Figure 15.4), for a total of 44,329 MW in those two countries alone. Assuming a continental average of twelve bats killed per installed megawatt (Arnett et al. 2008), and assuming these fatality rates are representative of all installed facilities and were consistent in 2010, then the number of bats killed by turbines in the United States and Canada in 2010 and 2011 could have been well into the hundreds of thousands given these assumptions. These fatality rates presumably are not sustainable, as bats are long-lived and have exceptionally low reproductive rates. Their population growth is relatively slow and their ability to recover from population declines is limited, thereby increasing the risk of local or regional extinctions (Barclay and Harder 2003; Racey and Entwistle 2003). Additionally, migratory tree-roosting species killed most frequently by turbines are not protected under federal law in the United States. Further, while bats may be protected under state laws pertaining to "nongame" animals, most states do not appear to enforce take of bats if killed by wind turbines.

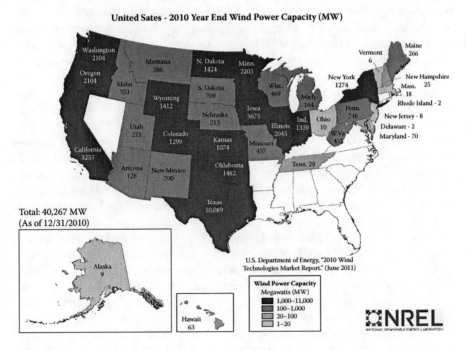

FIGURE 15.4 Installed wind energy capacity in the United States, by each state, as of December 31, 2010 (40,180 MW). (From http://www.windpoweringamerica.gov/wind_installed_capacity.asp)

The context of bird and bat fatalities remains a mystery, in part because little population data exist for several species of birds and most species of bats. Thus, understanding population-level impacts, as well as effectiveness of mitigation measures, has been seriously hindered. Perhaps the most challenging task in interpreting fatalities is in estimating exposure. For example, corvids are a common group of birds observed flying near the rotor swept area of wind turbines (e.g., Erickson et al. 2004; Smallwood and Thelander 2004), yet are seldom found during carcass surveys. Clearly, the role of abundance relative to exposure of birds and bats to collisions with wind turbines is modified by behavior within and among species and likely varies across locations. Estimating exposure, particularly for migrating passerines and bats, is problematic, but necessary to understand the context of fatalities. Studies using radar, thermal imaging, and other technologies simultaneously and concurrently with fatality studies would help determine the exposure risk of birds and bats. Model-based analysis of exposure risk also may be helpful, but empirical data generally are needed and lacking for developing such models (Arnett et al. 2007).

IMPACTS ON WILDLIFE HABITAT

Impacts of wind energy development on wildlife habitat can be considered direct (e.g., vegetation removal or modification and physical landscape alteration, habitat loss) or indirect (e.g., behavioral responses to wind facilities, hereinafter referred to as displacement or attraction) (Arnett et al. 2007; Kuvlesky et al. 2007). Impacts may be short-term (e.g., during construction and continuing through the period required for habitat restoration) and long-term (e.g., surface disturbance and chronic displacement effects for the life of the project). Duration of habitat impacts vary depending on the species of interest, the area impacted by the wind energy facility (including number of turbines), turbine size, vegetation and topography of the site, ancillary disturbances, and climatic conditions in a particular region. Road construction, turbine pad construction, construction staging areas, installation of electrical substations, housing for control facilities, and transmission lines connecting the wind energy facility to the power grid also are potential sources of negative habitat impacts. Presence of wind turbines can alter the landscape and has potential to change patterns of habitat use by wildlife, including behavioral avoidance (displacement) of wildlife from areas near turbines (Figure 15.5). Displacement effects of wind facilities may extend over large areas (e.g., several square kilometers) but have relatively low direct habitat loss. The Bureau of Land Management (BLM) Programmatic Environmental Impact Statement (BLM 2005) estimated that on average the permanent footprint of a facility is approximately 5% of the site, including turbines, roads, buildings, and transmission lines. Some direct impacts are short-term, depending on the length of time required to reclaim a site, which varies depending on the climate, vegetation, and reclamation objective. Ultimately, the greatest impact from habitat modification may be reduced wildlife use due to displacement. The degree to which this displacement results in impacts depends on the abundance and behavioral response of individual species to turbines and human activity within the wind energy facility.

FIGURE 15.5 Wind turbines located in native grasslands can have direct impacts due to loss of habitat or indirect impacts through negative behavioral response to turbines and avoidance of habitat. These impacts may be short- or long-term depending on several factors, including animal habituation and habitat rehabilitation. (© Ed Arnett, Bat Conservation International, www.batcon.org)

Relatively little research has been conducted to determine the effect of wind energy facilities on use of habitat by wildlife. Some studies have found that densities of birds along transects increase with distance from turbine strings (e.g., Leddy et al. 1999). Some prairie grouse (lesser and greater prairie chicken), which exhibit high site fidelity and require extensive grasslands, sagebrush, and open horizons (Giesen 1998; Fuhlendorf et al. 2002), may be especially vulnerable to wind energy development. Several studies indicate that prairie grouse avoid certain anthropogenic features and disturbances (e.g., roads, buildings, power lines) during key life-cycle phases, resulting in sizable areas of habitat rendered less suitable (e.g., Robel et al. 2004; Pitman et al. 2005). The long-term impacts of wind energy facilities on prairie grouse populations are not well understood, but are currently under investigation. Research on habitat fragmentation has demonstrated that several species of grassland birds are area-sensitive, prefer larger patches of grassland, and tend to avoid trees and other vertical structures. Area-sensitivity in grassland birds was reviewed by Johnson (2001); thirteen species were reported to favor larger patches of grassland in one or more studies. Other studies have reported an avoidance of trees by certain grassland bird species. Based on available information, it is probable that

FIGURE 15.6 Wind turbines located in agricultural settings like this facility in Illinois may have fewer habitat-related impacts as these areas often are less important to many species of wildlife. (© Ed Arnett, Bat Conservation International, www.batcon.org)

some disturbance or displacement effects may occur to grassland/shrub-steppe avian species occupying a site developed as a wind energy facility. The spatial extent of these effects and their significance will vary by species and habitat conditions, and may range from zero to several hundred meters.

Habitat impacts could be avoided by careful placement of wind facilities. For example, wind energy development in disturbed areas such as tilled agricultural lands likely will have fewer impacts because these areas tend to be less important to most species of wildlife (Kiesecker et al. 2011). Habitat impacts also can be mitigated with proper siting, habitat restoration at or near facilities, through on- or off-site habitat mitigation, purchases, or conservation easements. Developers also should attempt to reduce habitat loss and fragmentation by using existing roads when possible, limiting construction of new roads, and restoring disturbed areas to minimize a facility's footprint (Arnett et al. 2007). Placing turbines on disturbed lands (Figure 15.6) also may benefit expansion of transmission lines and associated infrastructure needed to facilitate wind development because disturbed lands are already in areas of high road and transmission line density (Kiesecker et al. 2011). Strategies for mitigating habitat impacts associated with wind facilities must be evaluated to determine their efficacy, as there currently is little empirical evidence.

OFFSHORE WIND ENERGY DEVELOPMENT AND WILDLIFE

Offshore wind facilities have been established throughout Europe, but few studies have been conducted to determine direct impacts on animals. European studies have been summarized by Winkelman (1994), Exo et al. (2003), and Morrison (2006). These authors conclude that offshore wind turbines may affect birds as follows: (1) risk of collision; (2) short-term habitat loss during construction; (3) long-term habitat loss due to disturbance by turbines, including disturbances from boating activities in connection with maintenance; (4) formation of barriers on migration routes; and (5) disconnection of ecological units, such as between resting and feeding sites for aquatic birds. A major concern with offshore developments has been loss of habitat from avoidance of turbines and the impact that boat and helicopter traffic to and from the wind facility may cause with regard to the behavior and movements of seabirds, although little is known about such effects. Some studies in Europe suggest that the development of wind energy facilities could displace migrating and breeding shorebirds and waterfowl, or disrupt daily movements and migration (e.g., Drewitt and Langston 2008). Resident seabirds and rafting (resting) waterbirds appear to be less at risk than migrating birds, as they may adapt better to offshore wind-energy facilities (Winkelman 1994; Exo et al. 2003). More empirical evidence is needed to document these potential impacts of offshore wind energy facilities and their effects on populations.

The effects of offshore wind energy facilities on marine mammals, fish, and benthic organisms are currently unknown, but warrant study and clarification. The potential impact of offshore wind energy development on bats also is poorly understood, although observations in Europe (Ahlen et al. 2009) and anecdotal accounts of bats occurring offshore suggest impacts are possible.

It is important to note that the actual impact of the first several offshore wind energy facilities, including those proposed on inland waters such as the Great Lakes, proposed and built in North America be evaluated extensively both for fatalities and displacement effects—although difficulties assessing fatalities at offshore facilities are expected to be greater than at onshore sites. While there is reason to believe that areas with high concentrations of birds (or bats) would present more risk than areas with lower densities of birds, this relationship has not yet been established on land or offshore, and it will be important for the first several offshore facilities to conduct studies needed to fill numerous information gaps. In addition, a method for estimating fatalities at existing and planned wind facilities offshore will be required to understand impacts and develop mitigation strategies, as retrieving dead birds and bats from water bodies will be a considerable challenge (Arnett et al. 2007).

SOME KEY CHALLENGES FOR RESOLVING WILDLIFE AND WIND ENERGY CONFLICTS

I believe there are a number of key challenges we currently face if wildlife and wind energy development conflicts are to be resolved so that responsible development that minimizes wildlife impacts can be achieved. This is not an exhaustive list of issues or casual factors, but a sample of a few I have experienced and developed opinions on and thus deem important for consideration.

MORE CREDIBLE SCIENTIFIC INFORMATION IS NEEDED

White (2001) suggested that all too often, we do not conduct appropriate research when we should and when we do conduct research, we often do it poorly. White (2001) also pointed out that "too often in wildlife management, [we] are asked to resolve conflicts that are impossible because the basic biological knowledge to understand the issue is lacking." With any such conflict "all stakeholders are right, under the assumptions each brings to the issue, but because the biological knowledge is inadequate to refute any of the assumptions, the conflict cannot be resolved in an objective fashion based on the biology of the problem." The insightful wisdom of White (2001) reflects exactly the situation in which we find ourselves regarding the science underpinning the development of wind energy facilities and their effects on wildlife impacts and decision making.

Far too often while reading study proposals, reports, and even manuscripts on wind-wildlife issues, both past and present, I am dumbfounded by a painfully obvious lack of rigor in the science. More often than not, it would appear that monitoring efforts or so-called "studies" are based on what budget level has been made available as opposed to asking a meaningful question and designing the effort to answer that question adequately. Many of the past, and unfortunately some of the present, wildlife monitoring efforts at proposed and operating wind energy facilities represent what I call "check-box monitoring," where monitoring represents the minimum effort to meet requirements of a requesting agency. Sometimes these efforts may be adequate, but more often they are not. In some cases, agency biologists must shoulder some of the blame as well when they require companies to conduct either trivial, redundant, or otherwise meaningless efforts or do not request the rigor needed to answer the question of interest, often which is never even clearly articulated by either party. Frequently, I hear the phrase uttered that "a little data is better than no data" and in almost all circumstances I could not possibly disagree more with this philosophy. This attitude fuels the fire to minimize effort and transform otherwise meaningful studies into a "check-box mentality." Inadequate or faulty information more often than not will crumble the foundation of the process for integrating science, management, and decision making, and in the end, the courts will be left to pick up the pieces (White 2001). All too often, the collection of information on wildlife impacts at wind energy facilities is inadequate to answer questions such that it cannot be upheld to scientific or legal scrutiny. This is not appropriate and is unacceptable. However, equally unacceptable is requesting a company or its consultants to collect inadequate information or data that ultimately will never be used or does not contribute to increasing our knowledge. While the quantity and quality of studies have improved over the past few years, scientists, management and regulatory biologists, and developers and their consultants must work more closely in the future to ensure meaningful questions are being asked and study designs and data collection are adequate to answer the questions of interest. Consistent implementation of guidelines (e.g., PGC 2007) could greatly improve the situation (Arnett et al. 2007). Forthcoming voluntary guidelines from the U.S. Fish and Wildlife Service (www.fws.gov/windenergy) represent a new minimum standard for all wind energy projects, and implementation of these guidelines will improve knowledge for decision making. We must ask ourselves if we would rather collect the data adequately up front to answer a question and resolve the issue, or pay for litigation to resolve the issue without the knowledge, all the while realizing that decisions will be made without reliable knowledge (White 2001). In the end, these situations cost far more in money, time, and consternation from professionals, politicians, and the public (Thomas and Burchfield 2000; White 2001), and I believe the management and policies surrounding wind and wildlife issues are far better served by gathering rigorous science-based information on the front end.

There also have been very few studies of wind energy and wildlife published in peer-reviewed scientific journals (Arnett et al. 2007), although this trend is changing. The shortage of scientific publication on wind-wildlife interactions (GAO 2005; Kunz et al. 2007a; NRC 2007) must be overcome to place the problem and decision making on a base of solid science. Most reports on wind-wildlife relationships are in the "gray literature," appear on the Internet, possibly accompanied by archived paper copies buried in file cabinets. However, rarely are these reports peer-reviewed in a credible, transparent fashion, and many other reports are retained by wind energy companies as proprietary material not available to outside parties, including regulatory agencies. While the vast majority of wind energy-wildlife data are gathered by consultants under confidentiality, my criticism is not necessarily directed at consultants, but rather toward their clients and the processes they choose to implement. However, a legitimate fear that wind energy companies have is that agencies may use data shared voluntarily as a basis for law enforcement, under the Migratory Bird Treaty Act (MBTA), for example (J. Anderson, American Wind Energy Association, personal communication). Solutions to this situation would seem to lie in better communication among stakeholders, trust, appropriate incentives, compliance with applicable laws, implementation of standard guidelines, and prosecutorial discretion by agencies when companies operate to minimize wildlife impacts.

Moving forward, more independent research via academic institutions and other organizations is needed. I personally have no issue with consultants conducting scientific studies. In fact, many have done outstanding work to advance our knowledge. However, their clients must allow them to operate with some independence and transparency, including freedom to publish without judicial rights over findings with which they disagree. To do otherwise taints the perception, and perhaps the reality, of the effort whether the researcher is an academician or a consultant. Greater transparency, credible peer-review, information sharing, and publication in reputable venues are essential if the trust and acceptance of diverse stakeholders and the public is to be gained. Proactive companies have recognized the situation and are responding, but many do not yet fully appreciate the situation. Research partnerships among diverse players will be helpful for generating common goals and providing adequate funding to conduct studies (e.g., Arnett and Haufler 2003; also see the Bats and Wind Energy Cooperative [www.batsandwind.org] and American Wind and Wildlife Institute [www.awwi.org]). Such partnerships garner support from state

FIGURE 15.7 Partners of the Bats and Wind Energy Cooperative (www.batsandwind.org) gather to discuss research priorities and work plans. Cooperative research programs like this garner support from diverse stakeholders, leverage dollars and logistical support, and provide transparency that yields credibility. (© Merlin D. Tuttle, Bat Conservation International, www.batcon.org)

and federal agencies, industry, NGOs, and academia that can leverage dollars and logistical support needed, provide peer-review, dissemination processes and transparency that yield credibility (Figure 15.7).

POLICY AND REGULATORY ISSUES

In this section, my goal is to discuss a few selected issues regarding policy and regulation of wind facilities as they relate to wildlife impacts and successful integration of science, policy, and management to improve siting that minimizes risk to wildlife. This discussion is by no means exhaustive or comprehensive, and useful reviews of policy and regulation regarding wind energy development are provided by the GAO (2005), the NRC (2007), and others.

The federal government's role in regulating wind power development is limited to projects occurring on federal lands, impacting federal trust species, or projects that have some form of federal involvement (e.g., interconnect with a federal transmission line) or require federal permits. While the Federal Energy Regulatory Commission (FERC) regulates the interstate transmission of electricity, natural gas, and oil, it does not approve the physical construction of electric generation, transmission, or distribution facilities. At present, such approval is left for state and local governments (GAO 2005). The primary federal regulatory framework for protecting wildlife from impacts from wind power includes three laws—the MBTA, the Bald and

Golden Eagle Protection Act, and the Endangered Species Act (GAO 2005; NRC 2007). Additionally, each state may enforce its laws regarding wind energy and wildlife impacts, or establish cooperative efforts to address impacts. For example, the Pennsylvania Game Commission (PGC 2007) developed a cooperative agreement with wind energy companies that allows for prosecutorial discretion for the take of state-listed species by the PGC when cooperating companies comply fully with the conditions of the agreement.

Because wind energy development has primarily occurred on nonfederal land, regulating such facilities is largely the responsibility of state and local governments (GAO 2005). The primary permitting jurisdiction for wind energy facilities in many instances is a local planning commission, zoning board, city council, or county board of supervisors or commissioners. Typically these local jurisdictional entities regulate wind projects under zoning ordinances and building codes (GAO 2005), often without the basic knowledge needed to make informed decisions. In some states, one or more state agencies play a role in regulating wind energy development, such as natural resource and environmental protection agencies, public utility commissions, or siting boards. Furthermore, some states have environmental laws that impose requirements on many types of construction and development, including wind power, that state and local agencies must follow (GAO 2005). It is apparent that most local jurisdictional entities lack experience in wildlife science and unless they coordinate with their state wildlife or natural resource agency, concerns about wildlife issues may fall by the wayside and never enter the discussion during decision making. While most developers proactively coordinate with state wildlife agencies to address concerns, a gap remains between the decision-making authority and those with expertise and jurisdiction over wildlife issues. The potential for miscommunication or no communication at all seems obvious and while policy and regulations vary among states, it seems prudent to suggest that stronger coordination and perhaps even policy and regulation may be warranted in many states, especially where wildlife agencies have little or no authority in decision making. The goal should not be to increase process, "red tape," or extend timelines, but rather to ensure environmental concerns are fully articulated and addressed throughout the permitting process. However, neither agencies nor local jurisdictional entities may have the financial wherewithal or resources necessary to act as an effective regulatory body. Merely having environmental regulations in place to carry out their mission sometimes may make for untenable situations whereby wind energy developers are forced to comply with the law without the agency achieving conservation goals it desires or the ability to effectively manage permitting programs, which in turn can have negative impacts on business.

Lack of regulations can sometimes create unique challenges for responsible and scientifically credible development of wind energy projects. Let us hypothetically envision a landscape that has a reasonable amount of commercial wind energy potential and five different developers are "prospecting" in the area to secure landowner leases and begin the permitting process. Along the route to permitting, some very important wildlife concerns are brought to the attention of each developer independently through consultation with local, state, and federal wildlife agencies, as well as academic and non-government experts. After careful review with the relevant agencies and their company officials, all five developers decide that this landscape is too environmentally sensitive for wind development and they all abandon their respective projects. Our landscape, which has now been unofficially deemed a high-risk site for development, is now safe from future development of a wind facility, right? No, it is not. There is absolutely no regulatory statute in any state of which I am aware that protects an area because a conscientious, proactive company (in our example, five of them) walked away due to environmental concerns. Another wind energy developer can proceed with prospecting and, within the existing laws, permit, build, and operate a facility right where our other five developers walked away, and in many instances with little or no consequence or requirements to assess or monitor impacts, depending on the state. For that matter, any other industry or land-use developer could do the same unless the land is put under conservation easement, for example, as landowners have the right to realize the highest and best economic use of their lands within the limits of the law. This obviously has nothing to do with the biology or science of the matter, but could have enormous implications for wind energy (and other land-use development) and wildlife impacts. Situations like these are not well documented, and most never reach public dissemination, but they happen and threaten wildlife and all stakeholders, including the industry. Leveling the playing field for all industry players is a challenge, as it may require more regulation than most industries wish to endure. However, I deem it critically important to establish that level playing field through reasonable regulation and compliance; otherwise, the lowest common denominator or industry behavior minimizes the incentive to abandon high-risk sites and rewards bad behavior.

However, regulations can also create unique challenges. Let's say, for example, we have two adjacent sites being developed by different companies that occur in occupied habitat by a federally listed threatened or endangered species. One company is proactive and decides to conduct extensive surveys for the listed species on its proposed site and it does so even though it is not legally required, but believes it is the right thing to do. Meanwhile, a neighboring developer chooses not to conduct such surveys and proceeds with development. Upon finding the listed species occupying its site, the proactive company reports to the U.S. Fish and Wildlife Service, as it should, but soon finds that the surveys and reporting now seriously jeopardize its project. This company may need to complete an expensive and time-consuming habitat conservation plan (HCP) and may never get to complete this project. While this is an appropriate course of action and possible outcome, our less conscientious company continues to develop unhindered without ever having spent a dime on surveys or conducted any consultation with the agencies. Its only risk at this point is likely to be if it were to actually take a listed species and then, even less likely, be caught doing so. With no consistent requirements in place regarding surveys for listed species, our proactive company has been penalized for good behavior and our less conscientious company is rewarded for bad behavior. This situation may be changing, with newly developing regional HCPs for whooping cranes (*Grus americana*; B. Tuggle, U.S. Fish and Wildlife Service, personal communication) and Indiana bats (*Myotis sodalis*; S. Pruitt, U.S. Fish and Wildlife Service, personal communication) that would create similar standards for developers. However, there is clearly a need for further policy discussions and decisions to level the playing field among wind companies. Creating incentives for responsible development and rewarding companies that adhere to guidelines, mitigate impacts, implement voluntary efforts, and support research will be critically important components for resolving wind energy and wildlife impacts in the future.

Another key issue is consistent application of regulations. In my discussions with wildlife professionals and industry developers, this appears to be an overarching problem with many federal agencies with multiple regions. A state agency colleague conveyed to me that four different U.S. Fish and Wildlife Service offices had different interpretations and applications of the Endangered Species Act given the same issue brought forth to each of them. This creates untenable situations, considerable uncertainty, consternation, and lack of trust among stakeholders that seems unnecessary.

DECISION-MAKING PROCESSES (OR LACK THEREOF)

A typical scenario for a wind developer after it has decided to pursue development of a site is to begin the process of addressing a wide range of factors potentially influencing the project (e.g., archeological sites, water quality issues, viewshed impacts, wildlife impacts). At some point, the developer will negotiate what is expected of it to address all of these types of issues and a pre-construction and post-construction plan will be agreed upon and, hopefully, implemented. Let's say bats are an issue, and the state wildlife agency and U.S. Fish and Wildlife Service request that pre-construction acoustic monitoring with detectors (see Kunz et al. 2007b) be conducted for at least one year. My question to the agencies is rather simple—what will the data being requested be used for and how will it influence decision making? Is there a decision-making framework upon which said data would inform and improve the process? While this seems fundamental, I would argue that in most cases, such decision-making frameworks are non-existent at present and rarely is there even a plan to synthesize the data in some meaningful way (although see efforts under way by the American Wind and Wildlife Institute; www.awwi.org). Part of the problem stems from companies, confidentiality with their consultants, and a general unwillingness to share information after it has been collected. However, scientists and biologists must share in some of the blame because it is our charge to develop these frameworks and demonstrate credibly how data that is demanded of the industry will actually be used. This task is not as easy as it may sound. If the project occurs in the range of a threatened or endangered species, such as the Indiana bat (Figure 15.8) and the aforementioned acoustic monitoring determines the species is present, there is a reasonable clear set of options for the developer, much of which may be dictated by the Endangered Species Act. The developer can choose to work with the agencies in pursuit of an HCP or abandon the project altogether, as two examples of how the data might be used. However, what if the acoustic monitoring does not indicate presence of a threatened or endangered species, but rather generates an average nightly call rate of 8.2 calls per night for the site? Our hypothetical agency also has received the same type of data collection from another proposed facility that found an average of 82.0 calls per night. What do these data tell us beyond the fact

FIGURE 15.8 Recent fatalities of the federally endangered Indiana bat (*Myotis sodalis*) will likely dictate future policy and regulatory decisions for wind energy development within the range of this species. (© Merlin D. Tuttle, Bat Conservation International, www.batcon.org)

that it seems plausible that one site has ten times more calls on average than the other? We currently do not have enough reliable data and analyses to yet know if pre-construction call data can predict risk, but let's say we do and the higher call rate indicates that high bat fatalities can be expected. Does this now trigger some action by the company or the agency? Perhaps it should, but I would submit that we simply do not yet have the conceptual framework for how new information will be used to improve decisions in most cases. Often, we see the words "adaptive management" smattered throughout the literature, in wind energy proposals, requests from agencies, guidelines, and in monitoring plans. I am not convinced that we currently have the capability to adaptively manage wind facilities, and we must strive to improve decision-making frameworks and develop new ones, for use of biological information, as this is fundamental and critically needed yet mostly absent. This huge gap must be filled in the coming years if we are to improve siting and permitting decisions to protect wildlife and develop wind energy responsibly.

COMMUNICATION AND INFORMATION EXCHANGE

Wildlife professionals have made great strides in communication with diverse audiences in recent decades, but I believe we still must improve our communications with the key audiences that really make a difference regarding wind energy and wildlife issues. Private landowners, the public, business/industry, and decision makers are groups that can make a real difference on the ground, in the political arena, and with decision making.

Landowners probably have more power and control over development on their private properties than they may realize. I contend that educating landowners about wildlife impacts, research needs, and mitigation measures is critical—knowledge is power, and by educating landowners, they can be empowered to work more closely with developers, demand the highest standards, and make the best decisions about their land when siting and operating a wind facility. As an example, the Texas Wildlife Association (http://www.texas-wildlife.org/) worked with the Texas Parks and Wildlife, NGOs, and others to conduct a series of wind energy and wildlife conferences targeting landowners and local decision makers to inform them of wildlife issues and wind energy development. Building and maintaining relationships with landowners and communicating the importance of conservation efforts needed to develop renewable energy responsibly must continue and expand in the future.

The public deserves to be better informed on the issues surrounding wind energy and wildlife issues, yet I cannot help but feel that wildlife professionals continually lose the public relations battle with sound bites on "green energy," "climate change," and "dependency on foreign oil." While these topics are important and are real aspects of energy development, wildlife impacts are equally important, yet remain poorly understood by the general public. Wildlife professionals must continue to communicate these issues so an informed public can appreciate discussions and decisions based on the full range of issues and trade-offs that include both the need for energy and minimization of environmental impacts.

Although wildlife professionals have excelled in educating the public and working with private landowners in recent years, in my opinion most lag behind in understanding and communicating with business, developers, and other industry officials and policy makers. I believe wildlife professionals should gain a basic understanding of all aspects of the wind industry (e.g., how it is financed, electricity markets, and policy matters) to better understand and communicate with developers and business professionals. Wildlife professionals must communicate the importance of wildlife, demand sound science, and require adequate mitigation measures where they are needed, but they need to articulate these in "business language." We must communicate our needs and desires in financial terms and demonstrate risks and rewards for doing "the right thing" as opposed to not doing so. We must emphasize the economic benefits realized from birding, hunting, and associated activities as part of these discussions.

I am of the opinion that the institutions financing wind projects may dictate the direction of how wildlife impacts are addressed long before new regulations are developed and widely implemented. Projects that are proposed in risky sites for wildlife are also risky for investors looking for quick returns on their investment rather than watching potential profits go toward years of litigation or large mitigation costs. Thus, educating the financial community and communicating wildlife issues, adherence to guidelines, and the risks of financing projects that do not comply will increase the odds of compliance with data collection standards, guidelines, mitigation measures and, I hope, minimize or eliminate projects that pose risks to wildlife. Another important audience influencing ecological siting decisions for wind facilities is the retail electricity suppliers (i.e., utility companies) that purchase wind-generated power. Like lenders, leaders in this industry are guided by inherent and consumer-driven pressures to be good environmental stewards; thus, many are reluctant to purchase electricity from wind energy facilities that pose notable wildlife detriments (R. Manes, The Nature Conservancy, personal communication). A foundational basis for the development and operation of wind energy facilities in most instances (except when it is operated as a merchant facility selling its power on the spot market) is a power purchase agreement (PPA) with an "off-taker." Without this agreement, most developers will not move forward with the construction and operation of a facility. It is incumbent, then, on wildlife professionals and advocates to ensure that utility companies understand the known and credibly suspected wildlife risks associated with wind energy development and work with developers (and eventually owner/operators) on effective mitigation measures without penalizing the operator for violating the terms of the PPA if unforeseen impacts necessitate the need for operational changes during certain times of the day and year. This will allow off-takers to condition requests for proposal and power purchase agreements to ensure proper siting and mitigation for wildlife and habitats. However, off-takers must be willing to work with developers/operators to allow for environmental-based operational changes the same way they do for other forms of curtailment and downtime for maintenance. It is unfair and difficult for operators to shoulder this burden on their own, and liability must be distributed in a more equitable fashion than what exists today. Similar outreach efforts also must be directed at companies that develop electrical transmission lines.

Government decision makers are perhaps the most challenging group with which to communicate, in part because many biologists and scientists shy away from the political process. This is due in part to unfamiliarity with the legislative and other political processes, concerns that their science will somehow be compromised, and the dissatisfaction that many Americans feel with the political system (Blockstein 2002). However, to not engage and communicate scientific evidence and concerns for wildlife may come with a high price, as we may not like the messenger or the message being communicated to decision makers if we fail to carry it ourselves. The broad goal, of course, is to ensure the full range of trade-offs for renewable energy development are analyzed and articulated clearly during planning and decision making, including development of legislation. I would be remiss, however, not to acknowledge the fact that agency biologists are simply not allowed to "lobby" per se on behalf of any legislation that might be beneficial to the fish and wildlife resources, and must be very careful on how they analyze legislation and respond when asked by citizens or legislative representatives.

CONCLUSIONS AND LOOKING TO THE FUTURE

Developing renewable energy sources is important for meeting future energy needs, but shifting energy production from fossil fuels to renewable energy that collects more energy that is diffuse from a broader spatial area will involve tradeoffs (Kiesecker et al. 2011). Wind power can contribute to renewable energy portfolios, but poorly planned developments will likely result in biologically significant cumulative impacts for some species of wildlife that must be avoided. Implementing consistent guidelines for siting, monitoring, and mitigation strategies among states and federal agencies would create greater certainty and equitable treatment from state-to-state and region-to-region, assist developers with compliance with relevant laws and regulations, and establish standards for conducting site-specific, scientifically sound biological evaluations (Arnett et al. 2007). Immediate unbiased research is needed to develop a solid scientific basis for decision making when siting wind facilities, evaluating their impacts on wildlife and their habitats, and testing efficacy of solutions.

Research priorities have been suggested for addressing wildlife impacts at wind energy facilities (e.g., Arnett et al. 2007; Kunz et al. 2007a; NRC 2007; Cryan and Barclay 2009). Wind energy developers should follow criteria and standards established within siting guidelines (e.g., PGC 2007; U.S. Fish and Wildlife Service [www.fws.gov/windenergy]) or regions that include avoidance of high-risk sites as determined by the best available information. Siting wind energy facilities in areas where habitat is of poor quality or already fragmented, for example, will likely result in fewer habitat-related impacts. The U.S. Department of Energy (2008) reported that an estimated 241 GW of land-based wind energy development on approximately 5 million hectares would be needed to reach 20% electricity production for the United States by 2030. However, Kiesecker et al. (2011) estimated that there are 3,500 GW of potential wind energy that could be developed across the United States on disturbed lands and concluded that their proposed disturbance-focused development strategy would avert development of 2.3 million hectares of undisturbed lands while generating the same amount of energy as development based solely on maximizing wind potential. Notwithstanding, many birds and bats may still fly over disturbed lands where wind energy facilities are sited and thus may still be at risk at such sites (Kiesecker et al. 2011).

Given projected increases in multiple sources of energy development, including biomass, wind, and oil and gas development, future conflicts surrounding land-use, mitigation, and conservation strategies should be anticipated. Habitat mitigation options, for example, when developing wind in open prairie may be compromised by development of other energy sources or urban planning. Regional assessments of existing and future land uses and planning regional conservation strategies among industries, agencies, and private landowners could reduce conflicts and increase options for mitigation (Arnett et al. 2007). There is an immediate need to insert wildlife impacts into the political dialogue so that all tradeoffs can be considered during decision making. Maintaining relationships with private landowners and communicating the importance of conservation efforts and their benefits will be critical toward responsibly developing renewable energy.

I began this chapter with a quote from President Obama's 2011 State of the Union Address that epitomizes the political support for wind energy development. Political and business perception and reality, as well as public demand, will drive renewable energy development forward, and I think all of us hope one day that renewables will become effective at slowing or reversing the effects of climate change. However, time may be running out, and the rush to develop renewable sources is on. Political and business timelines almost never match those of garnering reliable scientific knowledge upon which to make decisions. Aldo Leopold once said, "We shall never achieve perfect harmony with land, any more than we shall achieve absolute justice or liberty for people. In these higher aspirations, the important thing is not to achieve but to strive ...." I would submit that we will not likely achieve utopian science that renders perfect decisions to be made in resolving wind energy and wildlife conflicts, yet we should strive for a higher standard than we presently have. The old adage regarding one who "cannot see the forest for all the trees in front of them" offers a useful metaphor as we strive for high aspirations and attempt to balance energy demands, climate change, and human population growth—population growth, energy demands, and climate change, of course, representing the proverbial "forest" and wildlife impacts the "trees." Wildlife professionals, land managers, politicians, developers, and business officials are all challenged to think broadly about the forest, but they cannot forget about the trees along the way.

ACKNOWLEDGMENTS

My thoughts on wind and wildlife issues have been shaped and transformed by discussions with numerous colleagues, experiences while serving as the program coordinator for the Bats and Wind Energy Cooperative with Bat Conservation International, serving as technical editor for The Wildlife Society when developing the Society's technical review and position statement on wind energy and wildlife, and my service on the Department of Interior's Federal Advisory Committee to develop recommendations for wind energy and wildlife guidelines for the U.S. Fish and Wildlife Service. This manuscript was greatly enhanced by reviews from J. Anderson, K. Boydston, L. Brennan, C. Hein, N. Fascione, T. Kunz, J. Lindsay, R. Manes, and J. Sands.

REFERENCES

Ahlen, I., H. J. Baagoe, and L. Bach. 2009. Behavior of Scandinavian bats during migration and foraging at sea. *Journal of Mammalogy* 90:1318–1323.

Arnett, E. B., K. Brown, W. P. Erickson, J. Fiedler, T. H. Henry, G. D. Johnson, J. Kerns, R. R. Kolford, C. P. Nicholson, T. O'Connell, M. Piorkowski, and R. Tankersley, Jr. 2008. Patterns of fatality of bats at wind energy facilities in North America. *Journal of Wildlife Management* 72:61–78.

Arnett, E. B., and J. B. Haufler. 2003. A customer-based framework for funding priority research on bats and their habitats. *Wildlife Society Bulletin* 31:98–103.

Arnett, E. B., M. M. P. Huso, M. R. Schirmacher, and J. P. Hayes. 2011. Changing wind turbine cut-in speed reduces bat fatalities at wind facilities. *Frontiers in Ecology and the Environment* 9(4):209–214; doi:10.1890/100103 (published online November 1, 2010).

Arnett, E. B., D. B. Inkley, R. P. Larkin, S. Manes, A. M. Manville, J. R. Mason, M. L. Morrison, M. D. Strickland, and R. Thresher. 2007. Impacts of wind energy facilities on wildlife and wildlife habitat. *Wildlife Society Technical Review* 07-2. Bethesda, MD: The Wildlife Society.

Baerwald, E. F., J. Edworthy, M. Holder, and R. M. R. Barclay. 2009. A large-scale mitigation experiment to reduce bat fatalities at wind energy facilities. *Journal of Wildlife Management* 73:1077–1081.

Barclay R. M. R., and L. M. Harder. 2003. Life histories of bats: life in the slow lane. In: *Bat Ecology*, T. H. Kunz and M. B. Fenton, Eds. Chicago, IL: University of Chicago Press, pp. 209–253.

Bernstein, M. A., J. Griffin, and R. Lempert. 2006. Impacts on U.S. energy expenditures of increasing renewable energy use. Technical report prepared for the Energy Future Coalition. RAND Corporation, Santa Monica, CA.

Blockstein, D. E. 2002. How to lose your political virginity while keeping your scientific credibility. *BioScience* 52:91–96.

Bureau of Land Management (BLM). 2005. Final programmatic environmental impact statement on wind energy development on BLM administered land in the western United States. U.S. Department of the Interior Bureau of Land Management, Washington, DC.

Cryan, P. M. 2003. Seasonal distribution of migratory tree bats (*Lasiurus* and *Lasionycteris*) in North America. *Journal of Mammalogy* 84:579–593.

Cryan, P. M., and R. M. R. Barclay. 2009. Causes of bat fatalities at wind turbines: hypotheses and predictions. *Journal of Mammalogy* 90:1330–1340.

Doherty K. E, D. E. Naugle, B. L. Walker, and J. M. Graham. 2008. Greater sage-grouse winter habitat selection and energy development. *Journal of Wildlife Management* 72:187–195.

Drewitt, A. L., and R. H. W. Langston. 2008. Collision effects of wind-generators and other obstacles on birds. *Annals of the New York Academy of Science* 1134:233–266. doi: 10.1196/annals.1439.015

Dürr, T., and L. Bach. 2004. Bat deaths and wind turbines—a review of current knowledge, and of the information available in the database for Germany. *Bremer Beiträge für Naturkunde und Naturschutz* 7:253–264.

Energy Information Administration (EIA). 2011. Annual energy outlook 2011 with projections to 2035. U.S. Department of Energy, Energy Information Administration, Washington, DC. http://www.eia.gov/forecasts/aeo Accessed August 15, 2011.

Erickson, W. P., J. Jeffrey, K. Kronner, and K. Bay. 2004. Stateline wind project wildlife monitoring final report: July 2001–December 2003. Western EcoSystems Technology, Inc., Cheyenne, WY, and Northwest Wildlife Consultants, Inc., Pendleton, OR.

Erickson, W. P., G. D. Johnson, M. D. Stickland, D. P. Young Jr., K. J. Sernka, and R. E. Good. 2001. Avian collisions with wind turbines: a summary of existing studies and comparisons to other sources of avian collision mortality in the United States. National Wind Coordinating Committee, Washington, DC.

Exo, K. M., O. Hüppop, and S. Garthe. 2003. Birds and offshore wind facilities: a hot topic in marine ecology. *Wader Study Group Bulletin* 100:50–53.

Fleming, T. H., and P. Eby. 2003. Ecology of bat migration. In: *Bat Ecology*, T. H. Kunz and M. B. Fenton, Eds. Chicago, IL: University of Chicago Press, pp. 156–208.

Fuhlendorf, S. D., A. J. W. Woodward, D. M. Leslie, Jr., and J. S. Shackford. 2002. Multi-scale effects of habitat loss and fragmentation on lesser prairie-chicken populations. *Landscape Ecology* 17:601–615.

Gehring, J., P. Kerlinger, and A. M. Manville II. 2009. Communication towers, lights, and birds: successful methods of reducing frequency of avian collisions. *Ecological Applications* 19:505–514.

Giesen, K. M. 1998. Lesser prairie-chicken (*Tympanuchus pallidicinctus*). In: *The Birds of North America*, Number 364, A. Poole, and F. Gill, Eds. Philadelphia, PA: The Birds of North America, Inc.

Government Accountability Office (GAO). 2005. Wind power: impacts on wildlife and government responsibilities for regulating development and protecting wildlife. U.S. Government Accountability Office, Washington, DC, Report to Congressional Requesters, GAO-05-906.

Horn, J., E. B. Arnett, and T. H. Kunz. 2008. Behavioral responses of bats to operating wind turbines. *Journal of Wildlife Management* 72:123–132.

Howell, J. A. 1997. Bird mortality at rotor swept equivalents, Altamont Pass and Montezuma Hills, California. *Transactions of the Western Section of The Wildlife Society* 33:24–29.

Inkley, D. B., M. G. Anderson, A. R. Blaustein, V. R. Burkett, B. Felzer, B. Griffith, J. Price, and T. L. Root. 2004. Global climate change and wildlife in North America. *Wildlife Society Technical Review* 04-1. Bethesda, MD: The Wildlife Society.

Johnson, D. H. 2001. Habitat fragmentation effects on birds in grasslands and wetlands: a critique of our knowledge. *Great Plains Research* 11:211–231.

Johnson, G. D. 2005. A review of bat mortality at wind-energy developments in the United States. *Bat Research News* 46:45–49.

Kiesecker, J. M., J. S. Evans, J. Fargione, K. Doherty, K. R. Foresman, T. H. Kunz, D. Naugle, N. P. Nibbelink, and N. D. Niemuth. 2011. Win-win for wind and wildlife: a vision to facilitate sustainable development. *PLoS ONE* 6(4): e17566. doi:10.1371/journal.pone.0017566

Kunz, T. H., E. B. Arnett, W. P. Erickson, A. R. Hoar, G. D. Johnson, R. P. Larkin, M. D. Strickland, R. W. Thresher, and M. D. Tuttle. 2007a. Ecological impacts of wind energy development on bats: questions, research needs, and hypotheses. *Frontiers in Ecology and the Environment* 5:315–324.

Kunz, T. H., E. B. Arnett, B. M. Cooper, W. P. Erickson, R. P. Larkin, T. Mabee, M. L. Morrison, M. D. Strickland, and J. M. Szewczak. 2007b. Methods and metrics for studying impacts of wind energy development on nocturnal birds and bats. *Journal of Wildlife Management* 71:2449–2486.

Kuvlesky, W. P. Jr., L. A. Brennan, M. L. Morrison, K. K. Boydston, B. M. Ballard, and F. C. Bryant. 2007. Wind energy development and wildlife conservation: challenges and opportunities. *Journal of Wildlife Management* 71:2487–2498.

Leddy, K. L., K. F. Higgins, and D. E. Naugle. 1999. Effects of wind turbines on upland nesting birds in Conservation Reserve Program grasslands. *Wilson Bulletin* 111:100–104.

McLeish, T. 2002. Wind power. *Natural New England* 11:60–65.

Morrison, M. L. 2006. Bird movements and behaviors in the Gulf Coast region: relation to potential wind-energy developments. NREL/SR-500-39572. Golden, CO.

National Research Council (NRC). 2007. *Ecological Impacts of Wind-Energy Projects*. Washington DC: National Academies Press.

Naugle, D. E., Ed. 2011. *Energy Development and Wildlife Conservation in Western North America*. Washington DC: Island Press.

Orloff, S., and A. Flannery. 1992. Wind turbine effects on avian activity, habitat use and mortality in Altamont Pass and Solano County Wind Resource Areas. Report to the Planning Departments of Alameda, Contra Costa and Solano Counties and the California Energy Commission, Grant No. 990-89-003 to BioSystems Analysis, Inc., Tiburton, CA.

Pennsylvania Game Commission (PGC). 2007. Pennsylvania Game Commission wind energy voluntary cooperation agreement. http://www.pgc.state.pa.us/pgc/lib/pgc/programs/voluntary_agreement.pdf

Pitman, J. C., C. A. Hagen, R. J. Robel, T. M. Loughin, and R. D. Applegate. 2005. Location and success of lesser prairie-chicken nests in relation to vegetation and human disturbance. *Journal of Wildlife Management* 69:1259–1269.

Racey, P. A., and A. C. Entwistle. 2003. Conservation ecology of bats. In: *Bat Ecology*, T. H. Kunz and M. B. Fenton, Eds. Chicago, IL: University of Chicago Press, pp. 680–743.

Robel, R. J., J. A. Harrington, Jr., C. H. Hagen, J. C. Pittman, and R. R. Reker. 2004. Effect of energy development and human activity on the use of sand sagebrush habitat by lesser prairie-chickens in southwestern Kansas. *Transactions of the 69th North American Wildlife and Natural Resources Conference* 69:251–266.

Rydell, J., L. Bach, M. Dubourg-Savage, M. Green, L. Rodrigues, and A. Hedenstrom. 2010. Bat mortality at wind turbines in northwestern Europe. *Acta Chiropterologica* 12:261–274.

Smallwood, K. S., and B. Karas. 2009. Avian and bat fatality rates at old-generation and repowered wind turbines in California. *Journal of Wildlife Management* 73:1062–1071.

Smallwood, K. S., and C. Thelander. 2004. Developing methods to reduce bird mortality in the Altamont Pass Wind Resource Area. Final Report to the California Energy Commission, Public Interest Energy Research–Environmental Area, Contract No. 500-01-019, Sacramento, CA.

Smallwood, K. S., and C. Thelander. 2008. Bird mortality in the Altamont Pass Wind Resource Area in California. *Journal of Wildlife Management* 72:215–223.

Smallwood, K. S., C. G. Thelander, M. L. Morrison, and L. M. Rugge. 2007. Burrowing owl mortality in the Altamont Pass Wind Resource Area. *Journal of Wildlife Management* 71:1513–1524.

Strickland, M. D., E. B. Arnett, W. P. Erickson, D. H. Johnson, G. D. Johnson, M. L., Morrison, J. A. Shaffer, and W. Warren-Hicks. 2011. Comprehensive Guide to Studying Wind Energy/Wildlife Interactions. Prepared for the National Wind Coordinating Collaborative, Washington, DC.

Thomas, J. W., and J. Burchfield. 2000. Science, politics, and land management. *Rangelands* 22: 45–48.

U.S. Department of Energy. 2008. 20% Wind Energy by 2030: Increasing Wind Energy's Contribution to U.S. Electricity Supply. DOE/GO-102008-2567. U.S. Department of Energy, Office of Scientific and Technical Information, Oak Ridge, TN.

U.S. Department of Energy. 2011. Fossil fuels. http://www.energy.gov/energysources/fossilfuels.htm Accessed February 1, 2011.

White, G. C. 2001. Why take calculus? Rigor in wildlife management. *Wildlife Society Bulletin* 29:380–386.

The Wildlife Society. 2007. Impacts of wind energy facilities on wildlife and wildlife habitat. *Wildlife Society Technical Review* 07-2. Bethesda, MD: The Wildlife Society.

Winkelman, J. E. 1994. Bird/wind turbine investigations in Europe. *Proceedings of the National Avian-Wind Power Planning Meeting*, Lakewood, Colorado, July 20–21, 1994, pp. 43–47. Proceedings prepared by LGL Ltd., environmental research associates, King City, Ontario.

16 The Role of Joint Ventures in Bridging the Gap between Research and Management

James J. Giocomo
Oaks and Prairies Joint Venture
American Bird Conservancy
Temple, Texas

Mary Gustafson
Rio Grande Joint Venture
American Bird Conservancy
Mission, Texas

Jennifer N. Duberstein
Sonoran Joint Venture
U.S. Fish and Wildlife Service
Tucson, Arizona

Chad Boyd
Eastern Oregon Agricultural Research Station
U.S. Department of Agriculture
Burns, Oregon

CONTENTS

> Progress toward global conservation, if we accept the bargain, will pick up or falter depending on cooperation among three secular stanchions of civilized existence: government, the private sector, and science and technology.
>
> **—E. O. Wilson (2002)**

ABSTRACT

No single entity can effectively address conservation planning and actions for migratory bird species that move across continents annually to fulfill their life cycle needs. Successful landscape-level conservation requires cooperation and coordination of efforts among individual conservation entities. U.S. bird habitat joint ventures (JVs) are highly successful partnerships of public agencies, private organizations, corporations, and individual landowners that work cooperatively to meet shared goals. JVs identify and address strategic habitat conservation needs for priority bird populations through biological planning, conservation design, research, monitoring, communication, education, and outreach (CEO) that maximize the effectiveness of conservation delivery activities of the individual member agencies/organizations of the partnership. JVs have a greater impact than individual partners working independently. The highly successful model for JV bird conservation partnerships has been successfully copied for other taxa and issues, including Regional Alliances in Mexico, the Monarch Joint Venture, Fish Habitat Partnerships, and Landscape Conservation Cooperatives.

INTRODUCTION

During the early 1980s, waterfowl biologists and managers in central North America were alarmed by steeply declining populations of early-nesting waterfowl species such as Mallards (*Anas platyrhynchos*) and Northern Pintails (*Anas acuta*), which were at their lowest population levels since annual surveys began in 1955. Intensifying agricultural land use combined with a five-year drought in key breeding areas reduced wetland habitat availability and reproductive success of waterfowl. During 1985, waterfowl experts from Canada and the United States came together to design an international strategy for conservation of North American waterfowl with the aim of "maintaining an adequate habitat base to ensure the perpetuation of waterfowl populations." The resulting North American Waterfowl Management Plan was signed in 1986 by the Canadian Minister of the Environment and the U.S. Secretary of the Interior, and in 1994 Mexico became a signatory to the plan (USFWS 1986, 1994).

These waterfowl experts were representatives from concerned agencies and conservation organizations. They recognized that landscape-scale conservation would require long-term planning, close cooperation, and coordination of management activities. A key principle was that no single entity could effectively address complex landscape-level conservation issues to sustain migratory bird populations and their supporting habitats in perpetuity. "Joint ventures of private and governmental organizations [would] be considered as an approach to financing high priority research and management projects of international concern that can only be addressed through a pooling of resources" (USFWS 1986). Starting with five original waterfowl habitat-based joint ventures (JVs) and three species-based JVs, today there are eighteen bird habitat JVs covering most of the United States, four in Canada, and three species JVs (Figure 16.1). Our discussions throughout this chapter are centered on U.S. bird habitat JVs.

JVs work with many state and federal agencies to leverage public conservation programs with private corporation and conservation organization funding. Collectively, JVs have invested over $4.5 billion to protect, restore, and enhance more than 6.35 million hectares of wetland habitat (USFWS 2010). This investment was possible in part with the enactment of the North American Wetlands Conservation Act, a competitive grant program that provides federal funds to leverage state and private matching funds for wetland conservation projects at a minimum 1:1 ratio. More recently, the Neotropical Migratory Bird Conservation Act (NMBCA) was enacted to leverage federal funds with private conservation funding to promote the long-term conservation of Neotropical migratory birds and their habitats throughout the hemisphere. Between 2002 and 2010, the NMBCA provided over $175 million (matched 3:1) in support of 333 projects, with participation from forty-eight U.S. states and thirty-six countries. These funds benefited approximately 0.8 million hectares of bird habitat (USFWS 2010). JVs also work with the Natural Resources Conservation Service in the U.S. Department of Agriculture to help target U.S. Farm Bill conservation programs in high priority areas and species associated with agricultural lands (i.e., grassland and shrubland birds).

ORGANIZATIONAL STRUCTURE

JVs are regional, self-directed partnerships of government and non-governmental agencies and organizations as well as corporations and individual landowners. JV partners work across administrative boundaries to deliver landscape-level planning and science-based conservation, which results in linking on-the-ground management with national bird population goals. The administrative planning boundaries of many JVs are organized by Bird Conservation Regions (Figure 16.1). Bird Conservation Regions encompass landscapes with similar bird communities, habitats, and resource issues. All JVs are set up to cross traditional jurisdictional boundaries and encourage cooperation among agencies and organizations that traditionally focus their work within administrative, geographic, or political boundaries (e.g., state wildlife agencies or USFWS regional offices).

JVs focus on a broad spectrum of bird conservation activities including biological planning; "on-the-ground" conservation; communications, education, and outreach; initiating research and monitoring projects; creating decision support tools; and raising funds for these activities through partner contributions and grants. JVs

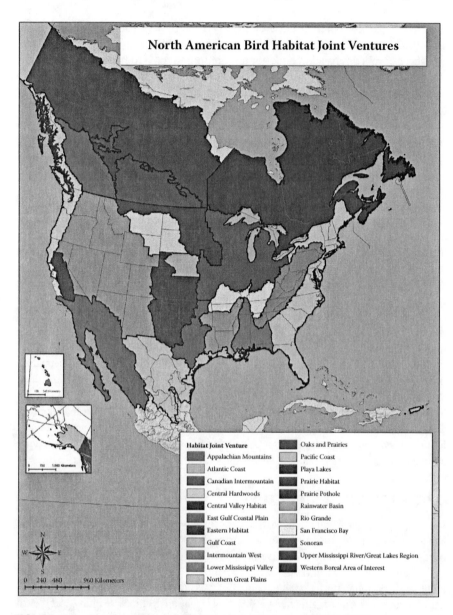

FIGURE 16.1   North American Bird Habitat Joint Ventures (USFWS 2010).

work to implement national and international bird conservation plans (i.e., water-fowl [North American Waterfowl Management Plan Committee 2004], landbird [Rich et al. 2004], waterbird [Kushlan et al. 2002], and shorebird [Brown et al. 2001]) by "stepping down" the population goals of these larger plans to regional or landscape habitat goals, while feeding local information (known as "rolling up") to the national and international planning groups. This process bridges the gap between continental-level planning and local-level actions by bringing national- and international-level priorities and resources to address local-level conservation issues. At the same time, the process ensures that local-level information is incorporated into national and international policy making. JVs document their proposed plans of action in an implementation plan, which provides guidance for the partnership on needed conservation actions and outcomes to further bird conservation in the region. These implementation plans are updated periodically to incorporate the best available knowledge and reflect the ever-changing nature of the interactions between human land use and the environment.

Although partnerships are common in wildlife conservation, one of the key distinctions in JV partnerships is that partners are invested in JV activities and share risks and costs as well as rewards. JV partners are expected to bring resources (personnel and funding) to the table to help identify and attain shared goals. These shared goals are broad enough to reach across the missions of the individual partners and are based upon managing bird populations and habitats at the landscape level. The shared goals are also broad enough so that they are not dependent on a single partner or program. The JV partnership model creates a shared motivation to continue to pursue shared goals even if one partner's circumstances change. The result of this process is continuity of conservation regardless of political or administrative changes. The partnership brings together the different abilities of the various partners to pursue common goals that some may be unwilling or unable to do alone. For example, one partner may lobby government officials for support, another may provide legal or tax expertise, while a third may provide scientific, technical, fundraising, or education and outreach abilities (Graziano 1993).

U.S. JVs receive funding from the U.S. Fish and Wildlife Service (USFWS), whose mission includes the conservation of birds and bird habitat. Funding for new JVs may initially provide funds for a coordinator and eventually funds may be available for additional staff. Some of the long-established JVs have annual budgets exceeding $1 million.

Although bird habitat JVs are designed to implement landscape-level science-based conservation of bird populations, a major principle is that local or regional priorities are informed by national objectives. With local control, each JV evolves as needed within its region, but national and international bird population objectives help to drive activities.

JV success was formerly evaluated by the USFWS using dollars and acres as the measures of progress—the total amount spent to put habitat in the desired condition and the number of acres affected. For the JV to be effective, these metrics have to link back to measurable changes in bird populations or vital rates (survival or productivity) for priority species. Recently, new metrics have been introduced to evaluate the success of JVs, including the number of USFWS Birds of Management Concern (USFWS 2008b) with habitat needs identified. For a given priority bird species, the habitat type, condition, and amount needed to support a population at a desired level (the population goal) has been determined. This gives the JV measurable goals to work toward as a part of its planning process. JV activities focus on several main themes, which have been outlined in a self-evaluation matrix used by each JV to assess its individual progress and next goals. These themes include organizational performance, biological planning, conservation design, habitat delivery, monitoring, research, communication, education, and outreach (CEO). Depending upon the conditions and needs within the geographic area of each JV, different JVs will invest resources in these themes in very different ways.

ORGANIZATIONAL PERFORMANCE

Organizational performance refers to the activities that make the partnership work from a personnel and funding management perspective. This includes coordinating the partners, creating a management board, organizing the technical community, and managing funding. At a minimum, each JV has a coordinator, a management board, and technical teams or advisors. JVs may have other staff to provide expertise in different areas including science; CEO; geographic information systems (GIS); modeling, etc.

The JV coordinator manages the day-to-day activities of the JV partnership as well as any staff who are assigned to the initiative. The coordinator works with the management board to determine priority staffing needs that address missing capabilities within the JV partnership. The JV coordinator provides leadership, coordinates operation of the partnership, and is accountable to the management board. The coordinator is responsible for providing organizational leadership by furthering the JV mission, vision, and implementation plan and provides programmatic, organizational, and financial management guidance, and maintaining communication among the partners as well as oversight of any additional JV staff. JV staff (i.e., coordinator and others) may work for or be funded by any of the partners.

The coordinator and additional staff, such as science coordinators, communication, education and outreach coordinators, and geospatial staff, work with the scientific, technical, and land management community to form technical teams of staff members from state and federal agencies, nonprofit conservation organizations, and universities to identify and address bird conservation science, management, and research needs.

JVs are overseen by management boards, which include representatives from the JVs' main partner organizations. Usually, management board members are top administrators within their agency/organization and have the authority to direct its resources to address the goals of the JV. The management board and the JV coordinator develop the vision for the future and establish and implement strategies to achieve that vision. The management board's responsibilities include:

1. Directing the activities of the partnership, including staff and JV technical teams.
2. Formulating strategies to further the JV's mission and periodically reviewing and updating the mission as necessary.
3. Providing oversight of organizational and programmatic planning and evaluation.
4. Ensuring legal and ethical integrity and maintaining accountability for the JV.
5. Promoting the activities of the JV and enhancing the JV's visibility among partner entities and the broader conservation community.

It is expected that management board members will:

1. Maintain commitments of time, focus, and financial support necessary to achieve the JV mission.
2. Consistently attend and engage fully in management board meetings, conference calls, and ad hoc working groups as needed.
3. Direct technical staff from their agency/organization to fully participate on JV technical teams and contribute to the development of technical documents.
4. Possess authority to represent their agency/organization in decision making on the JV management board.
5. Serve as active partners in the JV's planning and implementation activities.
6. Act as JV ambassadors to other public, private, and political leaders.
7. Be alert to opportunities and threats likely to be encountered by the JV.
8. Become familiar with JV finances and financial (or resource) needs.
9. Understand the policies and procedures of the JV.

JV technical teams serve as the forum for coordination and communication among JV partners in matters pertaining to implementing adaptive conservation (Figure 16.2; biological planning, conservation design, habitat delivery, and monitoring and research; also known as Strategic Habitat Conservation [USFWS 2008a]). Technical teams advise the management board and JV staff and ensure that the conservation actions by partners help to contribute to the overall goals of the JV. Most researchers interact with the JV by participating in technical teams that focus on developing models and decision support tools that link habitat to populations in ways that help predict the impacts of potential land management strategies. Investigators can then use hypothesis-driven research to identify and test the major assumptions used to create these models and outcome-based monitoring to evaluate the effectiveness of management strategies. The results of the research and monitoring before, during, and after implementation of management activities are then fed back into the next round of planning to improve future management and incorporate changes in the environment in a continuous cycle.

JV partners participating in the technical teams support the ongoing development and modification of JV conservation goals and objectives, and work to provide the biological science foundation for JV activities by using existing research to evaluate current management strategies. Responsibilities of the JV technical teams include the following:

1. Develop, refine, and integrate JV priority species and habitat objectives that contribute to range-wide bird conservation plan population objectives for all priority species (waterbird, shorebird, waterfowl, landbird, etc.).
2. Integrate state wildlife action plans into the JV planning process where applicable.
3. Implement an adaptive approach for bird conservation that includes habitat monitoring to evaluate impacts of JV partner conservation actions.
4. Design targeted and statistically rigorous monitoring programs to help test planning assumptions.

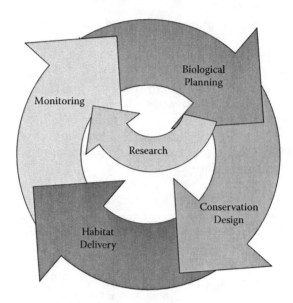

FIGURE 16.2   Joint ventures are generally organized around an adaptive management cycle, also known as Strategic Habitat Conservation (USFWS 2008a). Each JV may address any part of the cycle depending upon its own regional needs, and an adaptive conservation cycle can start at any point. Ideally, adaptive conservation is based on a plan that identifies what needs to be done (biological planning) and where the identified actions would have the highest probability of success (conservation design). Where uncertainties exist in the planning process, research needs are addressed and incorporated into future planning. For example, for new habitat projects, the project is a result of needs identified in the biological planning process. Through conservation design, the project is located on the land in the best available location. The Habitat Delivery is the on-the-ground conservation actions that most conservation organizations and agencies already are doing, but the adaptive conservation planning helps to focus individual actions to achieve larger landscape-level effects. Monitoring the project's success and the resulting impacts on the population or desired parameter (survival, productivity) allows for refinement of the process through research to ensure greater success in the future.

5. Provide technical support for biological planning including the development of population-habitat models and decision support tools, and the identification of basic research needs where insufficient information exists to build initial models.
6. Identify conservation actions and targeted research to test assumptions built into the JV biological science foundation.
7. Coordinate the implementation of research projects.
8. Provide technical support for conservation design by developing GIS tools and maps to identify strategic, biologically based locations for conservation actions.
9. Provide CEO to the public on the technical and scientific issues.
10. Organize ad hoc or standing sub-committees or working groups as necessary (e.g., focus areas, monitoring, etc.).

BIOLOGICAL PLANNING

Biological planning builds the conceptual biological science foundation that supports bird conservation needs. It is based on regional, national, and international conservation initiatives while feeding local information to inform the conservation planning at larger scales. JVs use the biological planning process to answer questions including, "Which species are a priority?", "How many birds can the region support?", and "How much habitat is needed?" Biological planning includes identifying and justifying priority bird species and suites of species and establishing abundance and demographic population objectives for those species to account for environmental or seasonal variability. Limiting factors for declining priority species are identified and population-habitat models are designed to link population objectives and habitat objectives. Assumptions built into these models then become targets for further research by JV partners.

CONSERVATION DESIGN

Conservation design refers to the process of assessing the existing landscape for conservation potential and providing spatially explicit targets for conservation activities. The conservation design process is used to answer questions such as, "How much habitat exists and how is the amount changing?", "How much more habitat is needed to meet objectives?", and "Where should conservation actions be focused?" The technical teams work with JV staff to create spatially implicit or explicit decision support tools to evaluate specific management actions to overcome identified limiting factors. Where possible, existing GIS data layers are used to create GIS-based decision support tools and maps to target conservation actions, and additional GIS layers are developed as needs are identified.

HABITAT DELIVERY

Habitat delivery refers to the major activities of JV partners, including state and federal agencies and many of the non-governmental organizations (NGOs) to protect, restore, and enhance bird habitat. JVs use the planning and decision support tools created through biological planning and conservation design to focus partner activities. Most JV partners already do a great deal of habitat delivery that is often effective at the local scale. Working with a landscape-level plan shifts conservation from a "shot-gun" pattern of unconnected projects, usually in areas where conservation efforts are politically or logistically convenient, to an integrated strategic approach that maximizes limited resources and improves conservation impacts by targeting areas that will have the largest positive impact for populations of priority species.

MONITORING AND RESEARCH

Monitoring allows JVs to measure bird population response to management activities and to evaluate "net" changes in habitat over time at the landscape scale. Monitoring the response of priority bird populations to direct management activities provides a local-scale check for the effectiveness of habitat conservation actions. Landscape-scale monitoring of habitat changes in conjunction with species-habitat relationship information allows JVs to estimate the total population of priority bird species within the JV. Ultimately, monitoring allows JV partners to adjust their conservation activities to provide the most benefit and helps the JV set and adjust bird and habitat conservation priorities and objectives. Finally, monitoring helps to justify continuing effective conservation actions or abandoning conservation actions that do not significantly increase priority bird populations.

Research is used as a tool to evaluate competing conservation strategies by testing the assumptions of biological planning and identifying factors that might minimize the uncertainty in these assumptions. The most basic assumptions of biological planning are that priority species are limited by available habitat, and that changing the breeding, migration, or wintering habitat structure and composition will promote increases in vital rates of these species. These assumptions are then evaluated, to the extent possible, through existing and future habitat manipulation projects, with local-scale experiments distributed to represent the landscape-scale.

Ideally, population-habitat models serve as a tool for generating questions to be answered with hypothesis-driven research. Often research is conducted without clear plans for linking what is learned to what is done. In order to address this shortcoming, whenever possible, JV conservation actions are conducted within an adaptive conservation framework formalized by the USFWS and the U.S. Geological Survey as Strategic Habitat Conservation (National Ecological Assessment Team 2006) that allows for statistical evaluation of conservation prescriptions (Figure 16.2). The basic idea is to implement management in replicated "blocks" and simultaneously monitor bird populations and habitat responses on untreated areas (control sites). The use of replicate treatments allows estimation of experimental error (i.e., variability in response to treatment), which then allows statistical comparisons either between competing treatments or treated versus non-treated areas.

COMBINING RESEARCH AND MANAGEMENT IN THE JOINT VENTURE SYSTEM

The adaptive conservation framework effectively combines and streamlines research and management into a single process, which allows research results to improve future decisions about conservation actions (i.e., adaptive management). Adaptive management provides a format for experimental learning based on statistical comparisons between competing alternatives, can be used to test assumptions and hypotheses associated with biological planning, and is carried out in an iterative fashion that allows for continual refinement of management approach. Although monitoring programs can be put in place at any time, monitoring protocols should be designed before the management action can be of maximum utility and generally should include pre- as well as post-treatment data. Overall, the adaptive management process is a key component of effectively linking research outcomes with future management actions.

Ultimately, research results in a biologically based foundation that forms the basis for decision support tools that link conservation actions with predicted species responses. Research can also be used to derive estimates of the amounts and conditions of habitat needed to provide sufficient habitat area and quality for reaching population objectives for priority species. For research to increase our understanding of the effects of our conservation actions, planning and funding of research activities must be considered as important as the on-the-ground habitat conservation and management.

COMMUNICATION, EDUCATION, AND OUTREACH

Human behavior underlies most of the threats facing bird populations throughout the world. JVs must account for the complex social and economic factors that influence ecological landscapes at various scales (Bogart, Duberstein, and Slobe 2009). As discussed earlier in this chapter, biological planning, conservation design, and monitoring and research all provide a strong scientific foundation for achieving conservation. In order to implement conservation goals and objectives on the ground, however, JVs must work with other agencies, NGOs, private landowners, homeowners, businesses, legislators, universities, land managers, and a wide range of other stakeholders. CEO is a critical tool for doing this.

Strategic CEO involves clearly identifying audiences that are critical for conservation success, determining desired behavioral outcomes, and developing the best tools to deliver a message that results in altered human behavior. JVs use strategic CEO to identify relevant stakeholders and provide them with the tools and information needed to ultimately achieve on-the-ground bird and habitat conservation. From the development of best management practices manuals and landowner guides to websites and listserves to videos and newsletters to training workshops and conferences to press releases and articles, there are many ways in which strategic CEO allows JVs to cultivate engaged, active partnerships, garner support, and deliver results. Researchers, for example, design stakeholder surveys to help land managers understand social or political barriers to conservation actions that may have significant biological support, but are not implemented due to misunderstandings of the possible actions (e.g., using prescribed fire for management in some areas).

SUMMARY

Every JV has different locally determined priorities and goals. Collectively, the JV model has had a huge impact on bird and habitat conservation in the New World. The JV model provides a bridge between research and management. The U.S. habitat JV model of a coordinator funded by USFWS (and employed by USFWS or an NGO) working for a management board of partner state and federal government agencies, conservation organizations, corporations, tribes, and individuals with land managers well represented working with researchers on the technical committees to determine the conservation needs in their region, and the partners implementing integrated and strategic conservation actions effectively marries management and research into strategic conservation.

Since 1986, JVs have invested more than $4.7 billion to protect, restore, and enhance habitat for birds on more than 7 million hectares. JVs continue to serve as a model for landscape-level conservation, including the emergence of the U.S. Department of the Interior's Landscape Conservation Cooperative Program to address conservation needs of all priority species—including birds.

JVs have become an incredibly successful and important strategy for science-based conservation of birds and bird habitat in a relatively short time (<30 years). Their success is partially rooted in a strategy that allows decision makers and land managers to work with researchers collaboratively and employ tactics that might otherwise not be possible if these people continued to work in isolation. The maintenance of this strong linkage between research and management is, and will continue to be, a major component of the continued success of JVs.

ACKNOWLEDGMENTS

Comments from several reviewers including Greg Esslinger greatly improved the manuscript.

REFERENCES

Bogart, R. E., J. N. Duberstein, and D. F. Slobe. 2009. Strategic communications and its critical role in bird habitat conservation: understanding the social-ecological landscape. In: *Tundra to Tropics: Connecting Birds, Habitats, and People*, T. D. Rich, C. Arizmendi, D. Demarest, and C. Thompson, Eds. Proceedings of the 4th International Partners in Flight Conference, February 13–16, 2008, McAllen, TX, pp. 441–452.

Brown, S., C. Hickey, B. Harrington, and R. Gill, Eds. 2001. *The U.S. Shorebird Conservation Plan*, 2nd ed. Manomet, MA: Manomet Center for Conservation Sciences.

Graziano, A. 1993. Preserving wildlife habitat: the U.S. Fish and Wildlife Service and the North American Waterfowl Management Plan. In: *Land Conservation Through Public/ Private Partnerships*, E. Endicott, Ed. Washington, DC: Island Press, pp. 85–103.

Kushlan, J. A., M. J. Steinkamp, K. C. Parsons, J. Capp, M. A. Cruz, M. Coulter, I. Davidson, L. Dickson, N. Edelson, R. Elliot, R. M. Erwin, S. Hatch, S. Kress, R. Milko, S. Miller, K. Mills, R. Paul, R. Phillips, J. E. Saliva, B. Sydeman, J. Trapp, J. Wheeler, and K. Wohl. 2002. *Waterbird Conservation for the Americas: The North American Waterbird Conservation Plan, Version 1*. Washington, DC: Waterbird Conservation for the Americas.

National Ecological Assessment Team. 2006. Strategic Habitat Conservation. U.S. Fish & Wildlife Service and the U.S. Geological Survey. http://www.fws.gov/science/doc/ SHC_FinalRpt.pdf (accessed January 2011).

North American Waterfowl Management Plan, Plan Committee. 2004. North American Waterfowl Management Plan 2004. Implementation Framework: Strengthening the Biological Foundation. Canadian Wildlife Service, U.S. Fish and Wildlife Service, Secretaria de Medio Ambiente y Recursos Naturales.

Rich, T. D., C. J. Beardmore, H. Berlanga, P. J. Blancher, M. S. W. Bradstreet, G. S. Butcher, D. W. Demarest, E. H. Dunn, W. C. Hunter, E. E. Inigo-Elias, J. A. Kennedy, A. M. Martell, A. O. Panjabi, D. N. Pashley, K V. Rosenberg, C. M. Rustay, J. S. Wendt, and T. C. Will. 2004. *Partners in Flight North American Landbird Conservation Plan*. Ithaca, NY: Cornell Lab of Ornithology.

USFWS (United States Fish & Wildlife Service). 1986. *North American Waterfowl Management Plan*. http://www.fws.gov/birdhabitat/NAWMP/files/NAWMP.pdf (accessed January 2011).

USFWS (United States Fish & Wildlife Service). 1994. *1994 Update to the North American Waterfowl Management Plan. Expanding the Commitment.* http://www.fws.gov/birdhabitat/NAWMP/files/NAWMP1994.pdf (accessed January 2011).

USFWS (United States Fish & Wildlife Service). 2008a. *Strategic Habitat Conservation Handbook: A Guide to Implementing the Technical Elements of Strategic Habitat Conservation (Version 1.0).* http://www.fws.gov/Science/doc/SHCTechnicalHandbook.pdf (accessed January 2011).

USFWS (United States Fish & Wildlife Service). 2008b. *Birds of Conservation Concern 2008.* Arlington, VA: U.S. Fish & Wildlife Service, Division of Migratory Bird Management. http://library.fws.gov/bird_publications/bcc2008.pdf (accessed January 2011).

USFWS (United States Fish & Wildlife Service). 2010. Division of Bird Habitat Conservation. http://www.fws.gov/birdhabitat/jointventures/index.shtm (accessed January 2011).

Wilson, E. O. 2002. *The Future of Life.* New York: Knopf.

17 Developing Management Strategies from Research
Pushmataha Forest Habitat Research Area, Oklahoma

Ronald E. Masters
Tall Timbers Research Station
Tallahassee, Florida
and
Oklahoma State University
Stillwater, Oklahoma

Jack Waymire
Oklahoma Department of Wildlife Conservation
Clayton, Oklahoma

CONTENTS

Land management is an art that builds on history and is based in science.

—Herbert L. Stoddard, Sr. (Crofton 2001:40)

ABSTRACT

Wildlife research typically generates descriptive knowledge about a species' biology, demographics, habitat, effects of changing land use, human dimensions, and other related aspects. Problem-oriented research is employed less often, particularly deductive management-oriented research that tests ecological assumptions. Ecological assumptions may be tested by combining the research tool of hypothesis testing, through appropriate experimental design using the practical experience of wildlife managers to propose relevant land management treatments. Managers and researchers have been reticent, however, to blend the two, leading to a perceived disconnect between research and management. This chapter provides a case study of a successful collaborative research project between management and research that has been ongoing since 1983.

The Pushmataha Wildlife Management Area (PWMA) in southeast Oklahoma was established in 1946 as a deer (*Odocoileus virginianus*) refuge. Harvest regulation and food plots were the primary management tools employed until the late 1970s. From 1969 to 1972, seventy-one elk (*Cervus elaphus*) were released on the area. Shortly thereafter distinct browse lines became apparent across PWMA and populations of both species declined precipitously. In 1982, the 30-ha Forest Habitat Research Area was established to examine various traditional and non-traditional methods of forest management to enhance forage production for deer and elk using different combinations of thinning and various prescribed fire regimes. The completely randomized and replicated experimental treatments employed were developed by both managers and researchers in a problem-oriented fashion. Both managers and researchers also wanted to understand plant succession trajectories, associated animal use, habitat quality issues, and ecosystem dynamics related to fire. Knowledge gained from this continuing long-term study has been used for ecosystem restoration projects and the development of different management strategies for deer, elk, and other species. We have learned much about the role of fire and plant succession trajectories following disturbance and animal community succession associated with plant community change. The ecosystem management practices developed through collaboration of researchers and managers on the small plot study area has been scaled up and applied on a landscape level to the 7,789-ha PWMA with benefits to nongame and game species alike. Additionally, it has impacted land management on public and private lands totaling over 1.83 million hectares.

INTRODUCTION

Wildlife research typically deals with a species' biology, population dynamics, habitat, or the related effects of current land use. Such studies give us descriptive knowledge based on inductive or scientific methods (Romesburg 1981). Although quantitative analyses are evident in wildlife science, actual testing of hypotheses is sorely lacking (Macnab 1983; Romesburg 1981; Hobbs and Hilborn 2006). In fact, many large mammal studies often generate a large mass of data and information with limited pragmatic application (Caughley 1980) and may be perceived by managers as irrelevant for addressing specific management issues.

While an argument can be made that a knowledge base is necessary (Gill 1985), wildlife professionals can benefit from designed research that tests ecological assumptions (Macnab 1983; Romesburg 1981) because managers often mimic ecological processes to achieve desired objectives. Sinclair (1979) stated, "Indeed, the principle of experimental management should be a basic concern in our understanding of ecosystem processes. Yet in most national parks and reserves, it is rare that management practices are designed so that biologists can compare differences with control areas and follow-up long term consequences. As a result, most wildlife management in North America, for example, still rests on untested myths." A partial solution to this disconnect is to use a problem-oriented approach similar to that used by medical researchers (Romesburg 1981). It is here wildlife managers can play a vital role in answering ecological questions through deductive management-oriented research (Macnab 1983). Managers and researchers have been reticent, however, to blend the two, leading to a perceived disconnect between research and management.

This management-oriented research approach is applicable to habitat manipulation practices commonly used, albeit in an inductive manner, by wildlife managers (Macnab 1983). Wildlife managers manipulate plant communities and plant succession to increase carrying capacity for a given species by providing essential life requisites (Yoakum et al. 1980), especially those that limit a given population. However, population responses to habitat manipulation are often difficult to adequately quantify (Ripley 1980).

Long-term studies in low productivity oak-pine (*Quercus* spp.–*Pinus* spp.) habitats have been suggested as an excellent place for an investigation on population response to habitat manipulation and other commonly employed management practices because population responses are presumably easier to quantify on such sites (Ripley 1980). For such a study to be successful, one must possess knowledge of how management techniques will affect carrying capacity or population status (Macnab 1985). Implicit assumptions include understanding ecological relationships of the system and how manipulation will impact those relationships (Macnab 1983). Many ecological assumptions may be tested by combining the research tool of hypothesis testing, through appropriate experimental design, and the practical experience of wildlife managers to devise a suitable array of potential management strategies to test. Realistically, managers often employ a cookbook approach to management with heavy dependence on conventional wisdom; a classic example is providing supplemental forage openings (food plots, feed patches) in the southeast United States. Researchers have tended to focus efforts on theoretical or descriptive research versus designing experiments with an applied perspective that incorporates manipulative experiments to explore ecological cause and effect relationships.

Much of the forested landbase in southeastern Oklahoma is in oak-pine or oak-hickory (*Quercus* spp.–*Carya* spp.) forest types (Hines and Bertelson 1987). The oak-pine forest is the largest forest type in the eastern United States (Lotan et al. 1978). Shortleaf pine (*P. echinata*) is a major constituent and has the widest geographic range of the southern pines (Lawson and Kitchens 1983). In spite of its prevalence and importance, a research void in the management of the oak-shortleaf pine type was evident at the beginning of this study (Komarek 1981a; Murphy and Farrar 1985).

The Ozark and Ouachita Mountain regions, similar to other interior highlands in the eastern United States, lack an adequate understory forage base and a suitable evergreen winter browse for white-tailed deer (*Odocoileus virginianus*) and Rocky Mountain elk (*Cervus elaphus*) compared to other physiographic regions of the United States (Segelquist and Pennington 1968; Segelquist et al. 1969, 1972; Masters et al. 1993b, 1997). Winter mortality of deer, decreased productivity, and summer fawn mortality have been related to oak mast failures (Segelquist et al. 1969, 1972; Logan 1972). The typical approach to this problem has been to establish supplemental forage openings (food plots) and honeysuckle (*Lonicera japonica*) or other plantings (Segelquist and Rogers 1974).

Cultivated forage openings have been criticized widely by wildlife managers (Larson 1967; Masters et al. 1997; Edwards et al. 2004). Managers' criticisms include cost inefficiencies, concentration of wildlife and therefore increased chances for disease transmission, inadequate testing as a management practice, limited duration of availability, and the perception that cultivated openings generally provided little benefit (Larson 1967; Masters et al. 1997). In addition, they are artificial and often rely on non-native and sometime invasive plantings (D. Elmore, personal communication).

However, the value of providing supplemental forage has been shown in forests of limited productivity (Segelquist 1974; Thompson et al. 1991). During late fall and winter, deer selected appreciable amounts of other foods (browse and herbage) only when hard mast was unavailable (Segelquist and Green 1968; Harlow et al. 1975). The greatest use of supplemental forage clearings occurred when mast was scarce. In deer enclosures with forage clearings, mortality was less in years of mast failure than in control areas (Segelquist and Rogers 1974). Although oak mast was substantially higher in nutritional quality than available forage, adequate nutrition could be provided without hard mast only in unique situations (Harlow et al. 1975). Goodrum et al. (1971) found that total mast failure never occurred during their twenty-year study, but their results may not be applicable to poor quality oak-pine habitats in mountainous terrain. Mast shortfalls have been observed in the Ozark Plateau region of Arkansas (Segelquist 1974) and in the Ouachita Highlands (Masters, unpublished data).

THE PRESENT STUDY

The 7,789-ha Pushmataha Wildlife Management Area (PWMA) lies in strongly dissected terrain with considerable topographic relief along the western edge of the Ouachita Highland Province in southeastern Oklahoma (Figure 17.1). Oak-pine, oak-hickory, and cross timbers vegetation types are interspersed and juxtaposed in close proximity along the western edge of the Ouachita Mountains. Areas of relict tallgrass prairie existed in small, interspersed pockets (Duck and Fletcher 1943) within 35 km of PWMA. Crandall and Tyrl (2006) have further described vegetation of PWMA.

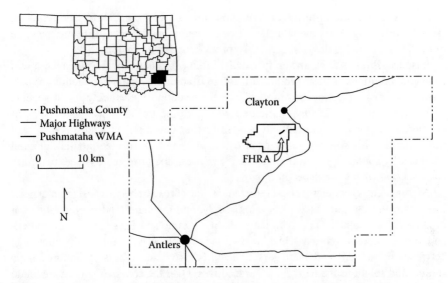

FIGURE 17.1 Location of the Pushmataha Wildlife Management Area and the Forest Habitat Research Area (FHRA), Pushmataha County, Oklahoma.

Before acquisition from 1946 to 1954, the Management Area was grazed, timber selectively harvested, and frequently burned. PWMA was initially established as a deer refuge in the 1940s. From 1969 to 1972, seventy-one elk were released in the area (Masters et al. 1993b, 1997). A total of nine elk (three males, six females) from a captive herd were released in March 2002.

Food plot development and maintenance, and regulation of harvest were the primary tools for management of deer on PWMA prior to the 1970s. Development of a distinct browse line was noted in the mid-1960s, despite the use of 221 ha of scattered food plots (0.2 to 1.6 ha in size) as a management tool on the area. Following dramatic declines of deer and elk, Oklahoma Department of Wildlife Conservation (ODWC) managers began to explore other ways to increase forage production and to improve habitat conditions.

At the time, much habitat research had been conducted on the effects of clearcutting across the Southeast United States and showed dramatic increase in forage production. Habitat quality of managed pine stands for white-tailed deer had been evaluated in the Ouachita Mountains (Segelquist and Pennington 1968; Fuller 1976; Reeb and Silker 1979; Fenwood et al. 1984). A five-year study on PWMA from 1965 to 1970 demonstrated that with overstory removal via aerial application of herbicide and application of prescribed fire, forage production was enhanced greatly (T. Silker, deceased, Oklahoma State University; Stillwater, unpublished data). Silker's study along with the body of research on deer habitat quality showed that aggressive overstory reduction could play a role in improving forage quality and quantity.

The area manager postulated that early succession forest openings (ESOs) and open woodlands could be developed through timber harvest and maintained with periodic prescribed fire as a strategy to improve the forage base for deer and elk. However, because of the importance of hard mast in fall and winter a proportion of the best oak trees, in terms of size and crown development should be retained. By fully releasing the crowns of these trees from competition by thinning, perhaps crowns would further develop and acorn production would increase.

Based on Silker's work and early forest thinning conducted by the manager, it was clear that additional forage produced through dramatic overstory reduction (>50%) was lost in a short period of time (~six to eight years) to advancing forest succession. Development of ESOs maintained with periodic fire as a deer and elk management strategy was untested and needed further development. Questions existed as to whether this strategy was cost effective or beneficial for deer and elk or even ecologically sound. The required frequency of prescribed fire necessary to maintain early successional stages was unknown.

At this point, the area manager enlisted the aid of researchers to design an experiment with the ultimate goal to develop, apply, and evaluate landscape application of appropriate management treatments on target game species such as deer and elk and later on breeding birds. Thus began a process of periodic interactive meetings between researchers and managers to identify problems versus symptoms, their source, and solutions from different perspectives. Over the course of a year, we made numerous field visits to previously thinned stands and Silker's study sites. We determined that a small plot study was needed to evaluate the wildlife management strategy of ESO development through timber harvest and periodic prescribed fire and to evaluate the effectiveness of periodic prescribed fire to manage plant succession on oak-pine sites. Once results were clear, then landscape application and evaluation would be the next step. Further application to similar oak-pine types was the ultimate end-point.

FOREST HABITAT RESEARCH AREA—TREATMENTS

The Forest Habitat Research Area (FHRA) was established in 1983 on a site similar to those where ESOs would be developed. Study area soils belong to the Carnasaw-Pirum-Clebit association with areas of rock outcrop. They were thin and drought prone, with a stony, fine, and sandy loam texture and a high proportion of surface rock (Bain and Watterson 1979). The FHRA was situated near a ridge top approximately 335 m in elevation, on a southeastern aspect slope. Soils and nutrient status were further described in Masters et al. (1993a).

The initial consensus was to examine fire intervals of two and four years on thinned areas to determine their efficacy to maintain ESOs. After consulting with a statistician and the Pittman-Robertson Federal Aid coordinator, additional treatments were suggested. We therefore incorporated a thinned but unburned control, an untreated control, and a series of annual burned treatments to determine how much fire hardwoods could tolerate. We added clearcut and hazard reduction burn (four-year interval fire) treatments because they had been studied extensively across the southeast and therefore would provide a basis of comparison between our study sites and others.

We applied ten treatments to thirty-six, 1.2- to 1.6-ha units in a completely randomized experimental design, beginning in summer 1984 (Table 17.1). Initially nine treatments were monitored on twenty-three units (Masters 1991b) and later ten

TABLE 17.1

Treatment Acronyms, Descriptions, and Number of Replications (N) for Monitored Treatments, 1983–Present, on the Pushmataha Forest Habitat Research Area, Pushmataha Wildlife Management Area, Oklahoma

Treatment	Description	n
Control	Control, no thinning, no burning	3
RRB	Rough reduction, late winter prescribed burn, 4-year interval	3
HNTI	Harvest pine timber only, late winter prescribed burn, 1-year interval; oak savanna, except 1995	3
HT	Harvest pine timber, thin hardwoods, no burn (natural regeneration to a mixed stand)	3
HT4	Harvest pine timber, thin hardwoods, late winter prescribed burn, 4-year interval	3
HT3	Harvest pine timber, thin hardwoods, late winter prescribed burn, 3-year interval	2
HT2	Harvest pine timber, thin hardwoods, late winter prescribed burn, 2-year interval, except 1995	3
HT1	Harvest pine timber, thin hardwoods, late winter prescribed burn, 1-year interval, except 1995	3
CCSP	Clearcut, summer burn-1985, contour rip 1986, planted to loblolly pine (*P. taeda*) 1986, winter burn 1998, thin summer 2001, January burn 2002, March burn 2007, 2009	3
PBS	Thin hardwoods, late winter prescribed burn, 1-year interval beginning in 1985-present, except 1995; pine-bluestem	1

treatments were monitored on twenty-seven of the treated units until the present because of time constraints and problems associated with burning on some replications. Merchantable pine timber was harvested in scheduled treatments, and hardwoods selectively thinned to approximately 9-m^2/ha basal area (BA) by single-stem injection using 2,4-D. Prescribed strip-head fires were applied on appropriate units in winter 1985 and in succeeding years at one-, two-, three-, and four-year intervals. Fireline intensity of prescribed burns has ranged from 33 to 3,100 kW/m (Masters and Engle 1994; Masters and Waymire 2010).

The clearcut site preparation treatment included windrowing of logging debris with a site preparation burn conducted during summer 1985. After contour ripping, genetically improved loblolly pine (*P. taeda*) seedlings were planted on a 2.1-m × 2.4-m spacing in early April 1986. In 2001, a post–salvage thinning was completed on these units for reducing average basal areas from about 26.4 m^2/ha to 20.7 m^2/ha to promote growth and salvage damaged crop trees from an ice storm in December 2000, approximately 20% of crop trees (Masters 2001).

Peripheral supplemental forage openings (1.2 to 4 ha) were included in 1988 to compare use of a traditional wildlife management technique with those under development. Inclusion of this additional treatment is valid in a completely randomized experimental design (Steel and Torrie 1980:126, 139). The food plot treatments were cultivated, fertilized, and planted to a mixture of fescue, rye, vetch, and Korean lespedeza; plots were mowed each fall and disked and fertilized periodically (~two- to three-year intervals) (N = 3).

Prior to treatment application, closed canopy post oak (*Q. stellata*), shortleaf pine, mockernut hickory (*C. alba*), and to a lesser extent blackjack oak (*Q. marilandica*) dominated the undisturbed overstory. Common understory species included tree sparkleberry (*Vaccinium arboreum*), greenbriar (*Smilax* spp.), grape (*Vitis* spp.), and sparse ground cover of little bluestem (*Schizachyrium scoparium*), panicums (*Panicum* spp., *Dicanthelium* spp.), and sedges (*Carex* spp., *Scleria* spp.) (Masters 1991a; Masters et al. 1993b; Crandall 2003).

HYPOTHESES

We were specifically interested in vegetation response with special attention to plant species suitable as deer and elk forage. We wanted to determine appropriate combinations of thinning and fire regimes to optimize habitat conditions for deer, turkey (*Meleagris galopavo*), Northern bobwhite (*Colinus virginianus*), and elk. Based on work primarily by Lay (1956, 1957) and Vogl (1974), we hypothesized that recurrent frequent fire would slow or halt forest succession (Vogl 1974), thus maintaining early seral openings and forage production. The experimental development and testing of ESOs as a deer management strategy offered an opportunity to fill a number of research voids. Assuredly, wildlife managers would benefit from additional management strategies, but our understanding of forest succession dynamics, fire ecology, and effects of forest management practices could be extended within a research setting. In order to develop management strategies from an ecosystem perspective, we needed to understand forest succession dynamics related to fire ecology in the Ouachita Mountain region. The point being that applied and basic research can be one and the same.

Based upon literature review, vegetation response to timber harvest and fire was postulated for our study. Initially, the harvested and winter-burned treatments would be similar to low intensity site preparation of clearcuts (Stransky and Richardson 1977; Stransky and Halls 1978). The clearcut treatment would be comparable to high intensity site preparation treatments commonly used in industrial forest settings (Stransky 1976). Rough reduction burns should cause a response similar to rough reduction burns elsewhere in mixed oak-pine habitats (Lay 1956, 1967; Wood 1988).

We postulated that the successional progression described by Hebb (1971) should be characteristic of clearcuts on the study area. The progression was: (1) the bare, soil-exposed site, (2) profusion of forbs and grasses, (3) dominance by relatively few species, and (4) shading out of understory by the developing overstory. We would expect plant species diversity and richness to increase initially then decline with increasing overstory and canopy closure (Masters 1991a).

Tallgrasses and other prairie constituents once prevalent in eastern Oklahoma (Nuttall 1980) should increase with overstory removal and fire. Fire should slow succession on seral stages dominated by scattered trees on harvested areas and stem girdle small <15 cm hardwoods on thinned and unthinned sites (Kucera 1978; Niering 1978). Frequent late-winter (February-March) prescribed burns (one- to two-year intervals) should increase grass production ten to fifteen times in ESOs and may control hardwood coppice to a degree depending upon fire frequency (Ferguson 1961; Kucera 1978). This fire season approximated part of the lightning fire regime

(Masters et al. 1995) and was also used by managers in the region. We did not know if the frequent fire return intervals would entirely halt secondary succession on oak-pine sites. Late-winter prescribed burns of three- to four-year intervals should increase grass production and allow woody plants to invade such as blackberry (*Rubus* sp.) and sumac (*Rhus* spp.) (Bragg and Hulbert 1976; Kucera 1978). Again, we did not know how long ESOs would persist under less frequent fire return intervals before forage production declined.

Woody browse species were important to deer and elk, but late-winter burns may lead to dominance by fire tolerant grasses that may not be utilized by deer (Stransky and Harlow 1981) or elk. Browse should increase in nutrient content and palatability on harvested and burned areas (Lay 1957; Halls and Epps 1969). Mast production on harvested areas would be impacted, but at that point, we did not know if forage production and nutrient increases would offset the loss of mast or if the hypothesized increase in mast production from selected trees would offset overall loss from the reduction within treated stands. These predictions were established *a priori* (Masters 1991a).

With expansion of selected treatments to the landscape level, we expected deer and elk populations to stabilize and possibly increase. Further, based on work by Wilson et al. (1995), we expected woodland-grassland obligate songbirds, many were species of special concern, to increase and closed-canopy forest dwelling species to decline.

FOREST HABITAT RESEARCH AREA OUTCOMES

TREE RESPONSE—BASAL AREA AND CANOPY COVER

Total basal area was similar on all units in 1983 before treatments with an average basal area of 25.3 m^2/ha (SE = 0.5). As expected, significant differences were evident following harvest and thinning (Figure 17.2a). Following thinning, the overstory on ESOs was composed of sparse (2 to 9 m^2/ha BA) hardwood stems >5 cm at 1.4-m in height (Figure 17.2b), mostly post oak and blackjack oak (Masters 1991a,b; Masters et al. 1993a). Canopy cover was closely correlated ($r^2 = 0.903$) with total basal area, but varied somewhat year to year with rainfall distribution and total rainfall.

On controls, and to a lesser extent, the rough reduction burn (RRB) treatment, there has been increased mortality to blackjack oaks from hypoxylon canker (*Hypoxylon atropunctatum*), and ice storm damage to pine in the past decade (Figure 17.2a and Figure 17.3a) (Masters 2001). Ice storm damage and wind-throw also lowered tree density on the pine-bluestem (PBS) treatment. Disease-related mortality on unthinned units was thought to be precipitated by a combination of competition in densely stocked stands and late summer drought stress. On treatments with partial or complete overstory retention, there has been a slight decline in basal area over time commensurate with fire frequency (Figures 17.2a,b and Figures 17.3a,b). The post oak-savanna treatment (HNT1) declined in total basal area because of fire-induced mortality of mockernut hickory, black hickory, and to a lesser extent blackjack oak (Figure 17.3a).

The HT and HT4 ESOs have developed into immature mixed pine and oak stands with the HT4 lower in density (Figures 17.2a,b and Figures 17.3a,b). Crown closure

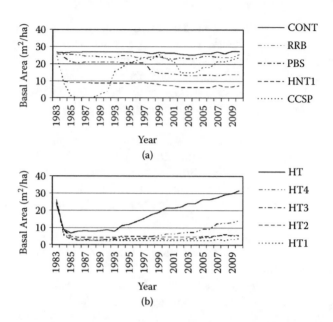

FIGURE 17.2   Change in basal area (m²/ha) prior to and following pine timber harvest, thinning of hardwoods and periodic fire on the Forest Habitat Research Area, Pushmataha Wildlife Management Area, Oklahoma from 1983 to 2009. (a) CONT = control, no treatment; RRB = rough reduction burn, 4-year burn interval; PBS = thin hardwood, 1-year burn interval; HNT1 = harvest pine, leave hardwood, 1-year burn interval; CCSP = clearcut, windrow, summer site preparation burn, plant loblolly pine. (b) HT = harvest pine, thin hardwood, no burn; HT4 = harvest pine, thin hardwood, 4-year burn interval; HT3 = harvest pine, thin hardwood, 3-year burn interval; HT2= harvest pine, thin hardwood, 2-year burn interval; and HT1 = harvest pine, thin hardwood, 1-year burn interval.

for the HT and CCSP occurred within six growing seasons post-cutting (Masters et al. 1993b). Although burn frequencies at three-, two-, and one-year intervals have held in check widespread hardwood and pine regeneration, the HT3 and HT2 units show signs of developing into savanna-like stands (Figure 17.2b and Figure 17.3b). It was clear that the four-year burn interval through twenty-five years could not maintain ESOs (Figure 17.2b and Figures 17.3a,b).

Eastern red cedar (*Juniperus virginianus*) development, a major invasive species in Oklahoma (Engle et al. 1996), has become evident on unburned control and HT (seed tree) treatments (Figures 17.2a,b and Figure 17.3a). Cedar only occurred as an understory constituent <1 m in height across all treatments at the beginning of the study but was eliminated in treatments incorporating fire. The only source of mortality for cedar in unburned treatments has been from antler rubbing by deer and elk.

MAST TREES

By 2009, total mortality among the selected 131 trees was 38 for 29.1%. Fire has caused mortality to 9.9% of all original trees. A total of twelve blackjack oak trees

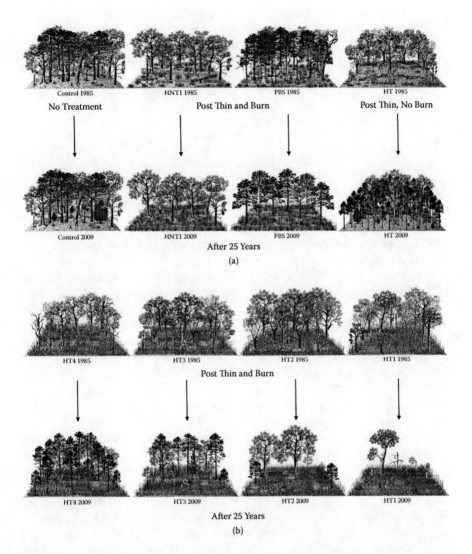

FIGURE 17.3 (a) Successional change in tree density and groundstory prior to and following pine timber harvest, thinning of hardwoods, and periodic fire on the Forest Habitat Research Area, Pushmataha Wildlife Management Area, Oklahoma, from 1983 to 2009. CONT = control; HNT1 = harvest pine, leave hardwood, 1-year burn interval; PBS = thin hardwood, 1-year burn interval; HT = harvest pine, thin hardwood, no burn. (b) Successional change in tree density and groundstory prior to and following pine timber harvest, thinning of hardwoods, and periodic fire on the Forest Habitat Research Area, Pushmataha Wildlife Management Area, Oklahoma, from 1983 to 2009. HT4 = harvest pine, thin hardwood, 4-year burn interval; HT3 = harvest pine, thin hardwood, 3-year burn interval; HT2 = harvest pine, thin hardwood, 2-year burn interval; and HT1 = harvest pine, thin hardwood, 1-year burn interval.

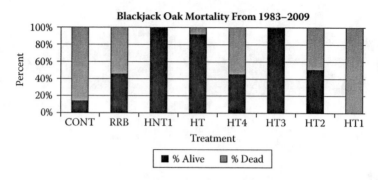

FIGURE 17.4 Percent mortality of selected blackjack oaks by treatment on Pushmataha Forest Habitat Research Area from 1983 to 2009, Pushmataha Wildlife Management Area, Oklahoma.

(9.2%) and one black oak have succumbed to hypoxylon canker all on unthinned treatments. Lightning struck three post oaks but has killed only two (1.5%).

Mortality of post and black jack oak was significant in the first and second year following introduction of fire on ESOs. This pulse of mortality was related to high amounts of residual logging slash, related fire behavior, and perhaps the shock of fire introduction after a long absence (Wade and Johansen 1986). Smaller trees were more susceptible (lower diameter, height, and crown area). Subsequent mortality for blackjack oaks has been almost entirely from hypoxylon canker in control units and RRB treatments. Unexpectedly control units experienced 87.5% mortality to blackjack oaks from hypoxylon canker (Figure 17.4). A higher proportion of total blackjack oak mortality was from disease rather than fire. Most likely, this was a result of competition and late summer drought stress (Conway and Olson 2004).

Post oak mortality was related to fire frequency but also influenced by lightning mortality (Figure 17.5). High blackjack oak mortality on HT4 and HT1 treatments was a result of the initial prescribed fires where flame lengths and fire intensity were greater and not a fire frequency effect. Wade and Johansen (1986) concluded that reintroduction of fire into stands long excluded from fire may cause delayed mortality from stem girdling because low intensity fires may smolder in the accumulation of sloughed bark and litter around the immediate bole of the tree. Tree mortality was related to bark thickness, moisture content of the tree (Martin 1963), diameter of the tree, fire intensity, and residence time (Wade and Johansen 1986). Large-stemmed oaks show a marked resistance to fire (Garren 1943; Kucera et al. 1963; Komarek 1981b; White 1986). The time for the cambium of trees to reach lethal temperatures increased with bark thickness (Hare 1965), and bark thickness increased with age (Davis 1953). Fire intensity determined largely the extent of bark char and crown damage (Cain 1984) and was primarily a function of fire type (backfire versus headfire) (Fahnestock and Hare 1964) and season of burn (Waldrop and Van Lear 1984). The height of crown scorch was a geometric function of fire intensity (Van Wagner 1973).

Aside from the high mortality rate from canker, the 100% blackjack mortality on the HT1 treatment but none on the HNT1 treatment over the course of the study also

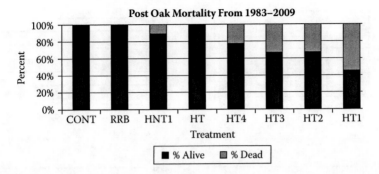

FIGURE 17.5 Percent mortality of selected post oaks by treatment on Pushmataha Forest Habitat Research Area from 1983 to 2009, Pushmataha Wildlife Management Area, Oklahoma.

was unexpected. Post oak also exhibited similar differential mortality (Figure 17.5). Because of the greater canopy cover on the HNT1 treatment versus other ESOs, substantially less fuel and thus lower fireline intensity occurred on this treatment. Lower fuel loads particularly adjacent to tree stems prevented stem girdling.

Implications for landscape implementation of fire treatments suggested less frequent fire regimes because of higher cumulative mortality to oaks from annual burning in low density stands (<10 m^2/ha BA). However, frequent fire might well be a suitable application in stands above 10 m^2/ha BA. A four-year frequency may allow oak regeneration to develop into small trees, which in time may replace those lost on these sites. However, this frequency would not maintain ESOs in an open state. Therefore, a variable fire regime was indicated.

FORAGE PRODUCTION

Post-treatment harvested and thinned openings were dominated by tallgrasses and woody sprouts of varying density, depending on fire frequency (Figures 17.3a,b). Grasses included big bluestem (*Andropogon gerardii*), little bluestem, and, to a lesser extent, Indiangrass (*Sorghastrum nutans*). Grass standing crop was the highest on treatments that annually or recently were burned and had canopy removal. These fire tolerant grasses composed from 60% to 80% of the standing crop on ESOs, but are rarely used by deer (Stransky and Harlow 1981) other than post-burn for the initial green-up period (Masters 1991a, b). However, they are used to some extent by elk (Schneider et al. 2006). The HT1 treatment had the highest grass standing crop with other treatments showing less, commensurate with lower fire frequency (Figure 17.6a).

Panicum standing crop was variable depending on panicum species present (Figure 17.6b) and apparently time since burned, rainfall distribution, and amount. The species composition was different in the panicum standing crop with the treatments with some overstory dominated by low panicum and harvested and thinned treatments tended to have switch grass (*P. vigatum*) as a dominant species associated

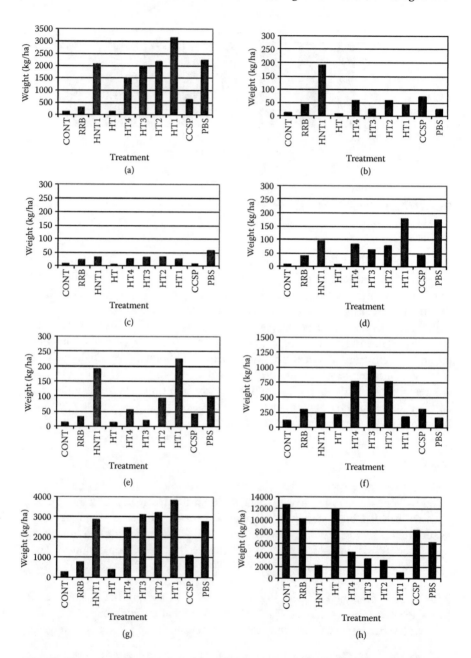

FIGURE 17.6 Standing crop (kg/ha) of various treatments by functional vegetation group, October 2008, on Pushmataha Forest Habitat Research Area, Pushmataha Wildlife Management Area, Oklahoma. (a) Grass, (b) panicum, (c) sedge, (d) legume, (e) forb (non-legume), (f) woody (current annual growth <1.5 m), (g) total, and (h) total litter. CONT = control, no treatment; RRB = rough reduction burn, 4-year burn interval; PBS = thin hardwood, 1-year burn interval; HNT1 = harvest pine, leave hardwood, 1-year burn interval; (*continued*)

with ephemeral drainages but also low panicum. This was particularly true of the oak-savanna treatment (HNT1). Sedge standing crop was generally low compared to other functional categories (Figure 17.6c). Sedge standing crop varied substantially from year to year and variation may be related to sampling intensity because this group tended to be less abundant and less evenly distributed across each unit. Sedges were distinctly benefited by fire over unburned treatments (Figure 17.6c). Sedge prevalence may also be related to time since fire and rainfall as for panicum.

Low panicum and sedge were utilized by deer (Masters 1991b) and important for deer in spring and winter because the green basal rosettes offer food material higher in nutrients and digestibility during a time when browse is of low quality (Jenks 1991; Short 1971). Elk also use switchgrass and sedges, but during summer (Schneider et al. 2006).

Legume standing crop (Figure 17.6d) was associated with the presence or absence of fire and fire frequency. Typically, more frequent or the most recent burn treatments had higher legume standing crop. The unburned treatments were an order of magnitude less in terms of legume standing crop. Non-legume forb standing crop (Figure 17.6e) also was associated with fire frequency or presence of fire. Forbs were highest with one- and two-year fire frequency intervals. In unburned treatments, production rates were similar and very low (Figure 17.6e). Forbs including legumes are important in the diet of both deer and elk. Forbs may compose almost 50% of the diet in spring for deer and just over 25% of elk diets year-round (Jenks 1991; Schneider et al. 2006).

Woody standing crop (<1.4 m in height) was current annual growth only. The mass was highest in burned treatments except for HT1 (Figure 17.6f). Woody sprouting is generally greatest in the growing season following fire and plant development generally increases as fire frequency decreases. The HT4 treatment has much greater canopy cover of woody shrub cover above the sampling zone than the more frequently burned treatments and some shading effect that lowered total forage production was exhibited.

Of the woody browse, approximately 40% by weight was considered preferred including winged sumac (*R. copallina*), muscadine (*V. rotundifolia*), winged elm (*Ulmus alata*), poison ivy (*Toxicodenron radicans*), and green briar. Post oak, black jack oak, and mockernut hickory, and, to a lesser extent, shortleaf pine were nonpreferred species prevalent in browse samples. On the thinned and burned treatments, woody plants valuable as deer forage have been promoted (Masters 1991b; Masters et al. 1993b). On the HT treatment, these species were much less prevalent as dense saplings have shaded out desirable deer forage species. Woody browse is particularly important as it is the major constituent of deer diets in the Ouachita Mountains for most of the year (Jenks 1991) except spring and when hard mast was

FIGURE 17.6 (*continued*) CCSP = clearcut, windrow, summer site preparation burn, plant loblolly pine; HT = harvest pine, thin hardwood, no burn; HT4 = harvest pine, thin hardwood, 4-year burn interval; HT3 = harvest pine, thin hardwood, 3-year burn interval; HT2 = harvest pine, thin hardwood, 2-year burn interval; and HT1 = harvest pine, thin hardwood, 1-year burn interval.

available in fall and winter (Fenwood et al. 1984) and may compose a third of the annual diet for elk (Schneider et al. 2006).

Total standing crop was highest in treatments that had both canopy removal and prescribed burning. The control and HT treatment had significantly lower total production because of fire exclusion, high litter, and canopy cover (Figure 17.2 and Figures 17.6g,h). The harvested and thinned treatments show an increasing amount of standing crop commensurate with increasing fire frequency as predicted. Fire stimulates an increase in all forage categories by removing litter, enhanced nutrient cycling, and scarification of some herbaceous seed, thus increasing overall production.

Fire aids in improving forage by increasing palatability, nutrient content, digestibility, productivity, and availability of grasses and forbs (Lay 1967; Komarek 1974; Reeves and Halls 1974; Lewis et al. 1982; Masters 1991a; Masters et al. 1993b, 1996). Others have reported similar dramatic increases in productivity (Oosting 1944; Lewis and Harshbarger 1976). Lay (1956) and Oosting (1944) have documented plant species composition change, as well as increased forage production after burning. The change in vegetation composition generally lasts two to three years. Exclusion of fire led to declines in ground cover herbaceous plants (Kucera and Koelling 1964; Lewis and Harshbarger 1976). Lewis and Harshbarger (1976) used seasonal and cyclic-fire treatments and found that in all cases forage production was increased over unburned controls. On annual and biennial summer burns, grasses became the dominant understory plants. Forage production on South Carolina loblolly pine sites was higher on annual winter burns than on unburned, periodically winter-burned, or any frequency of summer burning (Lewis and Harshbarger 1976).

WILDLIFE RESPONSE ON FHRA

Deer, elk, and rabbit (*Sylvalagus floridanus*) initially used the thinned units over unthinned units at a much higher rate as shown by browse transects and by pellet counts (Masters 1991a; Masters et al. 1997). All species' use of the various treatments including adjacent food plots showed some change through time as successional changes in woody vegetation occurred. As thinned treatments responded differentially to fire frequency, we saw little difference in use across the range of fire frequencies represented except that the unburned treatment declined dramatically in use by all species prior to canopy closure. We did not detect any difference between food plot use and use of thinned and burned treatments in 1988 or 1994 (Masters et al. 1997). Elk preferred more open grass dominated treatment units and deer treatment units with a greater shrub component. Rabbits showed a distinct preference in use of burned and thinned units over food plots (Masters et al. 1997).

Before this study began, fall bobwhite population density was estimated at less than one bird/50 ha. Following treatment application, fall density has ranged from a low of one bird/2.3 ha to a high of one bird/0.4 ha. The combination of thinning and burning has created usable space and suitable habitat where none existed before. However, useable space has declined on several of the treatments because of change in vegetation structure as plant succession proceeds in a different fashion based on fire frequency (Masters and Waymire 2010).

PWMA LANDSCAPE APPLICATION

MONITORING WILDLIFE RESPONSE

Deer surveys were conducted annually in March and April along twelve 1.6-km Hahn census lines from 1966 to 1992 scattered across the entire PWMA. This deer census technique was later dropped by the department in favor of the more accurate spotlight census technique. Both techniques were used concurrently on the area for fourteen years. Spotlight counts began in 1977 and continue to the present using ODWC spotlight protocols in August and September. Spotlight route length was 16.1 km for the first seven years, and then 25.6 km thereafter; surveys were repeated ten nights. Elk surveys were based on average total counts in February made during late afternoon by driving the deer census route. Turkey populations were indexed by incidental sightings during winter; however, lack of standardization precluded using that data. Because songbirds have been shown to be highly sensitive to change in forest structure and composition such as in ecosystem renewal treatments (Wilson et al. 1995; Masters et al. 2002), we used Breeding Bird Survey (BBS) (http://www.pwrc. usgs.gov/bbs/index.html) stop data from the Pushmataha County route from 1994 to 2009 to index songbird and bobwhite population change. Of the fifty stops on the route, we used thirty-seven that were located on the WMA along the deer spotlight census route to determine response to forest management treatment. Although the BBS route was established in 1992, individual stop data were not available for the first two years that the route was run.

LANDSCAPE APPLICATION OF TREATMENTS

Active landscape level forest management after initial acquisition was non-existent on PWMA until the late-1970s. Thinning to create ESOs averaged 132 ha each year from 1978 to 1994 before the area manager retired. Prescribed burning also increased during this time, particularly following establishment of the FHRA, with an annual average of 390 ha burned per year. With the change in personnel, hectares burned dropped and the thinning program declined (Figure 17.7). ESOs had already begun to show successional changes because of the four- to eight-year burning cycle then employed and thus the area of ESOs began to decline (Figure 17.7).

The change in personnel was a critical time during the study. The new manager heard criticisms about the timber cutting and burning from individuals who had used the area. Sentiment had been expressed that the FHRA had served its purpose and should be abandoned. The new manager and researchers met in 1995 to discuss criticisms, and review and evaluate mid-term FHRA results and ESO change on the area. We concluded that targeting appropriate areas within the landscape for low density stand development (ecosystem renewal) and setting burn frequency, based on FHRA results and historic burn frequency (Masters et al. 1995), to manage succession within these low density stands, that is, ESOs, was the appropriate course of action. As additional thinning on targeted areas was discussed between management and research and planned from an operational standpoint, burning efforts were broadened and frequency increased so that additional ESOs would not be lost to advancing

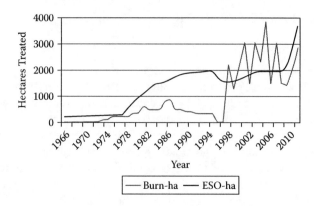

FIGURE 17.7 Management treatments (hectares treated) per year for prescribed burning and cumulative area (hectares) of early successional openings (ESOs) from 1966 to 2010 on Pushmataha Wildlife Management Area, Oklahoma.

forest succession. Initial thinning goals in targeted restoration management units had been confirmed by forage production results from the FHRA (Masters et al. 1993b).

A clearer picture of the proposed landscape developed as research on historical landscapes, fire history, and ecosystem restoration progressed in the Ouachita Mountain region (Foti and Glenn 1991; Kreiter 1994; Masters et al. 1995, 1996, 1998, 2007; Bukenhofer and Hedrick 1997; Sparks et al. 1998). In discussions of these findings, we observed that elements of our FHRA treatment regime approximated historic reference conditions including the fire regime (~three-year interval) (Masters et al. 1995). We used descriptions from historical accounts and analysis of General Land Office survey notes from the Ouachita Mountain region to provide guidance for landscape restoration of historic forest conditions on the area (Foti and Glenn 1991; Masters et al. 1995, 2007) through ESO development. We digitized existing habitat cover types in ARC-GIS (ESRI, Inc., Redlands, CA). We assessed the potential of each management compartment for ecosystem restoration to pine-bluestem, post oak savanna, or to be retained as north-slope hardwoods, mixed oak-pine woodlands, or protected drainages and riparian corridors.

Stands slated for restoration were chosen based on site characteristics, tree species composition, and topographic position (Johnson 1986; Foti and Glenn 1991; Kreiter 1994; Masters et al. 2007) (Table 17.2, Figure 17.8). These stands were scheduled for thinning to focus on reducing the hardwood component (pine-bluestem sites), the pine component (oak woodland and savanna restoration sites), or overall (mixed oak-pine woodlands or savanna). Stands were thinned to basal area ranges characteristic of presettlement forest density; oak-pine woodland and pine-bluestem woodland, 14 m^2/ha BA; oak savanna and oak-pine savanna 7 m^2/ ha BA (Foti and Glenn 1991; Masters et al. 2007). Riparian corridors, steep north-slope hardwoods, and drainages were not thinned. Major roads were day-lighted beginning in 1999 in preparation for increasing the spatial scale of burns through use of helicopter ignition. Thinning efforts were increased again in 2008.

TABLE 17.2
**Management Parameters for Early Successional Opening Development
and Ecosystem Restoration Treatments and Species of Management
Interest as Developed from the Small Plot Study on FHRA and Applied
on the Landscape of Pushmataha Wildlife Management Area, Oklahoma**

Management Type	Basal Area Target[a]	Burn Frequency	Season of Burn[b]	Emphasis Species
Oak savanna	6.9	2–3 year	D, G	Elk, bobwhite, woodland-grassland songbirds
Oak-pine savanna	6.9	2–3 year	D, G	Elk, bobwhite, woodland-grassland songbirds
Pine bluestem	9.2–13.8	1–2 year	D, G	Elk, bobwhite, woodland-grassland songbird
Oak pine woodland	13.8–16.1	3–4 year	D	Deer, turkey, woodland-songbird
North slope hardwood	No harvest	Limited	D	Deer, turkey, closed canopy songbirds
Riparian corridor and drainage	No harvest	Limited	D	Deer, turkey, closed canopy songbirds

[a] (m^2/ha)
[b] Season of burn: D = dormant season, January–March; G = growing season, April–July.

We segregated the target game species, deer, elk, turkey, and bobwhite into two management groups based on similarity of habitat requirements from the literature and their response to different research treatments on the FHRA. Elk and bobwhite preferred open savanna or woodland conditions with grass dominated understory. Thus, savanna-like ESOs (HNT1, HT1, HT2, HT3, and PBS treatments) maintained by high frequency fire would be targeted for these species recognizing that woodland–grassland songbirds would benefit as well (Table 17.2). Deer and turkey were more suited to woodland settings with less frequent fire, thus allowing a greater woody shrub component interspersed with grasses to provide browse for deer and nesting habitat for turkey (RRB, HT3, HT4, HT, and control) (Table 17.2). In these settings, forest dwelling songbirds with closed canopy requirements or with some midstory requirement would benefit. Because the mountainous terrain on PWMA has east-west trending ridges, the broad oak-pine woodland management units were dissected by north-slope hardwood stands or upland drainages and riparian corridors. These would also provide suitable habitat for songbirds with niches for closed canopy forest. Historical evidence suggests that these management types were all characteristic of the presettlement landscape structure, a structure where each of the above species or groups of species once thrived. From 1998 to the present, both hectares burned and ESOs created through heavy thinning have steadily increased (Figure 17.7).

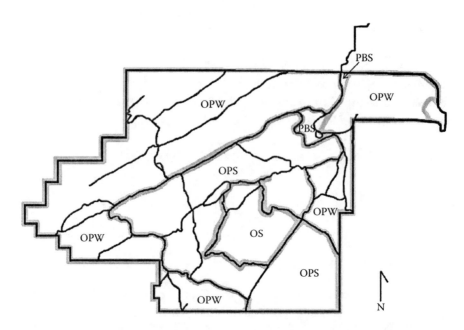

FIGURE 17.8 Target management types for the Pushmataha Wildlife Management Area, Oklahoma, based on research results from the Forest Habitat Research Area. OS = oak savannas, OPW = oak-pine woodland, OPS = oak-pine savannas, PBS = pine–bluestem woodland. Riparian corridors, steep north-slope hardwoods, and drainages within management types were not thinned.

PWMA WILDLIFE RESPONSE

ELK

By 1984, PWMA supported only six elk, after a long downward trend following release. These individuals localized on the FHRA and adjacent food plots after burning began on the study area. Mortality from the meningeal worm (*Parelaphostrongylus tenuis*) had been taking a toll (Stout et al. 1972), and some elk had begun moving off the area because of drought conditions and woody plant succession on the earliest ESOs (Figure 17.9). It seemed that elk would be gone from the area before enough research had been completed to recommend burn frequency for ESO maintenance. However, as burning and ESO development progressed on PWMA landscape, the population slowly increased to twenty-seven elk in 1994 (Masters et al. 1997) and fifty by 2001. Elk populations began rebounding once 21% of the PWMA was in ESOs (Masters et al. 1997). Following management area personnel retirement in 1994, it became evident that additional ESOs and a more aggressive approach to burning were needed as elk numbers begin fluctuating (Figure 17.9). From 1995 to 2000, two to three elk were observed annually exhibiting symptoms of meningeal worm infection. It was also noted that with a decrease in hectares burned, ESOs were being lost to succession. This caused seasonal dispersal of elk from the PWMA. With implementation of restoration thinning and an aggressive approach

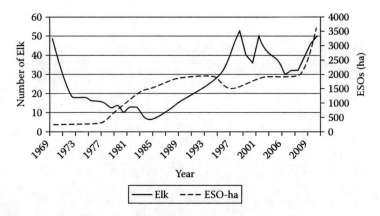

FIGURE 17.9 Rocky Mountain elk population estimates from driving route counts (1969–2010) and cumulative area (hectares) of early successional openings (ESOs) on Pushmataha Wildlife Management Area, Oklahoma.

to burning over the past decade, elk have responded once again. The number of elk observed exhibiting meningeal worm symptoms has declined and has ranged from zero to one annually since 2001. This is thought to be a result of high observed mortality to woodland gastropods, the intermediate host, from fire and the thinning efforts creating less suitable conditions for these snails. The movement of elk off the area to fescue pastures has raised depredation concern to private landowners in the vicinity; therefore, growing season burns are being explored as an avenue to provide a fresh green-up of forage during mid to late summer as a means of holding animals throughout the time when forage quality is declining.

DEER

Following the dramatic deer population decline on PWMA of the mid to late 1960s, management efforts continued to focus efforts on maintaining 221 ha of food plots, except that a few small stands were burned and fewer still were thinned. The timber management program began in earnest in the late 1970s and hectares burned began increasing (Figure 17.8). The PWMA supported an average density of 8.7 deer/km^2 (SE = 0.4) from 1986 to 1990 (Masters et al. 1993b) based on Hahn census lines. However, spotlight count data suggested more than double that number (Figure 17.10). An outbreak of epizootic hemorrhagic disease in 1993 lowered the deer population to <5 deer/km^2.

After management and research developed a consensus on FHRA treatments to use for landscape application, a more aggressive approach to ESO development through thinning and burning was adopted by the late 1990s, and the deer population began an increasing trend and has generally increased for over a decade (Figure 17.10). We recognize that day-lighting of roads along the deer census route likely biased initial deer counts upward. However, the magnitude of effect and current area thinned and burned suggests that some level of response is a direct result of increased available forage and improved habitat conditions on a landscape level.

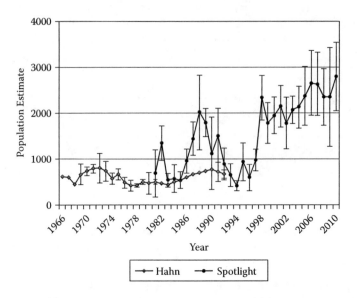

FIGURE 17.10 White-tailed deer population estimates (+SE) from Hahn (1966–1991) versus spotlight count (1977–2010) methods on Pushmataha Wildlife Management Area, Oklahoma.

As well, surrounding Pushmataha County deer populations have not increased in a similar fashion.

SONGBIRDS

The songbird data from BBS included an average of thirty-nine (SD = 3) species per year for the portion of the route that was run on PWMA. Therefore, we selected ten species that were representative of woodland grassland obligates (five) and closed canopy forest species (five) on which to focus (Figure 17.11). At the beginning of BBS surveys on PWMA in 1994, 11 of the 37 points had not been thinned within 400 m. Those points that had thinning within a 400-m radius typically had less than 50% of the surrounding area disturbed. By 2009, all points included stands that had been thinned and burned several times.

The five species chosen to represent closed canopy forest species were yellow-billed cuckoo (*Coccyzus americanus*), chuck-will's widow (*Caprimulgus carolinensis*), black and white warbler (*Mniotilta varia*), pileated woodpecker (*Dryocopus pileatus*), and red-eyed vireo (*Vireo olivaceus*). Regional trends assuredly influence local landscape populations of many species, particularly Neotropical migrants, and therefore must be taken into account when interpreting local (route level) trends. From 1994 to 2007, yellow-billed cuckoo and chuck-will's widow were experiencing a significant downward regional trend. Pileated woodpecker and black and white warbler showed stable regional populations, while the red-eyed vireo exhibited a significant increasing trend.

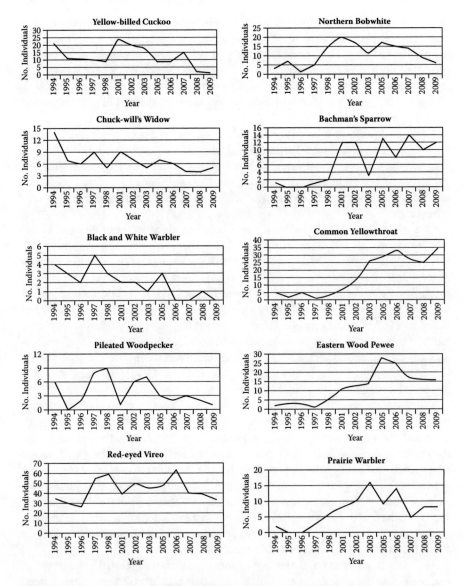

FIGURE 17.11 Breeding bird population trend estimates (BBS) for selected species from 1994 to 2009 on Pushmataha Wildlife Management Area, Oklahoma.

Based on previous studies, we expected certain forest interior birds such as chuck-will's widow, black and white warbler, pileated woodpecker, and red-eyed vireo to slightly decline because of our landscape level treatments to open up the forest canopy (Wilson et al. 1995). We found declines were most dramatic for the yellow-billed cuckoo and to some extent chuck-will's widow, but these were also experiencing significant regional declines (Figure 17.11). However, we did find declining numbers of black and white warbler and pileated woodpecker, which were stable on a regional basis, but these species were never of great abundance prior to landscape level treatments. The red-eyed vireo counts although variable did not exhibit the same general pattern of decline as the other species in this group. Other studies have shown little influence on these species because of thinning and burning (Wilson et al. 1995; Masters et al. 2002), but stands in those studies were not thinned to the same extent as in this study.

Early bird work in mixed pine-hardwood stands suggested that most breeding birds were associated with the hardwood component. Canopy stratification was often distinct in pine plantations with competing hardwood midstories (Dickson 1982). Noble and Hamilton (1975) found that as canopy strata increased so did the number and kinds of birds. However, these findings were from managed even-aged pine stands.

Fire may be the most important factor controlling abundance of forest and woodland birds through its influence on vertical and horizontal structure and thus availability of habitat niches (Masters et al. 2002). This knowledge ties research directly to management through determination of management practices that create suitable habitat structure for species of management concern. Aside from habitat structure, fire directly affects food availability for both seed-eating and insectivorous birds (Landers 1987; Komarek 1974). At ground level, litter dwelling invertebrates were reduced by fire in the short term. As succulent herbaceous regrowth occurs, herbivorous insects increased (Dickson 1982). These changes in the invertebrate community may affect breeding success of some birds because insects were a critical source of nutrients for many breeding birds (Landers 1987).

When fire reduced the midstory hardwood component in mixed pine-hardwood forests, structural complexity was reduced. Foliage gleaners tied to deciduous midstory and low shrubs were purportedly disadvantaged by periodic fire, but those that require pine stands or early successional habitats were favored. However, Wilson et al. (1995) found little effect on midstory foraging species. In general, those species dependent on heavy litter accumulations, vertical and horizontal structural diversity, and edge or plant species diversity were disadvantaged (Dickson 1982; Landers 1987). Michael and Thornburgh (1971) noted increased bird numbers within pine-hardwood stands subjected to partial hardwood removal (reduced by 11%) and fire similar to Wilson et al. (1995) and Masters et al. (2002).

We recognized from the outset that historical evidence suggested at least part of a given landscape may well have had denser forest conditions, such as along steep north slopes and associated with drainages and stream corridors (Masters et al. 2007). These areas were purposefully withheld from application of thinning, and burning was targeted at less frequent intervals because these topographic features tended to reduce historic fire intervals on those specific types of sites. These denser forest habitats were retained to meet the needs of forest interior and other forest-associated species.

The five species representing woodland-grassland birds were northern bobwhite, Bachman's sparrow (*Aimophila aestivalis*), common yellow throat (*Geothlypis trichas*), eastern wood pewee (*Contopus virens*), and prairie warbler (*Dendroica discolor*). Regional populations of the northern bobwhite have exhibited a significant downward trend from 1994 to 2007, with the common yellow throat showing a slight decline, the eastern wood pewee and prairie warbler were stable or exhibited no regional trend, and the Bachman's sparrow had no regional report but was listed as a species of special concern in most southern states (Cox and Widner 2008).

All of the representative woodland-grassland species showed dramatic increases at some point during the survey period and exhibited an overall increasing trend. The much higher counts of the common yellow throat and Bachman's sparrow was particularly notable in the face of regional declines and Oklahoma and Arkansas designation of the Bachman's sparrow as a species of special management concern (Figure 17.11). The common yellowthroat has literally exploded on the area following thinning and burning. Based on previous research in the Ouachita Mountains, these results were expected with landscape implementation (Wilson et al. 1995; Masters et al. 2002).

Although some have suggested that with intensive cutting, foraging guilds for various bird species would be virtually eliminated (Webb et al. 1977), we did not find this to be the case in landscape restructuring and burning. Specific habitat structure for woodland-grassland obligates is entirely missing in extensive closed canopy forests, which historic evidence clearly shows to have been in open woodland or savanna-like conditions in southeast Oklahoma. Further, it is not possible to meet the structural habitat requirements on every given hectare for the diverse array of bird species (and thus niches) historically represented across the southeast United States.

SUMMARY

This project began as problem-oriented research designed to address a management question and provide a practical management solution as an outcome of experimental results. It was also designed to be implemented and evaluated over the long-term on a landscape level. It is rare that wildlife habitat research or ecological restoration studies adequately quantify wildlife population response; therefore, this was a salient focal point.

In early project stages, management personnel had to make decisions based on implications of research results and ecological responses from other physiographic regions. As results from the FHRA became clearer through time, adjustments were made to landscape management in what we now term an adaptive resource management strategy. Discussions between management and research twice annually were critical for making necessary adjustments and ensuring positive outcomes. Further joint field days organized by Oklahoma State University Cooperative Extension Service and led by research and management project personnel, for natural resource professionals and private landowners, has broadened and cemented the partnership. This research has impacted land management on public and private lands totaling over 1.83 million hectares (Masters and Waymire 2010).

Based on small plot research about fire frequency and thinning we found that, in general, the advantages of fire and thinning in oak-pine forests included:

1. The ability to control and direct hardwood understory and midstory development to achieve specific wildlife and forest management objectives (habitat structure).
2. Removal of litter for enhanced nutrient cycling and plant growing conditions.
3. Increased forage palatability, nutrient content, and digestibility.
4. Increased herbage production, species richness, and availability.
5. The ability to restructure and manage stands in a historical context to favor both game and nongame species.
6. That disturbance is essential in order to maintain the wide array of plant and wildlife species that depend on varying habitat structure.

Most species of wildlife require specific habitat structure and without some form of successional redirection or method of disturbance (such as purposeful thinning and fire), these habitats progressively change without most people ever noticing (Komarek 1963). The result is a growing list of species of special management concern or, worse, an increasing number of species highlighted on threatened and endangered species lists.

Knowledge gained from this continuing long-term study has been used for ecosystem restoration projects, and the development of different management strategies for deer, elk, and nongame species of special management concern. We have learned much about the role of fire in plant succession trajectories following disturbance and animal community succession associated with plant community change. The ecosystem management practices developed through collaboration of researchers and managers on the small plot study area has been scaled up and successfully applied on a landscape level with benefits to nongame and game species alike. Land management treatments chosen for the small plot experiments from just a research or a management perspective would have eliminated specific treatments that are now incorporated on a landscape level such as the pine-bluestem or oak-savanna treatment. The value of long-term research is also seen because some recommendations based on short-term research (<10 years) initially overestimated the ability of the four-year interval fire to control woody plant succession. The short-term perspective also overlooked the value of the annual burn restoration treatments and failed to fully realize that fire is an essential process for landscape health and maintenance and to ensure retention of biodiversity.

We encourage the joint collaboration between managers and researchers to identify problems and work on them together in an experimental context. The perspective of either, when absent, leaves much room for error, therefore lowering the likelihood of achieving either management or research goals. In this case study management was not initially interested in various statistical controls or treatments that tied the results to similar treatments in other regions such as the clearcut and rough reduction burn treatments. Research was not interested in the array of annual burn treatments that later became the basis for restoration treatments to apply on the landscape. Through a great deal of discussion and the wise counsel of a statistician, consensus was achieved. As the process unfolded, a place was made early on for ongoing dialogue where all ideas were placed on the table and critically examined in light of the current state of knowledge. This open dialogue and determination to see the project through was essential when personnel changed and criticism was mounting. A key to successful implementation was the healthy mutual respect for the essential roles of both management and research throughout the process.

State agencies with land management responsibilities have the opportunity to participate in and direct replicate research projects on landscapes, and gain much from a variety of disciplines while achieving their management goals. When both wildlife researchers and wildlife managers work together, their roles are seen to blend rather than being disjointed in purpose and scope (Macnab 1983). Therein much will be accomplished to the benefit of natural resources.

ACKNOWLEDGMENTS

We kindly thank those funding this effort through the years. Funding has been provided in various phases through Oklahoma Federal Aid to Wildlife Restoration Project W-80R, Oklahoma Department of Wildlife Conservation (ODWC); Oklahoma State University (OSU), McIntire-Stennis funds; Oklahoma Department of Agriculture and Horticulture, Forestry Services, Forest Stewardship Wildlife Assistance Grant; and Tall Timbers Research Station, Wildlife Research Endowment. Many managers, biologists, foresters, and range scientists deserve our thanks for assistance with fieldwork, open discussions, and sharing of thought. Although we can't name them all, we would be remiss if we did not mention several important players in moving this project along. We thank Ray Robinson, retired Area Manager; Leon Johnson, Assistant Manager; Reggie Thackston, former Forest Wildlife Biologist, ODWC; and Bill Warde (deceased), Statistician, OSU, for their contributions in conceiving, designing, and getting the project off the ground and in the early stages of implementation. Ray conceived the idea and insisted that management and research working as a team should be part of the development and implementation. This chapter is dedicated to him.

We thank the sampling crew from the Kiamichi Research Station, OSU for over a decade of dependable field assistance including Randy Holeman, Dennis Wilson, Keith Anderson, Greg Campbell, and Bob Heinemann. John Weir and his students from OSU are acknowledged for assistance with prescribed burns for the past decade. We thank Mark Howery who conducted many of the breeding bird surveys. We gratefully acknowledge the excellent assistance of Kaye Gainey for word processing, formatting, and development of graphics for Figure 17.3a,b. Finally, we thank Dwayne Elmore, Theron Terhune, Terry Bidwell, and Bill Palmer for useful reviews of the draft manuscript.

REFERENCES

Bain, W. R. and A. Watterson, Jr. 1979. Soil survey of Pushmataha County, Oklahoma. U.S. Department of Agriculture, Soil Conservation Service, Washington, DC.

Bragg, T. B., and L. C. Hulbert. 1976. Woody plant invasion of unburned Kansas bluestem prairie. *Journal of Range Management* 29:19–24.

Bukenhofer, G. A., and L. D. Hedrick. 1997. Shortleaf pine/bluestem grass ecosystem renewal in the Ouachita Mountains. *Transactions of the North American Wildlife and Natural Resources Conference* 62:509–515.

Cain, M. D. 1984. Height of stem-bark char underestimates flame length in prescribed burns. *Fire Management Notes* 45:17–21.

Caughley, G. 1980. The George Reserve deer herd (book review). *Science* 207:1338–1339.

Conway, K. E., and B. Olson. 2004. Hypoxylon cancer of oaks. Oklahoma Cooperative Extension Service Fact Sheet EPP-7620, Oklahoma State University, Stillwater, OK.

Cox, J., and B. Widener. 2008. Lightning-season burning: friend or foe of breeding birds? Tall Timbers Miscellaneous Publication No. 17. Tall Timbers Research Station and Land Conservancy, Tallahassee, FL.

Crandall, R. M. 2003. Vegetation of the Pushmataha Wildlife Management Area, Pushmataha County, Oklahoma. Thesis, Oklahoma State University, Stillwater.

Crandall, R. M., and R. J. Tyrl. 2006. Vascular flora of the Pushmataha Wildlife Management Area, Pushmataha County, Oklahoma. *Castanea* 71:65–79.

Crofton, E. W. 2001. Herbert L. Stoddard Sr.: King of the Fire Forest. In: *The Fire Forest: Longleaf Pine-Wiregrass Ecosystem,* J. R. Wilson, Ed. Covington, GA: Georgia Wildlife, Natural Georgia Series, 8(2): 40–41

Davis, K. P. 1953. *Forest Fire Control and Use.* New York: McGraw Hill.

Dickson, J. G. 1982. Impact of forestry practices on wildlife in southern pine forests. In: *Increasing Forest Productivity, Proceedings of the 1981 Convention of the Society of American Foresters, Orlando, Florida.* Bethesda, MD: Society of American Foresters, pp. 224–230.

Duck, L. G., and J. B. Fletcher. 1943. A game type map of Oklahoma. Oklahoma Game and Fish Department, Oklahoma City, OK.

Edwards, S. L., S. Demarais, B. Watkins, and B. K. Strickland. 2004. White-tailed deer forage production in managed and unmanaged pine stands and summer food plots in Mississippi. *Wildlife Society Bulletin* 32:739–745.

Engle, D. M., T. G. Bidwell, and M. E. Mosley. 1996. Invasion of Oklahoma rangelands and forests by eastern red cedar and Ashe juniper. Oklahoma Cooperative Extension Service Circular E-947, Oklahoma State University, Stillwater, OK.

Fahnestock, G. P., and R. C. Hare. 1964. Heating of tree trunks in surface fires. *Journal of Forestry* 62:799–805.

Fenwood, J. D., D. F. Urbston, and R. F. Harlow. 1984. Determining deer habitat capability in Ouachita National Forest pine stands. *Proceedings of the Annual Conference of the Southeast Association of Fish and Wildlife Agencies* 38:13–22.

Ferguson, E. R. 1961. Effects of prescribed fires on understory stems in pine-hardwood stands of Texas. *Journal of Forestry* 59:356–359.

Foti, T. L., and Glenn S. M. 1991. The Ouachita Mountain landscape at the time of settlement. In: *Proceedings of the Conference: Restoration of Old-Growth Forests in the Interior Highlands of Arkansas and Oklahoma,* D. Henderson and L. D. Hedrick, Eds. September 19–20, 1990, Winrock International, Morrilton, AK, pp. 49–65.

Fuller, N. M. 1976. A nutrient analysis of plants potentially useful as deer forage on clear-cut and selective-cut pine sites in southeastern Oklahoma. Thesis, Oklahoma State University, Stillwater.

Garren, K. H. 1943. Effects of fire on vegetation of the southeastern United States. *Botanical Review* 9:617–654.

Gill, R. B. 1985. Wildlife research—an endangered species. *Wildlife Society Bulletin* 13:580–587.

Goodrum, P. D., V. H. Reid, and C. E. Boyd. 1971. Acorn yields, characteristics and management criteria of oaks for wildlife. *Journal of Wildlife Management* 35:520–532.

Halls, L. K., and E. A. Epps, Jr. 1969. Browse quality influenced by tree overstory in the South. *Journal of Wildlife Management* 33:1028–1031.

Hare, R. C. 1965. Contribution of bark to fire resistance of southern trees. *Journal of Forestry* 63:248–251.

Harlow, R. F., J. B. Whelan, H. S. Crawford, and J. E. Skeen. 1975. Deer foods during years of oak mast abundance and scarcity. *Journal of Wildlife Management* 39:330–336.

Hebb, E. A. 1971. Site preparation decreases game food plants in Florida sandhills. *Journal of Wildlife Management* 35:155–162.

Hines, F. D., and D. F. Bertelson. 1987. Forest statistics for east Oklahoma counties—1986. U.S. Department of Agriculture, Forest Service, Resource Bulletin SO-121, New Orleans, LA.

Hobbs, N. T., and R. Hilborn. 2006. Alternatives to statistical hypothesis testing in ecology: a guide to self teaching. *Ecological Applications* 16:5–19.

Jenks, J. A. 1991. Effect of cattle stocking rate on the nutritional exology of white-tailed deer in managed forests of southeastern Oklahoma and southwest Arkansas. Dissertation, Oklahoma State University, Stillwater.

Johnson, F. L. 1986. Woody vegetation of southeastern LeFlore County, Oklahoma, in relation to topography. *Proceedings of the Oklahoma Academy of Science* 66:1–6.

Komarek, Sr., E. V. 1974. Effects of fire on temperate forests and related ecosystems: southeastern United States. In: *Fire and Ecosystems*, T. T. Kozlowski and C. E. Ahlgren, Eds. New York: Academic Press, pp. 251–277.

Komarek, Sr., E. V. 1981a. Prescribed burning in hardwoods. Tall Timbers Research Station Management Note 1, Tallahassee, FL.

Komarek, Sr., E. V. 1981b. Scorch in pines. Tall Timbers Research Station Management Note 2, Tallahassee, FL.

Komarek, R. 1963. Fire and the changing wildlife habitat. *Proceedings of the Tall Timbers Fire Ecology Conference* 2:35–43.

Kreiter, S. D. 1994. Dynamics and spatial patterns of a virgin old-growth hardwood-pine forest in the Ouachita Mountains, Oklahoma, from 1896–1994. Thesis, Oklahoma State University, Stillwater.

Kucera, C. L. 1978. Grasslands and fire. In: *Proceedings of the Conference: Fire Regimes and Ecosystem Properties*, H. A. Mooney, T. M. Bonnicksen, N. L. Christensen, J. E. Lotan, and W. A. Reiners, technical coordinators. U.S. Department of Agriculture, Forest Service, General Technical Report WO-26, Washington, DC, pp. 90–111.

Kucera, C. L., J. H. Ehrenreich, and C. Brown. 1963. Some effects of fire on tree species in Missouri prairie. *Iowa State Journal of Science* 38:179–185.

Kucera, C. L., and M. Koelling. 1964. The influence of fire on composition of central Missouri prairie. *American Midland Naturalist* 72:142–147.

Landers, J. L. 1987. Prescribed burning for managing wildlife in southeastern pine forests. In: *Managing Southern Forests for Wildlife and Fish*, J. G. Dickson and O. E. Maughn, Eds. U.S. Department of Agriculture, Forest Service, General Technical Report SO-65, New Orleans, LA, pp. 19–27.

Larson, J. S. 1967. Forests, wildlife and habitat management—a critical examination of practice and need. U.S. Department of Agriculture, Forest Service, Research Paper SE-30, Asheville, NC.

Lawson, E. R., and R. N. Kitchens. 1983. Shortleaf pine. In: *Silvicultural Systems for the Major Forest Types of the United States*, R. M. Burns, technical compiler. U.S. Department of Agriculture, Forest Service, Agricultural Handbook 445, Washington, DC, pp. 157–161.

Lay, D. W. 1956. Effects of prescribed burning on forage and mast production in southern pine forests. *Journal of Forestry* 54:582–584.

Lay, D. W. 1957. Browse quality and the effects of prescribed burning in southern pine forests. *Journal of Forestry* 55:342–349.

Lay, D. W. 1967. Browse palatability and the effects of prescribed burning in southern pines. *Journal of Forestry* 65:826–828.

Lewis, C. E., H. E. Grelen, and G. E. Probasco. 1982. Prescribed burning in southern forest and rangeland improves forage and its use. *Southern Journal of Applied Forestry* 6:19–25.

Lewis, C. E., and T. J. Harshbarger. 1976. Shrub and herbaceous vegetation after 20 years of prescribed burning in the South Carolina Coastal Plain. *Journal of Range Management* 29:13–18.

Logan, T. 1972. Study of white-tailed deer fawn mortality on Cookson Hills deer refuge eastern Oklahoma. *Proceedings of the Annual Conference of the Southeastern Association of Game and Fish Commissioners* 26:27–35.

Lotan, J. E., M. E. Alexander, S. F. Arno, et al. 1978. Effects of fire on flora—A state-of-knowledge review. U.S. Department of Agriculture, Forest Service, General Technical Report WO-16, Washington, DC.

Macnab, J. 1983. Wildlife management as a scientific experimentation. *Wildlife Society Bulletin* 11:397–401.

Macnab, J. 1985. Carrying capacity and related slippery shibboleths. *Wildlife Society Bulletin* 13:403–410.

Martin, R. E. 1963. A basic approach to fire injury of tree stems. *Proceedings of the Tall Timbers Fire Ecology Conference* 2:151–162.

Masters, R. E. 1991a. Effects of timber harvest and prescribed fire on wildlife habitat and use in the Ouachita Mountains of eastern Oklahoma. Dissertation, Oklahoma State University, Stillwater.

Masters, R. E. 1991b. Effects of fire and timber harvest on vegetation and cervid use on oak-pine sites in Oklahoma Ouachita Mountains. In: Proceedings of an international symposium. *Fire and the Environment: Ecological and Cultural Perspectives*, S. C. Nodvin and T. A. Waldrop, Eds. U.S. Department of Agriculture, Forest Service, General Technical Report SE-69, Asheville, NC, pp. 168–176.

Masters, R. E. 2001. Ice damage and the loblolly/shortleaf debate. *Oklahoma Renewable Resources Bulletin* 16(3):1, 6–7.

Masters, R. E., and D. M. Engle. 1994. BEHAVE-evaluated for prescribed fire planning in mountainous oak-shortleaf pine habitats. *Wildlife Society Bulletin* 22:184–191.

Masters, R. E., D. M. Engle, and R. Robinson. 1993a. Effects of timber harvest and periodic fire on soil chemical properties in the Ouachita Mountains. *Southern Journal of Applied Forestry* 17:139–145.

Masters, R. E., S. D. Kreiter, and M. S. Gregory. 2007. Dynamics of an old-growth hardwood-*Pinus* forest over 98 years. *Proceedings of the Oklahoma Academy of Science* 87:15–30.

Masters, R. E., R. L. Lochmiller, and D. M. Engle. 1993b. Effects of timber harvest and periodic fire on white-tailed deer forage production. *Wildlife Society Bulletin* 21:401–411.

Masters, R. E., R. L. Lochmiller, S. T. McMurry, and G. A. Bukenhofer. 1998. Small mammal response to pine-grassland restoration for red-cockaded woodpeckers. *Wildlife Society Bulletin* 28:148–158.

Masters, R. E., J. E. Skeen, and J. Whitehead. 1995. Preliminary fire history of McCurtain County Wilderness Area and implications for red-cockaded woodpecker management. In: *Red-Cockaded Woodpecker: Species Recovery, Ecology and Management*, D. L. Kulhavy, R. G. Hooper, and R. Costa, Eds. Nacogdoches, TX: Center for Applied Studies, Stephen F. Austin University, pp. 290–302.

Masters, R. E., W. D. Warde, and R. L. Lochmiller. 1997. Herbivore response to alternative forest management practices. *Proceedings of the Annual Conference of the Southeastern Association of Fish and Wildlife Agencies* 51:225–237.

Masters, R. E., and J. Waymire. 2010. Progress Report 2009–2010. Forest Stewardship—Forest Wildlife Assistance. Tall Timbers Research Station, Tallahassee, FL.

Masters, R. E., C. W. Wilson, G. A. Bukenhofer and M. E. Payton. 1996. Effects of pine-grassland restoration for red-cockaded woodpeckers on white-tailed deer forage production. *Wildlife Society Bulletin* 24:77–84.

Masters, R. E., C. W. Wilson, D. S. Cram, G. A. Bukenhofer, and R. L. Lochmiller. 2002. Influence of ecosystem restoration for red-cockaded woodpeckers on breeding bird and small mammal communities. In: *The Role of Fire in Non-Game Wildlife Management and Community Restoration: Traditional Uses and New Directions: Proceedings of a Special Workshop*, W. M. Ford, K. R. Russell, and C. E. Moorman, Eds. Newtown Square, PA: U.S. Department of Agriculture, Forest Service, General Technical Report NE-288, pp. 73–90.

Michael, E. D., and P. I. Thornburgh. 1971. Immediate effects of hardwood removal and prescribed burning on bird populations. *Southwestern Naturalist* 15:359–370.

Murphy, P. A., and R. M. Farrar, Jr. 1985. Growth and yield of uneven-aged shortleaf pine stands in the Interior Highlands. U.S. Department of Agriculture, Forest Service, Research Paper SO-218, Southern Forest Experiment Station, New Orleans, LA.

Niering, W. A. 1978. The role of fire management in altering ecosystems. *Proceedings of the Conference: Fire Regimes and Ecosystem Properties*, H. A. Mooney, T. M. Bonnicksen, N. L. Christensen, J. E. Lotan, and W. A. Reiners, technical coordinators. U.S. Department of Agriculture, Forest Service, General Technical Report WO-26, Washington, DC, pp. 489–570.

Noble, R. E., and R. B. Hamilton. 1975. Bird populations in even-aged loblolly pine forests of southeastern Louisiana. *Proceedings of the Annual Conference of the Southeastern Association of Fish and Wildlife Agencies* 29:441–450.

Nuttall T. 1980. *A Journal of Travels into the Arkansas Territory during the Year 1819*. Norman, OK: University of Oklahoma Press.

Oosting, H. J. 1944. The comparative effect of surface and crown fire on the composition of a loblolly pine community. *Ecology* 25:61–69.

Reeb, J. E., and T. H. Silker. 1979. Nutrient analysis of selected forbs on clearcut areas in southeastern Oklahoma. *Proceedings of the Annual Conference of the Southeastern Association of Fish and Wildlife Agencies* 33:296–304.

Reeves, H. C., and L. K. Halls. 1974. Understory response to burning in forests of southeast Louisiana. *The Consultant* 19:73–77.

Ripley, T. H. 1980. Planning wildlife management investigations and projects. In: *Wildlife Management Techniques Manual*, S. D. Schemnitz, Ed. Washington, DC: The Wildlife Society, pp. 1–6.

Romesburg, H. C. 1981. Wildlife science: gaining reliable knowledge. *Journal of Wildlife Management* 45:293–313.

Schneider, J., D. S. Maehr, K. J. Alexy, J. J. Cox, J. L. Larkin, and B. C. Reeder. 2006. Food habits of reintroduced elk in southeastern Kentucky. *Southeastern Naturalist* 5:535–546.

Segelquist, C. A. 1974. Evaluation of wildlife forage clearings for white-tailed deer habitat in a 600-acre Arkansas Ozark enclosure. Dissertation, Oklahoma State University, Stillwater.

Segelquist, C. A., and W. E. Green. 1968. Deer food yields in four Ozark forest types. *Journal of Wildlife Management* 32:330–337.

Segelquist, C. A., and R. E. Pennington. 1968. Deer browse in the Ouachita National Forest in Oklahoma. *Journal of Wildlife Management* 32:623–626.

Segelquist, C. A., and M. Rogers. 1974. Use of wildlife forage clearings by white-tailed deer in Arkansas Ozarks. *Proceedings of the Annual Conference of the Southeastern Association of Game and Fish Commissioners* 28:568–573.

Segelquist, C. A., M. Rogers, F. D. Ward, and R. G. Leonard. 1972. Forest habitat and deer populations in an Arkansas Ozark enclosure. *Proceedings of the Annual Conference of the Southeastern Association of Game and Fish Commissioners* 26:15–35.

Segelquist, C. A., F. D. Ward, and R. G. Leonard. 1969. Habitat-deer relations in two Ozark enclosures. *Journal of Wildlife Management* 33:511–520.

Short, H. L. 1971. Forage digestibility and diet of deer on southern upland range. *Journal of Wildlife Management* 35:698–706.

Sinclair, A. R. E. 1979. Dynamics of the Serengeti ecosystem: process and pattern. In: *Serengeti: Dynamics of an Ecosystem,* A. R. E. Sinclair and M. Norton-Griffiths, Eds. Chicago, IL: Chicago University Press, pp. 1–30.

Sparks, J. C., R. E. Masters, D. M. Engle, M. W. Palmer, and G. A. Bukenhofer. 1998. Effects of growing-season and dormant-season prescribed fire on herbaceous vegetation in restored pine-grassland communities. *Journal of Vegetation Science* 9:133–142.

Steel, G. D. R., and J. H. Torrie. 1980. *Principles and Procedures of Statistics: A Biometrical Approach.* New York: McGraw-Hill.

Stout, G. G., F. C. Lowry, and F. Carlile. 1972. The status of elk transplants in eastern Oklahoma. *Proceedings of the Annual Conference of the Southeastern Association of Game and Fish Commissioners* 26:202–203.

Stransky, J. J. 1976. Vegetation and soil response to clearcutting and site preparation in east Texas. Dissertation, Texas A & M University, College Station.

Stransky, J. J., and L. K. Halls. 1978. Forage yields increased by clearcutting and site preparation. *Proceedings of the Annual Conference of the Southeastern Association of Fish and Wildlife Agencies* 34:476–481.

Stransky, J. J., and R. F. Harlow. 1981. Effects of fire in the southeast on deer habitat. In: *Prescribed Fire and Wildlife in Southern Forests*, G. W. Wood, Ed. Georgetown, SC: The Belle W. Baruch Forest Science Institute, Clemson University, pp. 135–142.

Stransky, J. J., and R. Richardson. 1977. Fruiting of browse plants affected by pine site preparation in east Texas. *Proceedings of the Annual Conference of the Southeastern Association of Fish and Wildlife Agencies* 31:5–7.

Thompson, M. W., M. G. Shaw, R. W. Umber, J. E. Skeen, and R. E. Thackston. 1991. Effects of herbicides and burning on overstory defoliation and deer forage production. *Wildlife Society Bulletin* 19:163–170.

Van Wagner, C. E. 1973. Height of crown scorch in forest fires. *Canadian Journal of Forest Research* 3:373–378.

Vogl, R. J. 1974. Effects of fire on grasslands. In: *Fire and Ecosystems*, T. T. Kozlowski, and C. E. Ahlgren, Eds. New York: Academic Press, pp. 139–194.

Wade, D. D., and R. W. Johansen. 1986. Effects of fire on southern pine: observations and recommendations. U.S. Department of Agriculture, Forest Service, General Technical Report SE-41, Asheville, NC.

Waldrop, T. A., and D. H. Van Lear. 1984. Effect of crown scorch on survival and growth of young loblolly pine. *Southern Journal of Applied Forestry* 8:35–40.

Webb, W. L., D. F. Behrend, and B. Saisorn. 1977. Effect of logging on song bird populations in a northern hardwood forest. The Wildlife Society, Wildlife Monographs Number 55, Ithaca, NY.

White, A. S. 1986. Prescribed burning for oak-savannah restoration in central Minnesota. U.S. Department of Agriculture, Forest Service, Research Paper NC-266, North Central Forest Experiment Station, St. Paul, MN.

Wilson C. W., R. E. Masters, and G. A. Buckenhofer. 1995. Breeding bird response to pine-grassland community restoration for red-cockaded woodpeckers. *Journal of Wildlife Management* 59:56–67.

Wood, G. W. 1988. Effects of prescribed fire on deer forage and nutrients. *Wildlife Society Bulletin* 16:180–186.

Yoakum, J., W. P. Dasmann, H. R. Sanderson, C. M. Nixon, and H. S. Crawford. 1980. Habitat improvement techniques. In: *Wildlife Management Techniques Manual*, S. D. Schemnitz, Ed. Washington, DC: The Wildlife Society, pp. 329–404.

Section V

Conclusions and Future Directions

18 Moving Forward
Connecting Wildlife Research and Management for Conservation in the Twenty-First Century

*Joseph P. Sands**
Caesar Kleberg Wildlife Research Institute
Texas A&M University–Kingsville
Kingsville, Texas

Leonard A. Brennan
Caesar Kleberg Wildlife Research Institute
Texas A&M University–Kingsville
Kingsville, Texas

Stephen J. DeMaso†
Caesar Kleberg Wildlife Research Institute
Texas A&M University–Kingsville
Kingsville, Texas

Matthew J. Schnupp
King Ranch, Inc.
Kingsville, Texas

* Current address: New Mexico Department of Fish and Game, Santa Fe, New Mexico.
† Current address: U.S. Fish and Wildlife Service, Gulf Coast Joint Venture, National Wetlands Research Center, Lafayette, Louisiana.

CONTENTS

When we run out of country we will run out of stories. When we run out of stories we will run out of sanity.

—Rick Bass (1994:12)

ABSTRACT

Research and management in wildlife ecology and conservation have undergone radical changes in the past seventy-five years, including shifts from traditional game management to conservation biology, and from single-species management to ecosystem management. The next fifteen to twenty years will represent a critical juncture for wildlife conservation and management, an arena that faces at least four long-term, ongoing challenges: habitat loss, invasive exotic species, climate change, and demographic changes in the wildlife constituency. We conceptualize the wildlife management-research paradigm as a system of reinforcing feedback loops of information transfer among three primary entities (private institutions, public institutions, and research institutions), each of which is charged with its own mission. We contend that the system in which this paradigm exists is beset by a series of balancing loops and delays that inhibit the flow of useful information through the system and disrupt the beneficial outcomes of cooperation among entities. Consequently, members of these three entities would benefit by taking a broad view of the wildlife management-research system in order to evaluate its potential to meet these challenges and capitalize on opportunities that exist to improve wildlife conservation.

INTRODUCTION

This book has presented discourse on the disconnect between research and management in wildlife conservation, with a special emphasis on North America. It has examined the practical, technical, and sociopolitical reasons for the existence of what Potts (2008:58) described as "a polarization of cultures" between managers and researchers in wildlife ecology, and it has illustrated the importance of collaboration between management and research through case studies of successes and failures in management. What this book has provided is a conceptual foundation for some of the mechanisms necessary for improving wildlife conservation in the twenty-first century. DeMaso (this volume) discussed the longstanding, innate disconnect between research and management, and Krausman (this volume) discussed the primary reasons for its continued existence. The remaining chapters described institutional roles within the wildlife management-research paradigm and presented case studies of where the relationships between research and management have experienced success and failure.

Our objective is to conclude this volume with a chapter that highlights elements of wildlife management that our profession must address if it is to achieve long-term success in wildlife conservation. We do so by presenting a series of challenges and opportunities for the wildlife management-research paradigm in the twenty-first century; describing a wildlife management-research system using causal loop diagrams; and presenting a set of potential solutions for overcoming inherent problems within the system.

CONNECTING WILDLIFE RESEARCH, MANAGEMENT, AND CONSERVATION IN THE TWENTY-FIRST CENTURY

THE HUMAN ELEMENT

Wildlife science differs from other biological sciences like zoology, and it is interesting to us because it is intrinsically linked with a sociological component. Its true nature is as an applied science. Wildlife management carries with it the need to produce a commodity for the public. In our view, roughly two categories of commodities exist. The first is the traditional wildlife-game management commodity of what Leopold (1933:1) called "sustained annual crops of wild game for recreational use"; essentially, the directly consumptive end of wildlife use. The second involves the more aesthetic commodity or recreational opportunities for wildlife viewing; essentially, the nonconsumptive end of wildlife use.

CHALLENGES

Wildlife conservation will face a myriad of challenges in the twenty-first century. Among these are four long-term, ongoing challenges that will require collaboration between researchers and managers in wildlife in order to be addressed effectively. These challenges are habitat loss, invasive exotic species, climate change, and demographic changes within the wildlife constituency.

Mitigating and Overcoming Habitat Loss

The destruction of habitat, decline of North American wildlife populations since the arrival of Europeans, and the development of policy in the twentieth century to counteract these declines are well documented (Matthiessen 1959; Trefethen 1975; Dunlap 1988). Habitat loss (along with degradation and fragmentation) remains the most significant threat to wildlife conservation and biodiversity throughout the world (Wilcove et al. 1998). Excluding Antarctica, anthropogenic factors have impacted at least 83% of the Earth's land surface (Sanderson et al. 2002). Given that biodiversity loss is predicted to continue on a global scale despite increases in the amount of protected areas over the past twenty years (Mora and Sale 2011), the impacts of continued habitat loss will be devastating to global biodiversity. Mitigating habitat loss via protected areas or restoration of degraded areas is primarily a challenge for management. However, research must contribute to this process by determining optimal strategies for achieving these objectives.

Managing Invasive Species

The impacts of invasive species have global implications (D'Antonio and Vitousek 1992). For instance, behind habitat loss, exotic plants and animals are the second greatest extinction threat to wildlife in the United States (Wilcove et al. 1998), and exotic species invasions have caused more extinctions worldwide than any other factor except for habitat loss (D'Antonio and Vitousek 1992; Simberloff and Rejmanek 2011). Invasive exotic species disrupt ecosystem processes (e.g., invasive grasses altering the frequency and intensity of the grass-fire cycle; D'Antonio and Vitousek 1992), create a net loss of habitat for native species, or impact the direct mortality of native species. The development of techniques designed to mitigate impacts of invasive species will require innovative research and strong collaboration with managers in order to evaluate the effectiveness of research results for implementation.

Dealing with Climate Change

A great deal of uncertainty exists regarding the long-term, broad scale impacts of climate change on wildlife. Quantifying the long-term broad scale impacts of climate change on wildlife will be very difficult because of the confounding impacts of other variables that cloud this relationship. Whether climate change will result in broad scale shifts in the geographic distributions of species remains to be seen, although the potential is certainly there. In addition to the potential shifting of species' geographic ranges, Brennan (2007) described numerous secondary effects of climate change that have the potential to be problems for wildlife populations. These include more frequent summer heat waves, prolonged droughts (observed in the Midwestern United States in 2009 and 2011), and increased intensity of weather events such as rainstorms and hurricanes (Gong and Wang 2000; Intergovernmental Panel on Climate Change 2007). Little is known about how the impacts of climate change coupled with the impacts of habitat loss and fragmentation (see previous text) will interact with respect to wildlife populations. However, it can be assumed that for many area and temperature sensitive species (e.g., the northern bobwhite, *Colinus virginianus*), these impacts could be quite severe (Brennan 2007).

Changing Demographics

Hunting and game management helped provide a foundation for wildlife conservation (Trefethen 1975), the result of which was the protection of millions of acres of habitat and generation of billions of dollars for conservation. This is largely the reason why hunting is a central component of the North American Model of Wildlife Management (NAMWM). Thus, while recreational hunting may not have been "the cradle of the conservation movement" as Dunlap (1988:9) suggested, it has at least been the carriage, or the person pushing the carriage, for the better part of a century.

The number of people who hunt in the United States is declining and the average age of people who hunt is increasing (Beucler and Servheen 2009). While non-consumptive uses of natural resources are becoming very popular, the hunting public in North America still funds the lion's share of wildlife management and conservation efforts through license and stamp fees despite the fact that they represent a decreasing proportion of resource uses (Beucler and Servheen 2009). Clearly, this status quo is not a sustainable business model for wildlife conservation, or for state agencies whose mission it is to manage wildlife resources. Additionally, many established, successful non-governmental conservation organizations will likely be negatively impacted by a decreasing proportion of hunters as well. Nevertheless, the question remains: How do we maintain the numbers that we currently have while making hunting an attractive recreational opportunity for youth and non-hunters? This is admittedly an uphill battle considering the changing demographics of the United States. If declines in the percentage of Americans who hunt are truly irreversible, then criticisms regarding the position of sport hunting as a central component in the NAMWM are valid (Nelson et al. 2011; Dratch and Kahn 2011). However, regardless of how one feels about recreational hunting, removing hunting from this position will require the development of strategies and programs designed to tap alternative and consistent sources of funding for conservation. It is one thing to recognize inadequacies of philosophical constructs, but it is quite another to find new sources of revenue.

OPPORTUNITIES

Public Lands: Ecosystem Management and Conservation Areas

The amount of protected areas across the globe began to increase substantially in the 1970s, a trend that reflected changes in global policies toward conservation. There are millions of hectares of public lands in the United States. Wildlife conservation is a priority on many state and federally managed lands, and where this is the case, these lands should serve as showcases for conducting habitat and, where applicable, population management for targeted species.

Private Lands: Where Economics and Conservation Meet

Conservation opportunities on private lands have long been recognized by wildlife professionals (Leopold 1930). Eighty percent of the wildlife habitat in the United States is contained on private lands (Benson 2001), and there is great potential for private lands to impact conservation in a positive way (Knight et al. 1995; Schnupp and DeLaney, this volume). Ranching, farming, timber production, and other land uses have the potential to provide significant acreages of quality wildlife habitat but only if wildlife habitat is incorporated into a direct management objective of an organization. With declines in biodiversity predicted to continue on a global scale (as mentioned previously) despite increases in the amount of protected areas (Mora and Sale 2011), the application of sustainable strategies to the management of privately owned lands that are still dedicated to the use of natural resources (e.g., cattle, timber production) will continue to be absolutely critical to wildlife conservation.

The Joint Venture Approach to Conservation

Organized conservation efforts that operate at the landscape scale will be necessary to make headway in protecting and restoring usable habitat space for wildlife. In the latter portion of the twentieth century, broad scale approaches to conservation began to appear. Perhaps the most prominent of these conservation efforts is the North American Waterfowl Management Plan (Williams and Castelli, this volume) as well as a network of joint venture (Giocomo et al., this volume) partnerships designed to implement bird conservation across the continent. Using (1) biological planning, (2) conservation design, (3) program delivery, (4) outcome-based monitoring, and (5) assumption-driven research, this approach of integrating research and management with an adaptive management framework. Joint ventures have potential as a vehicle for achieving conservation success.

UNDERSTANDING THE REASONS FOR THE DISCONNECT

THE SYSTEMS PERSPECTIVE

DeMaso (this volume) and Brennan (this volume) each pointed out the similarities between research-management disconnects in wildlife science and management and in the medical profession. It is a modest leap of logic to hypothesize that these types of disconnects are just common, but are also inherent in many human enterprises. For instance, in the field of management and organizational science, there remains long-standing debate among scholars concerning the rigor of research versus its relevance to managers (Gulati 2007). How can this be so? The systems-thinking paradigm suggests that the inability of organizations to address problems is intrinsic and is the result of the systemic structure itself (Richmond 1993; Meadows 2008). A characteristic of systems is that they are resistant to change; thus, external

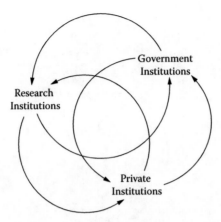

FIGURE 18.1 A conceptualized diagram of the wildlife management-research paradigm (see text for description of entities). In this diagram, each entity interacts with the other, and solid loops represent reinforcing flows of information among entities.

problems persist and often become more prevalent over time because the system is inherently unable to deal with them (Richmond 1993). The system's structure makes it resistant to change and unable to efficiently adapt to the impacts of external forces (Richmond 1993). The point is that public and private lands wildlife managers are not the problem, private lands wildlife managers are not the problem, and wildlife researchers are not the problem; *all of us* are the problem because we all represent entities within a system that is not operating under an optimized structure.

The challenges and opportunities presented in the previous sections of this chapter can be viewed as external forces with potential to impact the outcomes of the wildlife management-research paradigm. It is our opinion that this paradigm can be viewed as a system of reinforcing feedback loops of information transfer (Figure 18.1). We consider three primary entities within the system (research institutions, government institutions, and private institutions), which are separated by their respective missions.* Research institutions may represent either public or privately funded organizations, with the commonality between them being that their primary purpose is to conduct wildlife research and educate students. Government institutions can be state and federal wildlife agencies such as the Texas Parks and Wildlife Department or the U.S. Fish and Wildlife Service. Private institutions may be private landowners, or non-governmental organizations such as The Nature Conservancy.

Within the system, information flows either directly (e.g., from Government Institutions → Research Institutions or vice versa) or indirectly (e.g., Research Institutions → Government Institutions → Private Institutions) from one entity to another (Figure 18.1). Obviously, the system in Figure 18.1 is incomplete because each entity within the system is charged with its own purpose (see Section II, Chapters 3–8

---

* We recognize that there are many subgroups that could be incorporated within each entity; however, diagramming every possible element individually would reduce the utility and comprehensibility of these conceptual models (Meadows 2008). As it is not our intention to exclude a given group, we invite the reader to take a broad view of the system and use her or his knowledge to mentally place a particular group within an appropriate entity.

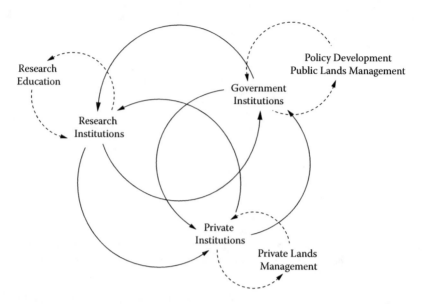

FIGURE 18.2   A conceptualized diagram of the wildlife management-research paradigm. Solid loops still represent a reinforcing flow of information among entities; however, the dashed loops represent actions undertaken by each entity to fulfill its specific purpose. These actions often occur independently of the information flowing from the other entities, and do not account for the purposes of other entities, which results in balancing process, slowing the flow of useful information through the system, and disrupting beneficial cooperation among entities.

of this volume). In the case of government institutions, a purpose may be setting rules and creating policies to comply with laws set forth by legislative bodies. For a research institution, the purpose may be the dissemination of new knowledge through research and publications. A private institution may be concerned primarily with providing quality deer hunts for lessees or some other targeted, focused service. Thus, with each entity there also exists a balancing action with respect to the reinforcing flow of information (Figure 18.2). These balancing actions inhibit the flow of useful information through the system and disrupt beneficial cooperation among entities.

The system in Figure 18.2 is incomplete as well. In reality, there is a natural delay between information disseminating from research institutions to management (Figure 18.3). In some cases, two delays may occur (e.g., Research Institutions → Government Institution → Private Institutions). Delays in information transfer strengthen the balancing actions, and do so because the activities conducted by each entity will continue in the absence of inflows of new information. When this happens, each entity may inadvertently work against the others. Ultimately, the outcome of these behaviors is that inappropriate actions are often taken by a particular entity (e.g., researchers ask the "wrong" questions; policies do not reflect best available science; management techniques are inappropriate for a given situation).

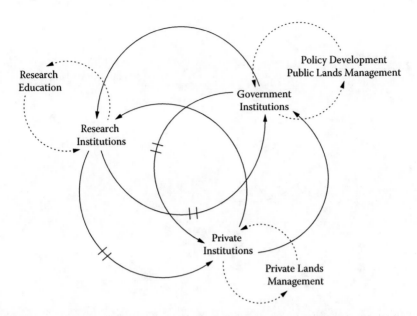

FIGURE 18.3 A conceptualized diagram of the wildlife management-research paradigm with delays (parallel dashes within solid loops) in information transfer from research institutions (see text for description of entities). Delays in information transfer result in the strengthening of balancing actions, which may result in inappropriate actions being taken by a particular entity (e.g., researchers ask the "wrong" questions; policies do not reflect best available science; management techniques are inappropriate for a given situation). See Figures 18.1 and 18.2 for description of the system.

A system where entity actions reinforced the flow of information from one to another would be ideal for connecting research with management in wildlife (Figure 18.4). Completely reversing the nature of these loops from balancing to reinforcing may be impossible because despite the fact that each of the entities in the system is linked, they are also bound by their individual purposes. In addition, because the system is circular, there are no defined starting and ending points for the flow of information. Finally, since the initiation of actions occurs independently among entities, one entity may undertake an action long before another considers it relevant, and consequently a given action taken as a solution at the present may create substantial problems in the future (Dowling et al. 1995). Schnupp and DeLaney (this volume) provided an example of this in discussing a grass introduction program in South Texas. Several species of grasses were introduced to the South Texas Plains ecoregion to solve a problem: a lack of forage for livestock during droughts. However, while these introductions were successful from a livestock production standpoint, they created unforeseen problems for future managers: exotic grasses negatively impact wildlife habitat. Research regarding the impacts of exotic grasses on organisms, habitats, and ecosystems has been a popular topic in rangeland and wildlife science for at least the past twenty years now (e.g., D'Antonio and Vitousek 1992; Williams and Baruch 2000; D'Antonio et al. 2011), but was non-existent at the time.

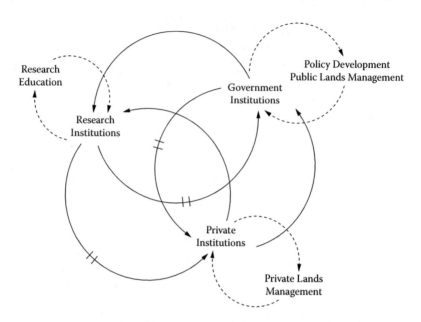

FIGURE 18.4 A conceptualized diagram of the wildlife management-research paradigm where actions of entities represent reinforcing loops, which improve information transfer among entities. Delays (parallel dashes within solid loops), while not optimal, have been retained for realism.

However, we believe that solutions exist that would help our profession confront the challenges and capitalize on the opportunities discussed previously. Next, we present a set of ideas and actions that would improve the transfer of information among these entities (researchers and managers). Certainly the set is not exhaustive, but it is designed to spark thought and debate and is based on the facts, thoughts, and opinions presented in the case studies in the previous chapters.

RECONNECTING WILDLIFE MANAGEMENT WITH RESEARCH

EDUCATION

Like most or all of our readers, each of us has been involved in university programs as students or educators (eight total institutions among us). The role of education within university settings is paramount to addressing the issues we face as a profession. This discussion is not new. Kevern (1973) suggested that the greatest contribution that universities could make to conservation is through education and that an undergraduate education should not be considered job training. We agree, although we suggest that providing students with some of the practical and technical skills necessary for job success is also important. Courses that teach these skills are easily integrated into curricula designed to provide a broad educational background. Perhaps the most important thing a university can teach a student is how to learn because no university can provide an exact set of skills necessary to perform every job in the wildlife profession. Thus, students must know how to be adaptable and to learn new skills as they advance through their career paths.

We question the necessity for non-thesis master's programs in wildlife and suggest instead that much of the curricula in these programs could be incorporated into undergraduate and graduate curricula alike (where relevant) based on a student's career goals. Web-based distance education has a place in wildlife education. Certain courses may have a significant value, especially for practicing professionals hoping to fill specific knowledge-gaps. Practicing professionals have presumably completed a degree program and presumably have a skill set that has allowed them to remain employed. Filling holes in this skill set via web-based courses seems practical and beneficial to these professionals. However, the ability of start-to-finish online education programs (e.g., Edge and Sanchez 2011) to provide an educational experience that is equivalent to a classroom-field-laboratory-based degree remains unknown. Oregon State University's online program will serve as an interesting case study regarding the long-term potential of this emergent learning paradigm.

RESEARCHER CONTRIBUTIONS TO POLICY AND CONSERVATION

Participation in Policymaking and the Political Process

Schenkel (2010) discussed the need to achieve evidence-based policy over policy-based evidence. This is not possible without direct participation from scientists in the policy-making process. Scientists should not be content to exist in what Schenkel (2010:1751) calls a "comfortable, parallel universe" with respect to the political process. Scientists should be part of the policy-making process (and especially so for controversial issues) because their participation is critical for basing conservation on the best available science. In doing so, however, researchers must remain objective when providing professional opinions, base opinions on the best science available, and be uncompromising about their data (DeStefano and Steidl 2001). Additionally, scientists should be part of the policy-making process because being involved in the process may provide a means for securing funding for research. Given the economic issues faced by governments in the United States and Europe, public funding for research will become more difficult to obtain. In order to maintain funding sources and promote a research-based management paradigm in an increasingly hostile environment, researchers must be involved in this process. How can a researcher convince managers and administrators that their work is (1) applicable to the public and (2) worth the investment if scientists are not present as conservation policy is developed and management decisions are made?

Applying Research to Solve Management Problems

Of course, asking the right questions and conducting research for the right reasons are also critical for strengthening the connection between research and management (Gavin 1989; Hunter 1989). Based on our collective experience, we often see lopsided examples of research and management priorities because both parties (researchers and managers) fail to distinguish between purposeful and cultural management actions. Purposeful management actions are those that provide resources needed by wildlife to meet their annual life-cycle needs. These resources typically are related to habitat features that provide cover, food, and other components needed for populations of wild vertebrates to persist, and perhaps even be elevated if possible and appropriate. Cultural management actions, on the other hand, typically have their motives rooted in what other people do and then they feel a need to implement similar actions. Providing supplemental feed, reducing nest predators, planting food plots, and so on are classic examples of cultural wildlife management actions. Typically, cultural management actions such as providing supplemental feed for game birds are a neutral practice. It is unusual to see demographic changes result from practices like supplemental feeding. Rather than any additional birds being produced, or living longer, existing birds are redistributed in relation to where the feed is placed.

Another aspect of the implications of neglecting purposeful management in favor of cultural management is related to understanding how prevailing land uses impact wildlife populations and their habitats. Being able to read a landscape and understand how past actions of humans have resulted in present-day configurations of wildlife habitat—or lack thereof—is a critical skill that seems to be overlooked or undervalued these days. The implications from understanding how to link land uses with wildlife habitat relationships are enormous. For example, the widespread lack of early successional forest habitat throughout much of the eastern third of the United States, while understood and appreciated by some (see Askins, this volume), is more often than not overlooked. Ditto for cattle grazing on public lands in the American West (see Herman, this volume). Not being able to see how prevailing land uses impact wildlife habitat and populations is a precursor to not being able to ask the right kinds of research questions that can provide a scientific basis for purposeful management actions that are so necessary to sustain, and elevate, species of wild vertebrates.

Manager Contributions to Research and Policy

Re-emergence of a Management Journal

The *Wildlife Society Bulletin* was resurrected in 2011 by The Wildlife Society after several years of debate and pleas from society membership who wanted a publication that contained information regarding "management and techniques and less information about statistical evaluations of research" (Ballard et al. 2011:1). The content that new *Bulletin* will provide for the membership in the future will depend in part on the variety of factors that go in to producing a scientific publication. However, for the *Bulletin* to truly be a management journal, it must consistently receive submissions from practicing wildlife managers. We consider practicing wildlife managers to be individuals who work in primarily non-research roles at state and federal agencies, or for private organizations (profit or nonprofit). There are many informative descriptive (and perhaps experimental) data available from federally and state-managed wildlife conservation lands. These data cannot remain locked in the filing cabinets or stored away on agency or company hard drives if they are to be useful to the wildlife community at large.

Staying Current

Wildlife practitioners in agency settings also need access to information that will enable them to remain current with the technical literature produced by researchers. One way of doing this is for agency administrators to encourage regular attendance and participation at professional society meetings. Our experience in Texas indicates that feedback from managers can be valuable for young (graduate students) and veteran (professors) researchers especially during the early phases of a project. The Texas Chapter of the Wildlife Society Annual Meeting is consistently well-attended by graduate students, tenure track researchers, management and extension agency staff, and administrative personnel from throughout the state, and promotes a strong interaction from both the research and management ends of the wildlife profession.

In addition to meeting attendance, managers should maintain memberships in professional societies and make an effort to read new articles relevant to their work. This is often challenging because direct access to a wide spectrum of journals is usually limited unless a person is located within an academic setting. Essentially, outside of a university with access to journal indexing services (e.g., JSTOR, BioOne, etc.) a manager could subscribe to several professional journals and still not have access to the full volume of literature necessary to remain current in his or her respective area of expertise. Agencies, professional societies, and publishers should explore potential avenues for making a broader volume of literature available to professional agency staff.

ADAPTIVE MANAGEMENT, LEARNING BY DOING

Adaptive management is an iterative process that involves the collaboration of a set of stakeholders to plan, act, monitor, and evaluate management actions (Stankey et al. 2005.) The concept has roots in a variety of professions including systems theory (Stankey et al. 2005), and is designed to improve natural resource management by using knowledge of the system to predict and learn from the outcomes of actions. Adaptive management acknowledges that managers must make decisions despite incomplete knowledge of a system and uncertainty regarding outcomes (Linkov et al. 2006). This concept has been applied to the development and waterfowl management to set harvest regulations since 1995 (Nichols et al. 1995, 2007). Using an adaptive management approach to conservation problems could foster the relationship between managers and researchers. Other benefits of an adaptive management approach to conservation include the following:

1. Efficient habitat management.
2. Transparent and defensible research and management.
3. Strategic allocation of limited resources.
4. More effective communication of conservation challenges and strategies to address them.
5. Accountability.

Without a doubt, adaptive management has the potential to more effectively and efficiently provide conservation when conducted properly.

CONCLUSIONS

KEEPING LANDSCAPES AND WILDLIFE WILD

Affecting change in the wildlife management-wildlife research disconnect requires close collaboration among all types of wildlife professionals (e.g., researchers and managers) and the stakeholders involved in specific wildlife management issues. We have a great responsibility to continue to improve the manner in which the practitioners of our profession interact. It is our job to collect, synthesize, and distribute the information necessary to management techniques. It is our job to properly advise our constituents on how best to manage private lands for wildlife. Wildlife management and science achieved some magnificent successes in the twentieth century, but must again rise to the challenge (those listed in this chapter and others) in the twenty-first century. These challenges will be great and at times daunting, but meeting them will be fundamental for achieving long-term success in wildlife conservation, and for keeping the last wild places, wild.

ACKNOWLEDGMENTS

This chapter benefited from support provided by the Richard M. Kleberg, Jr. Center for Quail Research, the South Texas Quail Associates Program, and the C.C. "Charlie" Winn Endowed Chair for Quail Research.

REFERENCES

Ballard, W., J. Wallace, and T. E. Boal. 2011. The road back for the Wildlife Society Bulletin. *Wildlife Society Bulletin* 35:1.

Bass, R. 1994. On Willow Creek. In: *Heart of the Land: Essays on the Last Great Places,* J. Barbato and L. Weinerman, Eds. New York: Pantheon, pp. 7–24.

Benson, D. E. 2001. Wildlife and recreation management on private lands in the United States. *Wildlife Society Bulletin* 29:359–371.

Beucler, M., and G. Servheen. 2009. Mirror, mirror, on the wall: reflections from a nonhunter. *Transactions of the North American Wildlife and Natural Resources Conference* 73:163–179.

Brennan, L. A. 2007. South Texas climate 2100: potential ecological and wildlife impacts. In: *The Changing Climate of South Texas 1900-2100*, J. Norwine and K. John, Eds. Kingsville, TX: National Science Foundation CREST-RESSACA Program, Texas A&M University–Kingsville, pp. 79–90.

D'Antonio, C., K. Stahlheber, and N. Molinari. 2011. Grasses and forbs. In: *Encyclopedia of Biological Invasions*, D. Simberloff and M. Rejmanek, Eds. Berkeley: University of California Press, pp. 280–290.

D'Antonio, C. M. and P. M. Vitousek. 1992. Biological invasions by exotic grasses, the grass/fire cycle, and global change. *Annual Review of Ecology and Systematics* 23:63–87.

DeStefano, S. and R. J. Steidl. 2001. The professional biologist and advocacy: what role do we play? *Human Dimensions of Wildlife* 6:11–19.

Dowling, A. M., R. H. MacDonald, and G. P. Richardson. 1995. Simulation of systems archetypes. *Proceedings of the International Conference of the System Dynamics Society* 13:454–463.

Dratch, P., and R. Kahn. 2011. Moving beyond the model. *The Wildlife Professional* 5:61–63.

Dunlap, T. R. 1988. *Saving America's Wildlife*. Princeton, NJ: Princeton University Press.

Edge, W. D., and D. Sanchez. 2011. An online fisheries and wildlife degree: can you really do that? *Wildlife Society Bulletin* 35:2–8.

Gavin, T. A. 1989. What's wrong with the questions we ask in wildlife research? *Wildlife Society Bulletin* 17:345–350.

Gong, D. Y., and S. W. Wang. 2000. Severe summer rainfall in China associated with enhanced global warming. *Climate Research* 16:51–59.

Gulati, R. 2007. Tent poles, tribalism, and boundary spanning: the rigor-relevance debate in management research. *Academy of Management Journal* 50:775–782.

Hunter, M. L., Jr. 1989. Aardvarks and Arcadia: two principles of wildlife research. *Wildlife Society Bulletin* 17:350–351.

Intergovernmental Panel on Climate Change. 2007. Climate change 2007: The physical science basis. Summary for policy makers. Geneva, Switzerland.

Kevern, N. R. 1973. The large university and the future of natural resources. *Wildlife Society Bulletin* 1:45–47.

Knight, R. L., G. N. Wallace, and W. Riebsame. 1995. Ranching the view: subdivisions versus agriculture. *Conservation Biology* 9:459–461.

Leopold, A. 1930. The American game policy. *Transactions of the American Game Conference* 17:284–307.

Leopold, A. 1933. *Game Management*. New York: Charles Scribner's Sons.

Linkov, I. F. K. Satterstrom, G. A. Kiker, T. S. Bridges, S. L. Benjamin, and D. A. Belluck. 2006. From optimization to adaptation: shifting paradigms in environmental management and their application to remedial decisions. *Integrated Environmental Assessment and Management* 2:92–98.

Matthiessen, P. 1959. *Wildlife in America*. New York: Viking Press.

Meadows, D. H. 2008. *Thinking in Systems: A Primer*. White River Junction, VT: Chelsea Green Publishing Company.

Mora, C., and P. F. Sale. 2011. Ongoing global biodiversity loss and the need to move beyond protected areas: a review of the technical and practical shortcomings of protected areas on land and sea. *Marine Ecology Progress Series* 434:251–266.

Nelson, M. P., J. A. Vucetich, P. C. Paquet, and J. K. Bump. 2011. An inadequate construct? North American model: what's flawed, what's missing, what's needed. *The Wildlife Professional* 5:58–60.

Nichols, J. D., F. A. Johnson, and B. K. Williams. 1995. Managing North American Waterfowl in the face of uncertainty. *Annual Review of Ecology and Systematics* 26:177–199.

Nichols, J. D., M. C. Runge, F. A. Johnson and B. K. Williams. 2007. Adaptive harvest management of North American waterfowl populations: a brief history and future prospects.

Potts, G. R. 2008. Global biodiversity conservation: we need more managers and better theorists. In: *Wildlife Science: Linking Ecological Theory and Management Applications*, T. E. Fulbright and D. G. Hewitt, Eds. Boca Raton, FL: CRC Press, pp. 43–63.

Richmond, B. 1993. Systems thinking: critical thinking skills for the 1990s and beyond. *System Dynamics Review* 9:113–133.

Sanderson, E. W., M. Jaiteh, M. A. Levy, K. H. Redford, A. V. Wannebo, and G. Woolmer. 2002. The human footprint and the last of the wild. *Bioscience* 52:891–904.

Schenkel, R. 2010. The challenge of feeding scientific advice into policy-making. *Science* 330:1749–1751.

Simberloff, D., and M. Rejmanek, Eds. 2011. *Encyclopedia of Biological Invasions*. Berkeley: University of California Press.

Stankey, G. H., R. N. Clark, and B. T. Bormann. 2005. Adaptive management of natural resources: theory concepts and management institutions. General Technical Report PNW-GTR-654. USDA Forest Service, Pacific Northwest Research Station, Portland, Oregon.

Trefethen, J. B. 1975. *An American Crusade for Wildlife*. New York: Winchester Press.

Wilcove, D. S., D. Rothstein, J. Dubow, A. Phillips, and E. Losos. 1998. Quantifying threats to imperiled species in the United States. *Bioscience* 48:607–615.

Williams, D. G., and Z. Baruch. 2000. African grass invasion in the Americas: ecosystem consequences and the role of ecophysiology. *Biological Invasions* 2:123–140.

Index

Page references in **bold** refer to tables.